AF598367

BIOREACTORS AND BIOTRANSFORMATIONS

This volume is the collection of the papers presented at the International Conference on Bioreactors and Biotransformations held at Gleneagles, Scotland, on 9–12 November 1987, organised by the National Engineering Laboratory, East Kilbride, Glasgow G75 0QU, Scotland, UK, in co-operation with the Laboratory of the Government Chemist, Cornwall House, Waterloo Road, London SE1 8XY, UK.

Sponsored by the National Engineering Laboratory, the Laboratory of the Government Chemist, the Scottish Development Agency and the East Kilbride Development Corporation.

Acknowledgements

The assistance of the Local Organising Committee, the International Committee and referees is most gratefully acknowledged.

Local Organising Committee

Dr T. J. S. Brain *(Chairman)*	National Engineering Laboratory
Dr P. B. Baker	Laboratory of the Government Chemist
Dr R. P. M. Bond	Shell Research Centre
A. N. Emery	University of Birmingham
Dr J. Ingham	University of Bradford
Dr J. C. Middleton	ICI plc
G. W. Moody	National Engineering Laboratory
Professor A. W. Nienow	University of Birmingham
Dr P. J. Rodgers	ICI plc
Professor P. Richmond	AFRC Institute of Food Research, Norwich
Professor J. E. Smith	University of Strathclyde

International Committee

Professor A. M. Klibanov	Massachusetts Institute of Technology, USA
Professor T. Kobayashi	Nagoya University, Japan
Professor B. Kristiansen	Center for Industrial Research, Norway
Professor M. Moo-Young	University of Waterloo, Canada
Dr J. C. van Suijdam	Gist Brocades, The Netherlands

BIOREACTORS AND BIOTRANSFORMATIONS

Edited by

G. W. MOODY

National Engineering Laboratory, East Kilbride, Glasgow, UK

and

P. B. BAKER

Laboratory of the Government Chemist, Cornwall House, London, UK

Published on behalf of
THE NATIONAL ENGINEERING LABORATORY
by
ELSEVIER APPLIED SCIENCE PUBLISHERS
LONDON and NEW YORK

ELSEVIER APPLIED SCIENCE PUBLISHERS LTD
Crown House, Linton Road, Barking, Essex IG11 8JU, England

Sole Distributor in the USA and Canada
ELSEVIER SCIENCE PUBLISHING CO., INC.
52 Vanderbilt Avenue, New York, NY 10017, USA

WITH 67 TABLES AND 190 ILLUSTRATIONS

British Library Cataloguing in Publication Data

International Conference on Bioreactors and
Biotransformations *(1987: Gleneagles).*
Bioreactors and biotransformations
1. Biotechnology.
I. Title II. Moody, G. W. III. Baker, P. B.
IV. National Engineering Laboratory
660'.6 TP248.2

ISBN 1-85166-162-X

Library of Congress CIP data applied for

Printed in Great Britain by Galliard (Printers) Ltd, Great Yarmouth

CONTENTS

Bioreactors and Plants

Bioreactor Control

Bioreactors and Animal Cells

Combined Bioreaction and Separation

Bioconversions in Novel Media

Biotransformations (Products)

Biotransformations (Processes)

Novel Bioreactors

Bioreactor Performance and Scale-up

Immobilisation

INTERNATIONAL CONFERENCE ON BIOREACTORS AND BIOTRANSFORMATIONS
GLENEAGLES, SCOTLAND, UK: 9-12 NOVEMBER 1987

Paper A1

OPTIMISATION OF PRODUCT YIELD IN IMMOBILISED PLANT CELL CULTURES

M.A. Holden and M.M. Yeoman
Department of Botany, University of Edinburgh,
King's Buildings, Mayfield Road, Edinburgh EH9 3JH

ABSTRACT

Product yield is critical if plant cell cultures are to be widely used in the production of valuable secondary products *in vitro*. The immobilisation of freely suspended cells results in advantages in bioreactor configuration and in the physiological state of the cells in some systems. Different matrices can be used in immobilisation including gels of calcium alginate, agarose, agar, polyacrylamide as well as inert support systems such as the hollow fibre reactor and immobilisation in reticulate polyurethane foam. Examples are presented from the production of capsaicin by cultures of *Capsicum frutescens* of techniques which have resulted in increased product yield. These are: 1. Cell line selection, 2. Immobilisation in foam, 3. Depression of primary metabolism, 4. Alteration of primary metabolism, 5. Induction of activity of rate limiting enzymes, 6. Inhibition of competing reactions, 7. Exploitation of biotransformations, 8. Continuous product removal, 9. Inhibition of product breakdown or further metabolism. Most of these approaches have been attempted and proved successful in *Capsicum* cultures. Which, if any, of these techniques could prove successful in other species depends on a good understanding of the regulation of the biosynthesis of the desired product.

INTRODUCTION

Between 25% and 30% of prescribed pharmaceuticals are extracted from, or contain ingredients extracted from plant sources [1]. Plants are also used as sources of food colourings and flavourings, and also in the production of agrochemicals. These products supply a market worth billions of dollars per annum and it is not surprising that the possibility of using plant tissue culture systems in order to produce some of these plant secondary products *in vitro* has received considerable attention over the past decade.

There are several putative advantages in the use of tissue culture over the cultivation of the whole plant in the production of natural products. These include the continuity of, and control over, supply due to the independence of production from environmental factors such as climate and season. The product composition and yield may also be more predictable, and product recovery may be easier in cell culture systems. However, the use of mass cell culture as a means of producing natural products will only find wide acceptance if it offers a cost advantage over conventional systems. In what is a capital intensive system the fundamental problem of this technology is therefore cost of production. This paper discusses product yield which is one critical aspect of production cost.

Immobilisation

It is now recognised that plant cells in mass culture do not behave like microbial cells and different cultural techniques would be needed to successfully obtain product formation from plant cells in vitro. Plant cells are more susceptible to shear forces than micro- organisms [2] although the importance of this factor is being increasingly questioned (Scragg, pers. comm.). Furthermore, conditions that favour rapid growth tend not to favour product accumulation [3-5]. One solution to these problems is the immobilisation of cells within a solid matrix. Immobilisation binds cells together and protects them from shear and in some instances has beneficial physiological effects which result in increased product yield.

The simplest system of immobilisation is "self-immobilised" cultures which are simply highly aggregated cultures with the cells growing in compact clumps [6]. Manipulation of the culture medium can result in almost 100% of cells growing as aggregates, which in a number of species leads to enhanced secondary metabolite production [7].

Plant cells can also be attached to the surface of inert surfaces with no loss of biosynthetic ability. Cells of Solanum aviculare have been covalently linked to glutaraldehyde activated beads of polyphenylene oxide [8] and Humulus lupulus cells have been attached to a man-made matrix (nylon or polypropylene) [9]. Bornmann and Zachrisson [10] have shown that protoplasts can also be attached to inert supports such as cytodex micro carriers and maintain viability.

By far the most common method of immobilisation is by entrapment, either within a polymeric gel or within a solid support (for reviews see 11 and 12). The gels which have been employed for plant cell immobilisation include calcium alginate, agarose, agar, polyacrylamide, gelatin and carrageenan, of which the first two have generally been found to retain the greatest degree of viability and respiratory and biochemical activity of the cells [13-15]. Because of the mild conditions employed alginate has received most attention, however, high concentrations of Ca^{2+} are required for polymerisation and low phosphate must also be employed. Furthermore, the long term stability of the matrix is poor as cell division leads to break up of the alginate beads.

Techniques for immobilisation which eliminate the need for a gel matrix and depend on an inert mechanical support for the growing culture have recently been developed. The first of these systems which is similar to that employed in some animal cell cultures uses a hollow-fibre reactor in which the cells are trapped in a chamber which is transversed by a large number of semi-permeable membrane tubes through which the medium is pumped [16,17].

The approach used in this laboratory for immobilisation of plant cells in a supporting matrix involves passive entrapment within reticulate polyurethane foam [18]. Blocks of sterilised foam 1 cm^3 are added to suspension cultures and as the cells grow they invade the foam, divide and become entrapped within the pores of the matrix. The advantages of the technique are that the matrix is stable, non-phytotoxic, autoclavable and cheap, and as it is 97% (v/v) void it presents a negligible barrier to permeability. Scale-up of this culture

Species	Matrix	Product of interest	Primary location	Reference
Amaranthus tricolor	Ch	oxalate	M	Knorr & Teutonico (1986)
Asclepias syriaca	Ch	proteases	M	Knorr *et al*. (1985)
Beta vulgaris	My	betacyanin	C	Rhodes *et al*. (1985)
Capsicum frutescens	Pf	capsaicin	M	Lindsey & Yeoman (1984)
Catharanthus roseus	Ag	cathenamine	C	Felix *et al*. (1981)
Coffea arabica	Al	methylxanthines	-	Haldimann & Brodelius (1986)
Daucus carota	Al	phenolics	M	Hamilton *et al*. (1984)
Digitalis lanata	Al	β-methyldigoxin	M	Alfermann *et al*. (1983)
Glycine max	Hf	phenolics	-	Schuler (1981)
Glycyrrhiza echinata	Al	echinatin	M	Ayabe *et al*. (1986)
Hyoscyamus muticus	Pf	atropine	C	Authors (unpub. observ.)
Lavandula vera	Al	pigments	M	Nakajima *et al*. (1985)
Mentha sp.	Pa	neomenthol	M	Galun *et al*. (1983)
Morinda citrifolia	Al	anthraquinones	C	Brodelius *et al*. (1979)
Mucuna pruriens	Al	L-DOPA	M	Wichers *et al*. (1983)
Nicotiana sylvestris	Pa	monoterpene alcohols	M	Galun *et al*. (1985)
Nicotiana sp.	Al	cinnamoyl putrescines	C	Berlin (1985)
Papaver somniferum	Al	codein	M	Furuya *et al*. (1984)
Salvia miltiorrhiza	Al	diterpines	M/C	Miyasaka *et al*. (1986)
Solanum aviculare	Sb	steroid glyco-alkaloids	M	Jirku *et al*. (1981)

Table 1. The production of phytochemicals by immobilised plant cell systems. Al (alginate); Ag (agarose); Pa (polyacrylamide); Ch (chitosan); Ny (nylon); Pf (polyurethane foam); Hf (hollow fibre); Sb (surface binding).

C = cells M = medium

system is also possible and bioreactors have been developed which are suitable for foam immobilised cells [19].

Immobilised cell systems in the production of plant secondary products

Immobilised plant cell cultures have a number of advantages over freely suspended cells in the practical field of bioreactor configuration [6,19] and in the physiological state of the cells during the culture period [12]. The high degree of cell aggregation which arises in immobilised systems, results in greater cell-cell contact and possibly communication which act to reduce the rate of cell division and permit the establishment of gradients of nutrients and growth substances which are important in the regulation of secondary metabolic activity. The use of fixed bed reactors which is facilitated by the use of immobilised cells [19] also lends practical advantages to the employment of immobilised cells. Cultural conditions can be changed at will (e.g. from full growth, to nutrient limited, media) in order to shift the balance from conditions favouring biomass production to those favouring secondary product accumulation. The use of circulating medium in a fixed bed reactor also allows the re-use of biomass [12] as well as making possible continuous product removal which in itself leads to increased yields in some systems (see below).

Constraints to the employment of immobilised cultures

The major constraint and limitation to the use of immobilised cultures in the production of secondary metabolites is the requirement for the product to be released into the medium. If the product is retained within the cells the advantages of the fixed bed reactor with circulating medium and continuous product removal no longer apply. However, the limitation to the use of immobilisation technology may prove to be less of a problem than at first anticipated, as Hall *et al* (1987) have shown that the primary location of the product in the majority of reported examples of immobilised cultures is in the medium (Table 1). Furthermore, in several instances the immobilisation of the cells results in a preferential partitioning of the product into the culture medium (Brodelius, pers. comm.).

Permeabilising agents which can cause product release with limited loss of viability have also been considered as part of a culture process [20]. The use of these agents is, however, outwith the scope of this article but the reader is referred to the review of Parr *et al*. [20].

Increased product yield from immobilised cultures

Given the constraints of the need for product release, several techniques have been developed which can lead to increased yields of desired secondary products in cultures. All of the following examples are taken from the experience in this laboratory in the production of capsaicin in cultures of *Capsicum frutescens* but the authors believe that with a good understanding of biosynthetic pathways, these general principles can be applied to any desired plant secondary product. The selection of high yielding cell lines and their stability over prolonged periods of culture is obviously of great importance in successful process development but discussion of these subjects is, however, outwith the scope of this article.

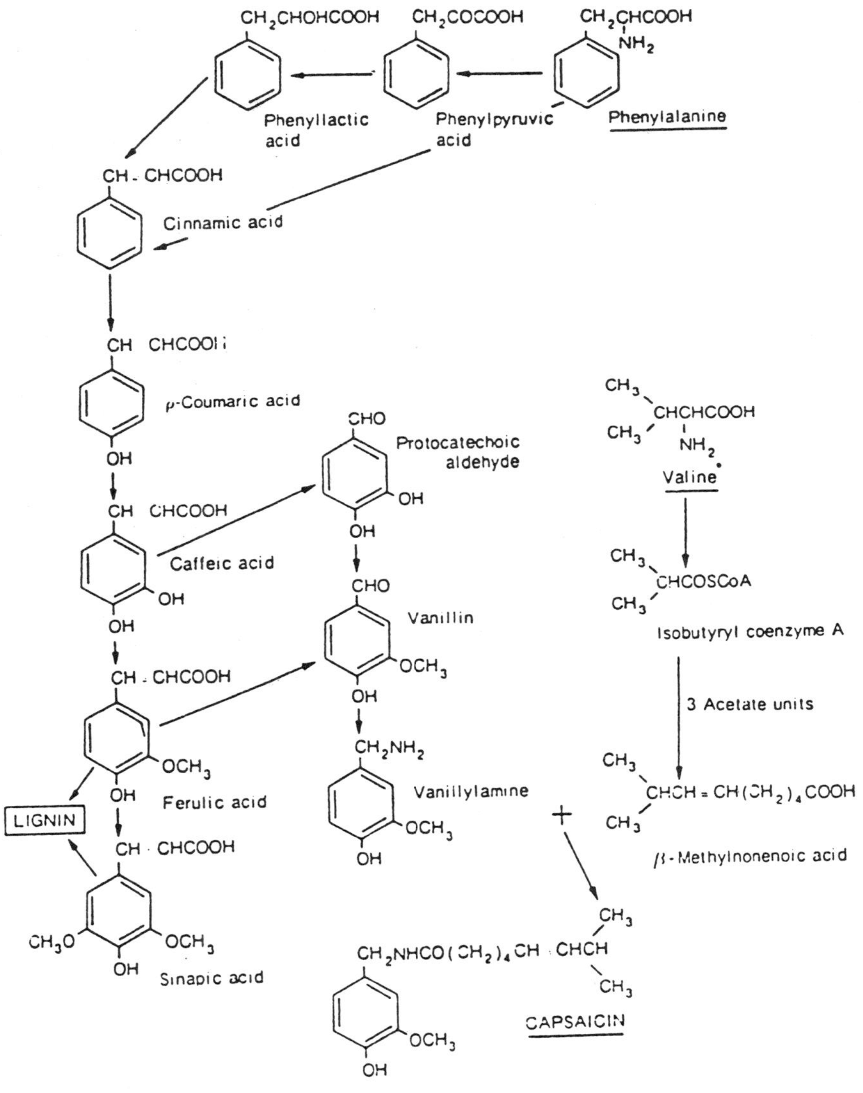

CH2CHOHCOOH
CH2COCOOH
CH2CHCOOH
NH2
Phenyllactic acid
Phenylpyruvic acid
Phenylalanine
CH - CHCOOH
Cinnamic acid
CH CHCOOH
p-Coumaric acid
OH
CHO
Protocatechoic aldehyde
OH
OH
CH CHCOOH
Caffeic acid
OH
OH
CHO
Vanillin
OCH3
OH
CH - CHCOOH
CH2NH2
OCH3
OH
Ferulic acid
LIGNIN
Vanillylamine
OCH3
OH
CH - CHCOOH
CH3O
OCH3
OH
Sinapic acid
CH3
CHCHCOOH
CH3
NH2
Valine
CH3
CHCOSCoA
CH3
Isobutyryl coenzyme A
3 Acetate units
CH3
CHCH = CH(CH2)4COOH
CH3
β-Methylnonenoic acid
+
CH3
CH2NHCO(CH2)4CH CHCH
CH3
CAPSAICIN
OCH3
OH

Figure 1. Proposed biosynthetic pathway for capsaicin from phenylalanine and valine. (After Yeoman et al. 1980)

Reduction in, or alteration to primary metabolism

From Fig. 1 it can be seen that one branch of the capsaicin biosynthetic pathway is derived from phenylalanine and one from valine. As these compounds are amino acids, they are largely incorporated into proteins when cultured under conditions which favour rapid cell growth [5,22]. If, however, cultures are grown under conditions of nutrient limitation which reduce growth rate, larger pools of these precursors are available for secondary metabolite production. Conditions which limit growth (i.e. absence of nitrate, phosphate and growth regulators) can therefore be used in order to increase product accumulation. Table 2 shows that incorporation of radioactivity from ^{14}C phenylalanine is directly proportional to the initial nitrate concentration in the media. Thus at high nitrate levels incorporation into protein is high and incorporation into capsaicin is low and *vice versa*. The preferential production of the desired secondary product can therefore be achieved by using conditions of nutrient limitation [23]. Although not a universal observation, the inverse relationship between cell growth and product formation has been observed in several systems [12].

Table 2. The effect of initial nitrate concentration on the incorporation of [^{14}C]Phe into soluble protein and capsaicin in cultures of immobilized pepper cells after 9 d. Cultures (20 ml) were supplied with 74 kBq of L-[U-^{14}C]Phe for 24 h. Results represent the mean of three replicates $\pm$ SE.

Initial NO_3 (mM)	Radioactivity in soluble protein		Radioactivity in capsaicin	
	mean (Bq g^{-1} DW)	% Full	mean (Bq g^{-1} DW)	% Full
0	4481 $\pm$ 695	34	484 $\pm$ 68	272
5	5892 $\pm$ 655	45	375 $\pm$ 83	210
10	8646 $\pm$ 374	66	166 $\pm$ 45	93
25	13044 $\pm$2784	100	178 $\pm$ 33	100
50	23870 $\pm$1814	183	111 $\pm$ 19	62

After Lindsey [23]

As the supply of phenylalanine has been shown to limit capsaicin accumulation in immobilised cultures, other techniques can be developed to increase the supply of this key precursor. Cell lines have spontaneously arisen in cultures of different species which produce greatly elevated levels of certain amino acids including phenylalanine. These over-producers can be conveniently selected by growing treated or non-treated calli on media containing synthetic analogues of phenylalanine. As the analogues are toxic, cultures which survive may have elevated levels of endogenous phenylalanine. The selection of phenylalanine over-producers may therefore result in cell lines with increased capsaicin accumulation [24].

Increased activity of an enzymic step between primary and secondary metabolism

Phenylalanine ammonia-lyase (PAL) converts phenylalanine to cinnamic acid in the first step of the phenylpropanoid branch of the capsaicin biosynthetic pathway. This enzyme is considered to be a key switchpoint between primary metabolism and the phenolic pathways of secondary metabolism, and several examples can be cited of increased enzymic activity leading to increased product synthesis [25]. PAL activity can be induced by many environmental factors, including low temperature, illumination with UV light, dilution of the biomass and the addition of fungal elicitors [25]. Of these treatments, the use of fungal elicitors has found favour with an increasing number of workers.

The use of elicitors is especially relevant if the product of interest is a phytoalexin such as phaseolin, but other compounds which are not thought to be phytoalexins can also show increased yields in response to elicitor treatment.

In the example of *Capsicum* cultures, it has been shown that spores of *Monilinia fructicola* can increase both capsaicin accumulation and incorporation of radioactive precursors into the product (Lindsey and Holden, unpublished observations). Present studies are under way to determine if this effect is mediated by increasing PAL activity.

Inhibition of competing secondary reactions

It has been shown that a high proportion (up to 63%) of supplied phenylalanine is not incorporated into capsaicin but is instead utilised in cell wall synthesis, either as esterified hydroxy-cinnamic derivatives or into lignin [22]. It may prove possible to divert intermediates into capsaicin synthesis by a number of treatments.

If excess product of a competing pathway is supplied to cultures, the equilibrium of the reaction can be shifted in favour of substrate accumulation. Sinapic acid which is derived from ferulic acid (an intermediate in the capsaicin pathway [Fig. 1]) can be fed to *Capsicum* cultures and results in increased capsaicin accumulation [19]. Feedback inhibition probably results in reduced demand for ferulic acid which then becomes available for capsaicin biosynthesis.

Exploitation of biotransformations

One or two step reactions may result in the synthesis of a high-value compound from a low-value precursor. Simple biotransformations such as the glycosylation of digitoxin to purpurea glycoside A and the hydroxylation of β-methyldigitoxin to β-methyldigoxin can be performed over long culture periods in immobilised cells [26]. The L-tyrosine to L-Dopa biotransformation can also be successfully achieved with cultured immobilised cells of *Mucuna puriens* [27]. This use of immobilised cultures is obviously only economically viable if the product is substantially more expensive than the supplied substrate and if the product is released into the culture medium. In the capsaicin pathway the ferulic acid to vanillin biotransformation (Fig. 1) meets these criteria. Vanillin is a high cost flavouring used extensively in the food and drinks industries and therefore has the potential to be exploited in culture systems. In our laboratory cell lines have been established which preferentially release this product into the medium and the full biosynthetic potential of these cultures is presently under investigation.

Continuous product removal

Continuous removal of capsaicin from the circulating medium which is possible in the fixed bed bioreactor described by Lindsey and Yeoman [28] and scaled up by Mavituna *et al*. ([19] and this volume) can increase product yield in two ways.

Firstly, removal of the product reduces the possibility of feedback inhibition on the final step(s) of the pathway and maintains the equilibrium of the final enzyme(s) in the direction of product synthesis. This is consistent with the finding that capsaicin fed to *Capsicum* cultures results in decreased product synthesis (Hall, unpublished results). Therefore it appears that the partitioning of product from the circulating medium can result in an increased yield per unit biomass in a set time [21].

The second advantage of continuous product removal is that further metabolism of the desired compound can be prevented. Capsaicin is not subject to degradation in culture but certain products of interest are broken down by the cultures that synthesise them. For example nicotine is further metabolised to nornicotine and then to other alkaloids [29]. Therefore, continuous product removal may reduce the possibility of further metabolism of the desired product and therefore result in increased yields.

CONCLUSIONS

Product yield from suspension cultures is usually disappointingly low in a few cases where empirical investigations have resulted in significantly increased accumulation of desired secondary products. However, if successful and repeatable procedures are to be developed which make the commercial exploitation of tissue culture viable on a large scale, a thorough knowledge of the particular biosynthetic pathway of interest is vital. In this article we have presented a series of approaches, which although it is by no means exhaustive, has proved successful in increasing capsaicin production in *Capsicum* cultures. Several or all of these approaches may be applicable to the production of secondary products in cultures of different species. However, in deciding which, if any, method should be adopted in order to optimise product yield, an understanding of the regulation of the desired pathway is essential. This information can only be obtained by basic biochemical and molecular biological research on the particular biosynthetic pathway of interest.

Acknowledgements

The authors wish to thank Mrs. E.A. Raeburn and Mrs. J. Summers for typing and processing this manuscript. We would also like to thank Albright & Wilson Ltd. and Glaxo Group Research Ltd. for financial support.

REFERENCES

1. Fowler, M.W., Commercial applications and economic aspects of mass plant cell culture. In Plant Biotechnology, eds. S.H. Mantell and H. Smith. Society for Experimental Biology Seminar Series 18, Cambridge University Press, Cambridge, 1983, pp. 3-37.

2. Tanaka, H., Technological problems in cultivation of plant cells at high density. Biotechnol. Bioeng., 1981, 23, 1203-1218.

3. Zenk, M.H., El-Shagi, H. and Stockigt, J., Formation of the indole alkaloids serpentine and ajmalicine in cell suspension cultures of Catharanthus roseus. In Plant Tissue Culture and its Biotechnological Applications, eds. W. Barz, E. Reinhard and M.H. Zenk, Springer-Verlag, 1977, pp. 27-43.

4. Knobloch, K.H. and Berlin, J., Influence of the medium composition on the formation of secondary compounds in cell suspension cultures of Catharanthus roseus (L.) G. Don. Z. Naturforsch, 1980, 35, 551-556.

5. Lindsey, K. and Yeoman, M.M., The relationship between growth rate, differentiation and alkaloid accumulation in cell cultures. J. Exp. Bot., 1983, 34, 1055-1065.

6. Berlin, J., The use of immobilised plant cells - an evaluation. IAPTC Newsletter, 1985, 46, 8-14.

7. Fuller, K.W. and Bartlett, D.J., The chemosynthetic potential of plants and its realisation by immobilised systems. Ann. Proc. Phytochem Eur., 1985, 26, 229-247.

8. Jirku, V., Macek, T., Vanek, T., Krumphanzl, V. and Kubauez, V., Continuous production of steroid glycoalkaloids by immobilised plant cells. Biotechnol. Lett., 1981, 3, 447-450.

9. Rhodes, M.J.C. and Kirsop, B.H., Plant cell cultures as sources of valuable secondary products. Biologist, 1982, 29, pp. 134-140.

10. Bornmann, C.H. and Zachrisson, A., Immobilisation of protoplasts by anchoring to microcarriers. Pl. Cell. Rep., 1982, 1, 151-153.

11. Hall, R.D., Holden, M.A. and Yeoman, M.M., The immobilisation of higher plant cells and protoplasts. In Biotechnology in Agriculture and Forestry, ed. Y.P.S. Bajaj, Springer-Verlag, Berlin, 1987.

12. Yeoman, M.M., Techniques, characteristics, properties and commercial potential of immobilised plant cells. In Cell Culture and Somatic Cell Genetics, ed. F. Constabel, Academic Press, 1987, in press.

13. Brodelius, P. and Nilsson, K., Entrapment of plant cells in different matrices. FEBS Letters, 1980, 322, 312-316.

14. Brodelius, P., Production of biochemicals with immobilised plant cells. Ann. New York Acad. Sci., 1983, 413, 383-393.

15. Nakajima, H., Sonomoto, K., Usui, N., Sato, F., Yamada, Y., Tanaka, A. and Fukui, S., Entrapment of Lavandula vera cells and production of pigments by entrapped cells. J. Biotechnol., 1985, 2, 107-112.

16. Prenosil, J.E. and Pedersson, H., Immobilised plant cell reactors. Enzyme and Microbial Technol., 1983, 5, 323-331.

17. Schuler, M.L., Sahai, O.P. and Hallsby, G.A., Entrapped plant cell cultures. Ann. New York Acad. Sci., 1983, 413, 373-412.

18. Lindsey, K., Yeoman, M.M., Black, G.M. and Mavituna, F., A novel method for the immobilisation and culture of plant cells. FEBS Lett., 1983, 155, 143-149.

19. Mavituna, F., Park, J.M., Wilkinson, A.K. and Williams, P.D., Characteristics of immobilised plant cell reactors. In Plant and Animal Cells, Process Possibilities, eds. C. Webb and F. Mavituna, Ellis Horwood, Chichester, 1987, pp. 92-114.

20. Felix, H., Permeabilised cells. Anal. Biochem., 1982, 120, 211-234.

21. Parr, A.J., Robins, R.J. and Rhodes, M.J.C., Release of secondary metabolites by plant cell cultures. In Plant and Animal Cells, Process Possibilities, eds. C. Webb and F. Mavituna, Ellis Horwood, Chichester, 1987, pp. 229-237.

22. Holden, M.A., Hall, R.D. and Yeoman, M.M., Limitations to product yield in rapidly growing cultures of Capsicum frutescens. Biochem. Soc. Trans., 1987, in press.

23. Lindsey, K., Manipulation by nutrient limitation of the biosynthetic activity of immobilised cells of Capsicum frutescens Mill. cv. annuum. Planta, 1985, 165, 126-133.

24. Lindsey, K. and Yeoman, M.M., The synthetic potential of immobilised cells of Capsicum frutescens Mill. cv. annuum. Planta, 1984, 162, 495-501.

25. Jones, D.H., Phenylalanine ammonia-lyase: regulation of its induction and role in plant development. Phytochem., 1984, 23, 1349-1359.

26. Alfermann, A.W., Schuller, I. and Reinhard, E., Biotransformation of cardiac glycosides by immobilised cells of Digitalis lanata. Planta Med., 1980, 40, 218-223.

27. Wichers, H.J., Malingre, T.M. and Huizing, H.J., Optimisation of the biotransformation of L-tyrosine into L-dihydroxyphenylalanine (DOPA) by alginate-entrapped cells of Mucuna pruriens. Planta, 1985, 166, 421-428.

28. Lindsey, K. and Yeoman, M.M., Novel experimental systems for studying the production of secondary metabolites by plant tissue cultures. In Plant Biotechnology Soc. Exp. Biol. Seminar Series, **18**, eds. S.H. Mantell and H. Smith, Cambridge University Press, 1983, pp. 39-66.

INTERNATIONAL CONFERENCE ON BIOREACTORS AND BIOTRANSFORMATIONS
GLENEAGLES, SCOTLAND, UK: 9-12 NOVEMBER 1987

Paper A2

GROWTH AND PRODUCT FORMATION BY PLANT CELL SUSPENSIONS CULTIVATED IN BIOREACTORS

Alan H. Scragg, Pamela Bond, Fiona Leckie, Ray Cresswell, and Michael W. Fowler
Wolfson Institute of Biotechnology
University of Sheffield
Sheffield. S10 2TN
U.K.

and

Eunice J. Allan
Bacteriology Division
School of Agriculture
University of Aberdeen
Aberdeen. AB9 1UD
U.K.

ABSTRACT

Plant cell culture has been proposed as an alternative supply of phytochemicals to the normal plantation system. This alternative requires the growth of plant cell suspensions in bioreactors. Plant cell suspensions differ from microbial suspensions in a number of characteristics which effect their growth in bioreactors. These differences are illustrated by reference to cultures of *Catharanthus roseus*, *Helianthus annuus* and *Picrasma quassioides*.

INTRODUCTION

Plant cell culture has been proposed as an alternative supply to plantation crops for a number of plant-derived chemicals such as quinine and morphine [1-3]. Such a system could provide a product of consistant quality and quantity, free from the constraints of season, climate, disease or political considerations. Any new process will be initially capital expensive, at least initially, such that any first generation plant cell culture process will necessarily concentrate on low volume high value products. No process will be adopted unless it can compete on cost, or confer some other real advantage over existing sources. A further potential is that plant cell culture could act as a supply of completely novel substances opening up new product and market areas. Whatever the case, either situation would require the cultivation of plant cell suspensions in bioreactors, and with as high a productivity of the required compound as possible. Productivity measured as grams product/litre culture/day is influenced by the growth rate of the organism, the product yield and overall bioreactor run time. The highest productivityso far recorded for a plant cell culture, is the production of rosmarinic acid by Coleus cells at 0.91 g/l/day [4] which compares quite well with microbial products such as penicillin (1.4-2.1 g l^{-1}, d^{-1}) but is lower than that for citric acid (30-37.5 g l^{-1}, d^{-1}) and monosodium glutamate (53-66 g.l^{-1}, d^{-1}) [5].

This paper is specifically concerned with the difference between plant and microbial cells and their growth and productivity performance in bioreactors.

MATERIALS AND METHODS

The cultures of Catharanthus roseus C87 and ID1, Helianthus annuus, and Picrasma quassioides were maintained and grown as described [6,7]. The cultures were grown in air-lift bioreactors of 7 and 30 litre working volume and a stirred-tank of 3 litre as described [8]. Cultural parameters such as dry weight and pH were determined as described by Stafford et al. [9].

Size distribution : The distribution of aggregate sizes in cell suspensions were determined by gently washing 100 ml of culture through a range of nylon meshes of known pore size (Simon Ltd., Stockport). The meshes were dried, and the results presented as the percentage of the total dry weight.

Shear sensitivity : Shear sensitivity of a cell culture was determined by exposing the culture to shear in a 3 litre stirred-tank bioreactor of a standard design, fitted with a six blades turbine impeller. The bioreactor was sterilised empty, and the cells added later. The culture was exposed to a fixed impeller speed (400-1,000 rpm) at 25°C, and with an aeration rate of 100 ml. min^{-1}. At intervals sterile samples were removed, and 3 x 20 ml aliquots used to inoculate 100 ml of fresh medium. After 14 days incubation, growth was estimated by wet and dry weight measurements.

Settling rate : The rate of settling of plant cells was determined by following the decrease in optical density at 600 nm.

RESULTS

The main differences between plant cell suspensions and microbial cells and some consequences for their growth in bioreactors are outlined in Table 1. These are described as follows:

TABLE 1

The differences between plant cell suspensions and microbial cultures

Feature	Microbial culture	Plant Cell suspension	Consequence for plant cell culture
Size	1-10 μm in length	40-100 μm in length	rapid sedimentation; possibly shear sensitivity
Individual cells	often	infrequently, mostly aggregates	rapid sedimentation; microenvironment; low viscosity
Growth rate	rapid, doubling times of hours	slow, doubling times of days	long culture runs; increased danger of infection
Aeration	high requirement for aerobic processes	low requirement	oxygen supply not critical, can use low Kla bioreactors
Shear sensitivity	mostly insensitive	sensitive	slow stirrer speeds; alternative stirrer designs; use of air-lift bioreactors
Product formation	often into medium	into vacuole	cells require harvesting and extraction

Size

One of the most obvious differences between plant cell suspensions and microbial cells is the much larger size of plant cells. Microbial cells are typically 1-10 µm in length whereas plant cells can be from 40 to 100 µm in length. The microbial cell often occurs as a single cell, but it is rare to find a single viable plant cell in suspension. Instead plant cells in suspension occur as aggregates of various sizes, probably formed by the non-separation of the cells after division. These cell aggregates can contain from 10 to many thousand cells and range in size from 140 µm in diameter to above 1,000 µm (Figure 1). The consequence of

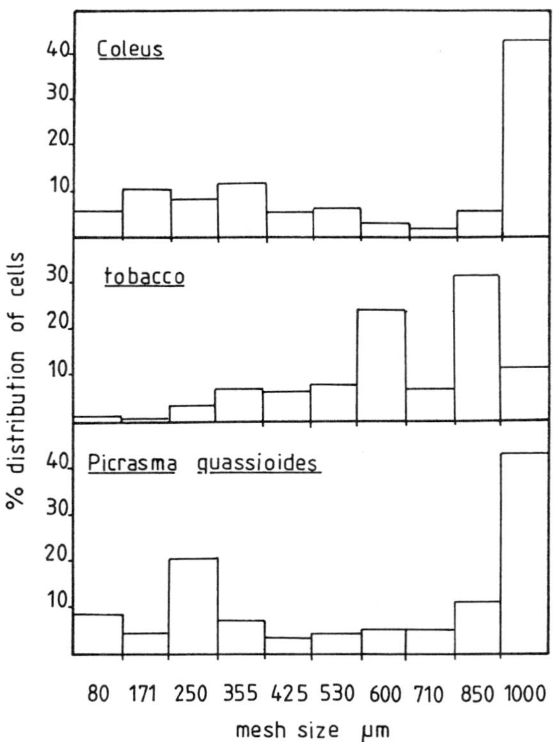

Figure 1 The size distribution of cell aggregates in cell suspensions of *Coleus*, tobacco and *Picrasma quassioides*

this aggregate formation is that plant cells will settle out of suspension very rapidly if mixing is stopped (Figure 2), and the aggregates make it difficult to take representative samples. In addition, the very large

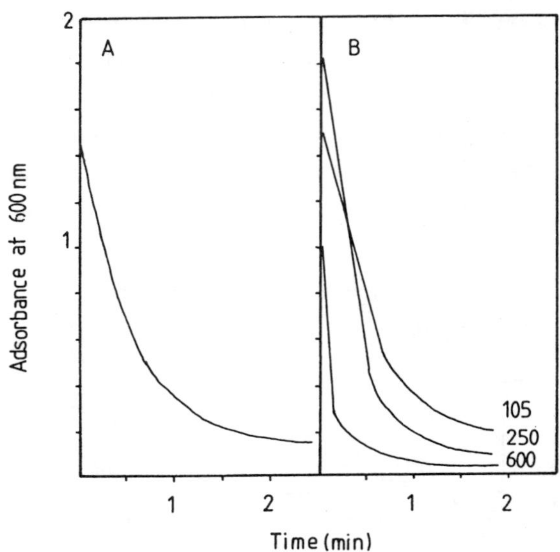

Figure 2 The settling rate of cell aggregates of *Picrasma quassioides* A, total suspension; B, aggregates of above 600 um, between 600 and 250 um, and between 250 and 105 um.

aggregates can block sampling ports or restrict the transfer of suspension from one bioreactor to another. Schuler [10] has proposed that the presence of aggregates in a plant cell suspension means that a proportion of the cells, those in the centre of the aggregate will be exposed to a different environment. The environment will differ due to a diffusional barrier to the import of nutrients such as oxygen and sucrose and export of metabolic by-products. It has been suggested that these types of environmental differences may be responsible for alterations in secondary product formation. The size distribution of cell aggregates can alter during the growth cycle (Figure 3) or can be altered depending upon the bioreactor used to cultivate the plant cells. (Figure 4). The change in size distribution of cell aggregates of Helianthus annuus suspension cultures cultivated in 2 litre flasks, a 7 litre and 80 litre (working volume) air-lift bioreactor, and a 3 litre stirred-tank bioreactor are shown in Figure 4. The stirred-tank reduced the mean aggregate size but did not alter the growth rate.

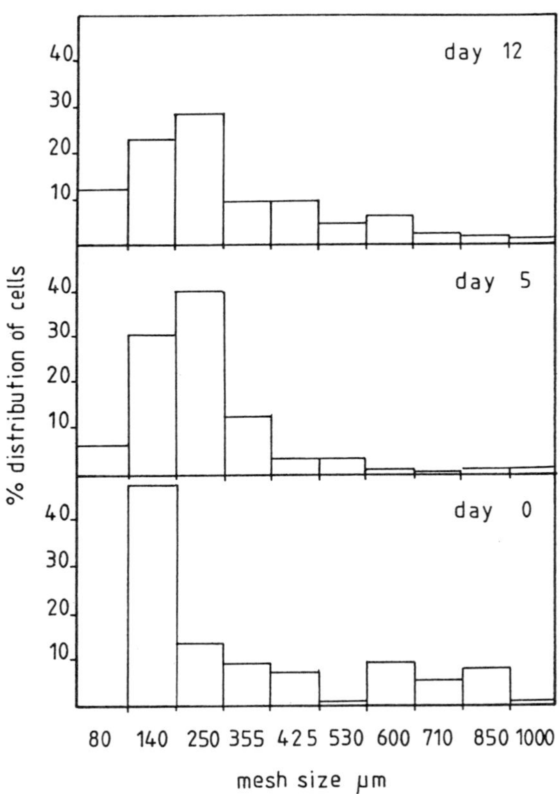

Figure 3 The size distribution of cell aggregates in a suspension of Catharanthus roseus grown in a 250 ml shake flask at various times after initiation.

Growth Rate

In general microbial cells have rapid growth rates with doubling times measured in hours, whereas in plant cell suspensions this is measured in days. One of the fastest growing cultures has been tobacco with a doubling time of 18 hours but a more typical time would be from 2-3 days [11]. This means that a bioreactor run of a plant cell suspension

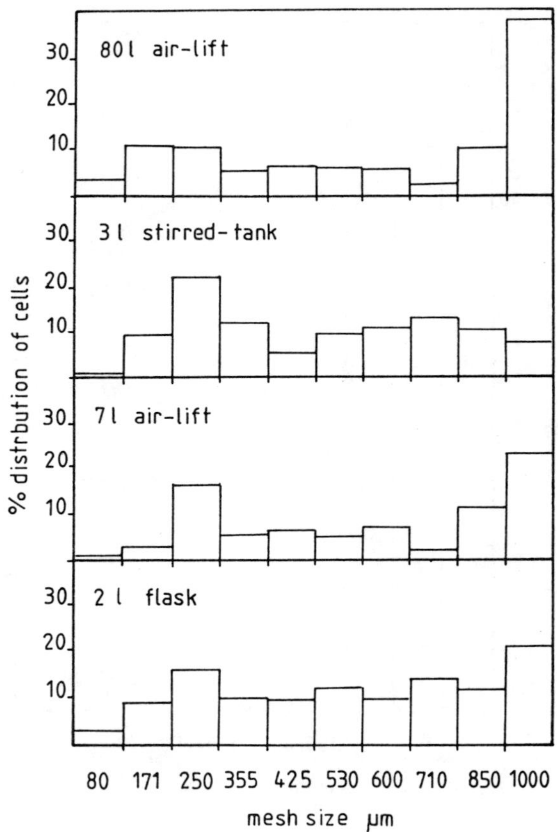

Figure 4 The size distribution of cell aggregates of a suspension of *Helianthus annuus* grown in various bioreactors.

can be from 1-4 weeks which means a much greater difficulty in maintaining sterility as the slow growth of plant cells means that almost any contamination will outgrow the culture. Water loss may be a significant problem with these long runs. In the development of a plant cell culture process the long bioreactor runs mean a considerable reduction in the number of runs possible per year and hence a reduction in overall productivity.

Aeration

The slow growth of plant cells is a reflection of their lower overall metabolic activity. This low activity is not due to slow metabolic processes but is probably a consequence of the effect of a very large vacuole found in most plant cells which reduces the cytoplasm to a very small percentage of the total volume of any cell. The lower overall metabolic activity means a reduced oxygen demand. Plant cells require 1-10mM O_2 l^{-1}. hr^{-1} compared with values of 10-300mMO_2 l^{-1}. hr^{-1} for microbial cells. A typical growth profile is shown in Figure 5. The reduced oxygen demand means that aeration is not as critical a feature in the bioreactor cultivation of plant cells, indeed with C.roseus cultures little inhibition of growth is seen until the dissolved oxygen drops below 25% saturation. The lower oxygen demand allows alternative bioreactor designs to be considered, for the cultivation of plant cells, for example air-lift designs. A note of caution, however, although the oxygen supply for plant cells appears not to be as critical, other gaseous components may be. It has been shown that high aeration rates causes inhibition of the growth of C.roseus by stripping off carbon dioxide rather than supplying excess oxygen [12].

The supply of gaseous components can not only affect growth, there is evidence for an effect on product formation. A cell line of C.roseus C87 was developed [6] which was capable of growth-linked serpentine production. This cell line was capable of producing high levels of serpentine in shake flasks after 21 days growth. However, when the cell line was grown in a low aspect ratio (1.6:1) airlift bioreactor of 7 litre working volume or a high aspect ratio L.H.E. (12:1) air-lift bioreactor of 30 litre working volume it lost the ability to produce serpentine (Figure

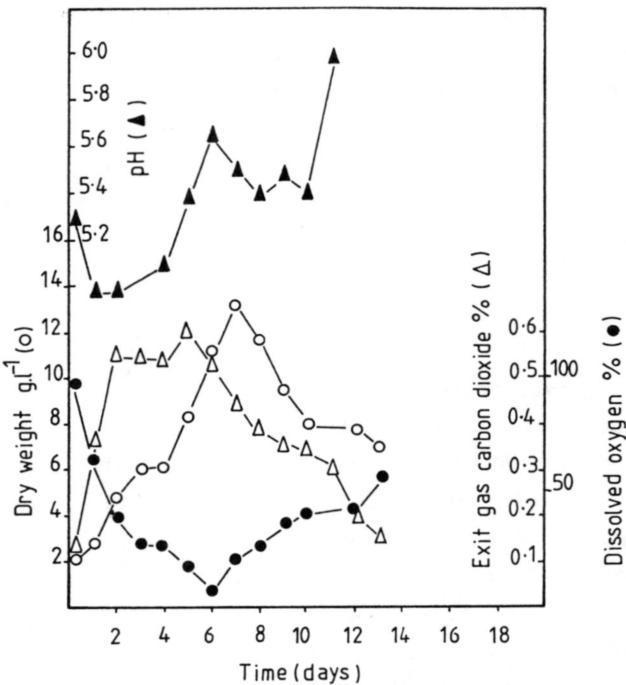

Figure 5 The growth of Catharanthus roseus suspension culture in a 30 litre L.H. Engineering air-lift bioreactor.

6). The loss of serpentine producing ability was clearly not related to the bioreactor design or due a permanent loss of the ability, as cells removed from the bioreactors of various time were able to form serpentine in shake flasks. The loss is perhaps a feature of particular cell lines as a second C.roseus line C87N also lost serpentine production when grown in bioreactors, but a related selected line C.roseus ID1 [13] did not show this phenomenon (Figure 6). In all cases the growth of the C.roseus lines in bioreactors was as good if not better than that in shake flasks. The loss of serpentine formation must be related to the differences in cultural conditions between shake flasks and bioreactors, most probably involving gaseous components such as oxygen, carbon-dioxide or even ethylene. The effects of oxygen and carbon dioxide levels on secondary product formation of C.roseus suspension cultures is at present under investigation.

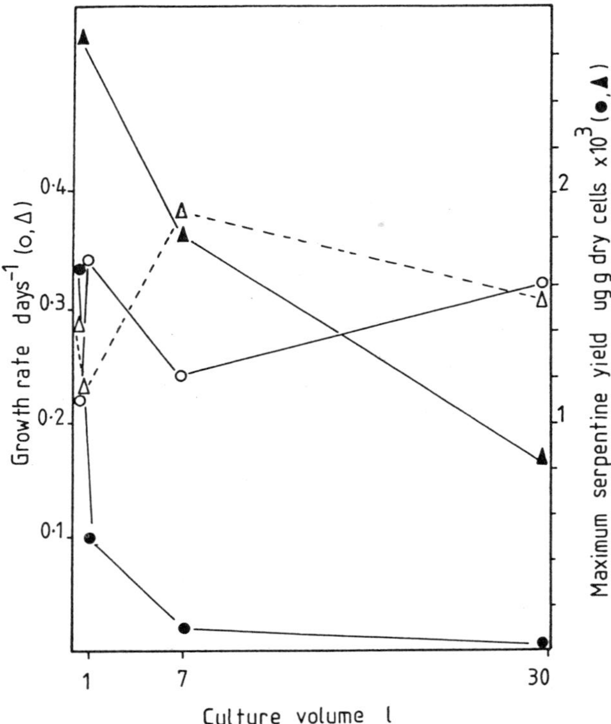

Figure 6 The effect of culture size on growth rate and serpentine formation by Catharanthus roseus suspension cultures. The volumes of 0.1 and 1 litre represent shake flasks, whereas 7 l. and 30 l. are low and high aspect ratio airlift bioreactors respectively. (O,Δ) C.roseus C87N ; (●,▲) C.roseus ID1

Shear Sensitivity

The large size, rigid cell wall, and large vacuole have been proposed as the reason that plant cells in suspension are sensitive to shear [4]. The sensitivity of plant cells to shear was suggested as the reason why plant cells, other than tobacco cells, proved difficult to grow in stirred-tank bioreactors. We have been trying to determine the shear sensitivity of a number of plant cell lines by exposing them to shear and subsequently following their ability to grow [6]. A summary of the data is shown in Table 2. It is clear that the plant cell suspensions are more resistant that was at first expected. The C.roseus lines C87 and ID1 and H. annuus having been shown to survive 5 hours of shear (shear rate 167

TABLE 2

The effect of shear on plant cell suspensions

Cell line	Shear rate average	maximum (sec^{-1})	Loss of viability after 5 hours exposure	Ability to grow a stirred-tank bioreactor
Catharanthus roseus C87N	167	1500	NIL	+
Catharanthus roseus ID1	167	1500	NIL	+
Helianthus annuus	167	1500	NIL	+
Picrasma quassioides				
(7.6.85)	73	660	lost after 2.5 hrs	-
(21.5.86)	167	1500	NIL	+

sec^{-1}) were subsequently grown in a small stirred-tank bioreactor (3 litre working volume) at 150 rpm. It was perhaps expected that the C.roseus C87 line would be more resistant to shear as it had been in culture for at least 5 years. However, the C.roseus ID1 line and the H.annuus were relatively new cultures,(H.annuus 1 year 36 subcultures and C.roseus ID1six 2 week subcultures) and therefore would not be expected to be resistant. The resistance to shear does appear to be related to the cultural characteristics such as growth rate, aggregate formation of the cell line. This can be seen with the Picrasma quassioides culture, which was initially sensitive but some 1 year later was resistant. This resistance appeared associated with an improvement in growth rate of the P.quassioides culture and the reduction in aggregate size distribution. Methods of achieving good mixing while having low shear conditions have involved the reduction of stirrer speed or to adopt the air-lift bioreactor [13,14]. Alternative designs of stirrer such as a helical screw have also been used [4].

Product Formation

Secondary products formed by plant cell cultures are normally stored in the vacuole and not exported into the medium. Thus plant cells require harvesting and extraction to obtain the product. The kinetics of product formation also greatly influence the nature of a plant cell culture process [15]. Secondary product formation in plant cell suspension generally occurs after growth has ceased or requires a change in cultural conditions. Some of the most successful plant cell systems that have been developed using a two stage process and some of these are shown in Table 3. Although successful these systems have the disadvantages of very long run times because of the need to change medium.

TABLE 3

Types of system using plant cell culture processes

Cell line	Product	Type of Process	First stage	Second stage
Lithospermum erythrorhizon	shikonin	two-stage	batch stirred tank	batch stirred tank
Coleus blumei	rosmarinic acid	two-stage	fed-batch culture spiral stirred tank	batch spiral stirred tank
Nicotiana tabacum	geraniol	two-stage	batch stirred tank	batch stirred tank
Panax ginseng	ginseng biomass	batch	batch stirred tank	-

CONCLUSIONS

The nature of plant cell suspensions is such that their growth in bioreactors has different requirements from those of microbial suspensions. The plant cell suspensions are slow growing and aggregated and therefore the supply of oxygen appears to be less critical than mixing. The sensitivity of plant cell suspensions to shear appears to be less of a problem than was initially thought which should allow the use of conventional stirred-tank bioreactors for their cultivation. Prior to this observation low shear bioreactors such as the airlift design were regarded as the most suitable for plant cell suspensions.

ACKNOWLEDGEMENTS

The authors would like to thank the SERC Biotechnology Directorate and Plant Science Limited for their financial support.

REFERENCES

1. Fowler, M.W., Commercial applications and economic aspects of mass plant cell culture. In Plant Biotechnology, eds. S.H. Mantell and H. Smith, Cambridge University Press, London, 1983, pp. 3-37.

2. Curtin, M.E., Harvesting profitable products from plant tissue culture. Biotechnology, 1983, 1, 649-657.

3. Berlin, J., Plant cell culture - a future source of natural products. Endeavour, 1984, 8, 5-8.

4. Ulbrich, B., Wiesner, W. and Arens, H., Large scale production of rosmarinic acid from plant cell culture of Coleus blumei. In Primary and Secondary Metabolism of Plant Cell Cultures, eds. K.H. Neumann, W. Barz and E. Reinhard, Springer-Verlag, Berlin, 1985,, pp. 293-301.

5. Sahai, O. and Knuth, M., Commercialising plant tissue processes : Economics, problems and prospects. Biotechnology Progress, 1985, 1, 1-9.

6. Morris, P., Regulation of product synthesis in cell cultures of Catharanthus roseus, Planta Medica, 1986, 121-126.

7. Scragg, A.H., and Allan, E., The effect of shear on the viability of plant cell suspensions. Submitted to Enzyme and Microbial Technology

8. Scragg, A.H., Morris, P., Allan, E.J., Bond P. and Fowler, M.W., The effect of scale up on serpentine formation by Catharanthus roseus suspension cultures. Accepted Enzyme and Microbial Technology.

9. Stafford, A., Smith, L. and Fowler, M.W., Regulation of product synthesis in cell cultures of Catharanthus roseus (L) G.Don. Plant Cell Tissue Organ Culture, 1985 4, 83-94.

10. Schuler, M.L., Production of Secondary metabolites from plant tissue culture - problems and prospects. Ann. N.Y. Acad. Sci, 1981, 369, 65-79.

11. Noguchi, M., Matsumoto, T., Hirata, Y., Yamamoto, K., Katsuyama, A., Kato, A., Azechi, S. and Katoh, K. Improvement of growth rates of plant cell cultures. In Plant Tissue Culture and Its Biotechnological Applications, eds. W. Barz, E. Reinhard and M.H. Zenk, Springer-Verlag, Berlin, 1977, pp. 85-94.

12. Hegarty, P.K., Smart, N.J., Scragg, A.H. and Fowler, M.W., The aeration of *Catharanthus roseus* (L.) G.Don suspension cultures in airlift bioreactors : The inhibitory effect at high aeration rates on culture growth. *J. Expt. Bot.*, 1986, **37**, 1911-1920.

13. Cresswell, R., Selection studies on *Catharanthus roseus*. In *Secondary Metabolism in plant cell cultures*, eds. P. Morris, A.H. Scragg, A. Stafford and M.W. Fowler, Cambridge University Press, Cambridge, 1986, pp. 230-236.

14. Mandels, M., The culture of plant cells. *Advances in Biochem Engineering*, **2**, 201-215.

15. Wagner, F. and Vogelmann, H., Cultivation of plant tissue culture in bioreactors and formation of secondary metabolites. In *Plant Tissue Culture and Its Biotechnological Applications*, eds. W. Barz, E. Reinhard and M.H. Zenk, Springer-Verlag, Berlin, 1977, pp. 130-146.

INTERNATIONAL CONFERENCE ON BIOREACTORS AND BIOTRANSFORMATIONS
GLENEAGLES, SCOTLAND, UK: 9-12 NOVEMBER 1987

Paper A3

PRODUCTION OF SECONDARY METABOLITES BY IMMOBILISED PLANT CELLS IN NOVEL BIOREACTORS

F. Mavituna, A.K. Wilkinson, P.D. Williams
Department of Chemical Engineering,
U.M.I.S.T, P.O. Box 88,
Sackville Street, Manchester, M60 1QD,
U.K.

and

J.M. Park,
Department of Chemical Engineering,
University of Minnesota,
Minneapolis, U.S.A.

ABSTRACT

There is evidence that the production of secondary metabolites by plant cell cultures can increase upon immobilisation. Two types of bioreactors have been developed for the *in situ* immobilisation of plant cell aggregates in reticulate polyurethane foam matrices. These bioreactors have been used for the production of capsaicin, the hot chilli flavour, by immobilised *Capsicum frutescens* cells.

INTRODUCTION

Plant cell cultures have a significant potential as a source of many important compounds [1]. Immobilisation of plant cells can lead to increased production levels of some valuable secondary metabolites when compared with suspension cultures [2]. Some of the reasons for this increased production may be attributed to the cell-to-cell contact [3,4], differentiation [5], and induction of stress due to the concentration gradients in the immobilised cell matrices. Immobilisation also offers advantages from a process engineering point of view, especially in providing an easily manipulated, low shear environment.

Several methods are available for the immobilisation of plant cells [6]. The system used here involves physical entrapment of plant cell aggregates in the open pore network of polyurethane foam matrices [7]. The advantages of this method compared with others include the simplicity of natural entrapment, the ease of aseptic manipulation, the ease of scale-up, the good mechanical strength and the stability of immobilised cell matrices and the fact that the technique is biologically harmless.

The production of capsaicin, the pungent flavour of chilli peppers, by immobilised *Capsicum frutescens* cells, which excrete it into the medium, was used as a model system to evaluate the bioreactor performance.

REQUIREMENTS FOR AN IMMOBILISED PLANT CELL BIOREACTOR

The combination of rich growth media and slow growth rates of plant cell cultures means stringent asepsis is required to prevent the development of microbial contamination. Carrying out the immobilisation of the plant cells *in situ* in the reactor should reduce the risk of contamination by eliminating handling and transport of the immobilised cells. Since immobilisation in reticulate polyurethane foam matrices is a method based on physical entrapment through filtration it is easily carried out *in situ*.

In order to immobilise the plant cell aggregates in the reticulated polyurethane foam matrix, a suspension culture of cells is introduced aseptically into a sterile liquid medium containing the empty foam matrix. The plant cell aggregates are carried from the bulk liquid into the porous foam matrix by the liquid elements flowing through the open pore structure of the initially empty foam matrices. Once physically entrapped, the cell aggregates grow and/or adhere to each other to fill the rest of the available space within the foam.

The most important factors affecting the physical entrapment are the size distribution of the cell aggregates compared with that of the pores of the foam and the relative movement between the cell aggregates and the matrix. A typical size distribution of cell aggregates of *C.frutescens* suspension cultures is given in Fig.1. Equal amounts of this culture were contacted with foams of 10, 30 and 40 pores per inch in an apparatus designed to simulate the physical entrapment conditions prevailed in the

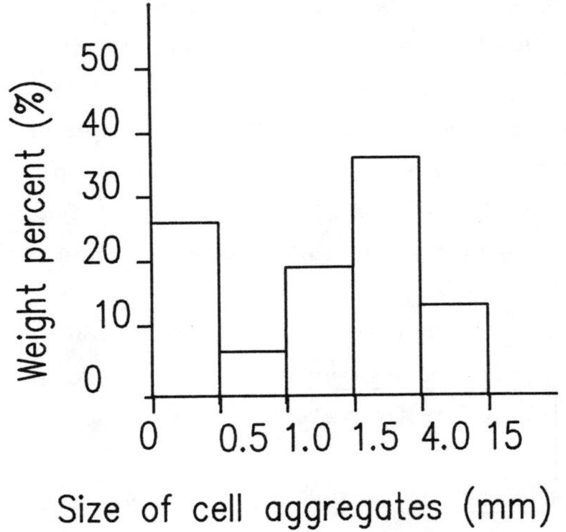

Figure 1: The size distribution of plant cell aggregates

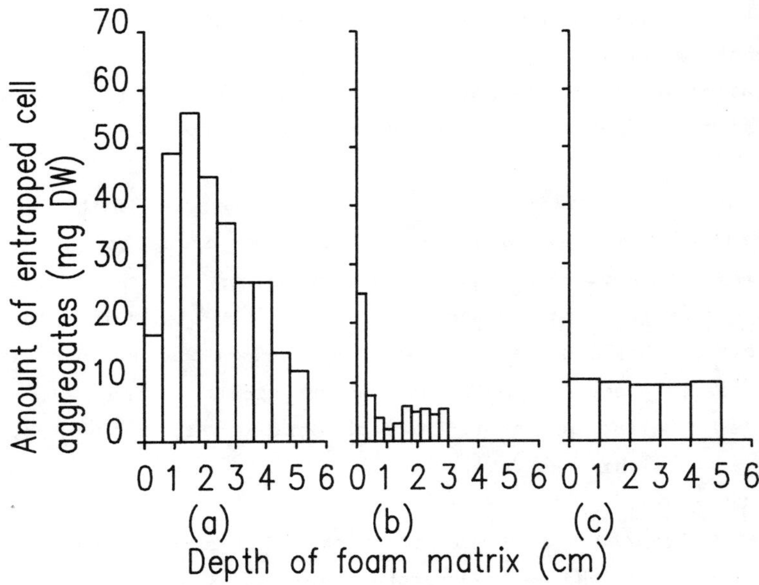

Figure 2: Entrapped-weight profiles: amount of plant cell aggregates from a suspension of cells with natural size distribution against depth of foam matrix. Average pore size of the foam: (a) 10 pores per inch, (b) 30 pores per inch, (c) 40 pores per inch

bioreactors. The results are shown in Fig.2 which indicates that the careful choice of the pore size is necessary for effective immobilisation.

BIOREACTOR CONFIGURATIONS

Two bioreactor configurations have been developed, one using reticulated polyurethane foam cubes in a circulating bed mode and the other using fixed sheets of foam.

Circulating Bed Bioreactor

This bioreactor is a 7 litre nominal capacity vessel fitted with an off-centre air sparger to induce circulation of typically one thousand 1 cm^3 foam particles in 5-6 litres of medium (Fig.3). As mentioned previously, in order to achieve efficient immobilisation it is necessary to hold the foam particles stationary in the bioreactor whilst circulating the plant cell aggregates. Therefore, initially the bioreactor is operated as a packed bed with the foam particles held stationary between two retractable stainless steel grids. Once the cell aggregates are fully immobilised the grids are retracted and the bioreactor is operated as a circulating bed.

The volumetric production rate depends on the total amount of biomass in the reactor. For an immobilised cell system such as this, where the cells are present in discrete foam particles, the volumetric rate of reaction, r_v, is given by:

$$r_v = (r_p\ m_p\ N_p)V_R$$

where

r_p = mean specific rate of reaction based on unit amount of immobilised biomass

m_p = mean biomass hold-up per particle

N_p = total number of particles

V_R = reactor volume

The percentage particle hold-up in the reactor, ϕ , is

$$\phi = 100(V_p\ N_p/V_R)$$

where,

V_p = volume of one particle

so,

$$r_v = \frac{\phi}{100} \cdot \frac{r_p m_p}{V_p} .$$

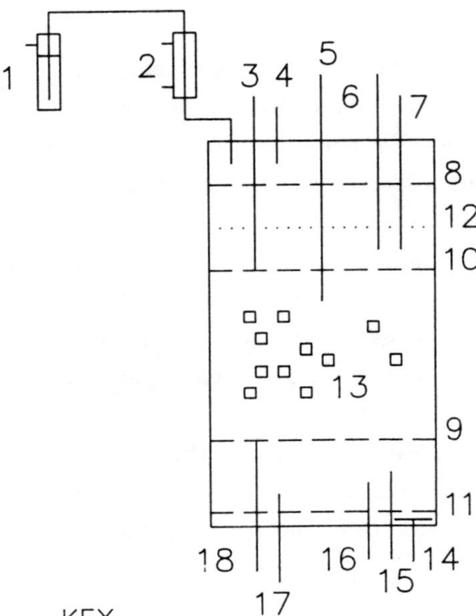

Figure 3: Circulating bed bioreactor

The terms r_p and m_p are biological parameters which depend both on the plant-cell system and on the mode of operation of the reactor. m_p is also dependent on the efficiency of immobilisation and in the present case on the efficiency of entrapment of plant-cell aggregates in the foam matrix. Also, if m_p and V_p are fixed the reactor productivity is directly proportional to , the particle hold-up. Hence in a reactor configuration like this one it is desirable to achieve high values for m_p and , in order that high volumetric production rates can be achieved.

However, for this type of reactor configuration there is a limiting maximum value for the particle hold-up, ϕ, in order to attain complete mixing. Some experiments were carried out with agar-filled foam particles to study the minimum superficial air velocity required to mix a given number of particles in reactors with different aspect ratios using different sparger designs. Figure 4 represents typical results. This indicates that maximum possible hold-up for particles to attain complete mixing is about 30%. This can be compared with the static packed bed voidage for the foam particles used, which is 42%.

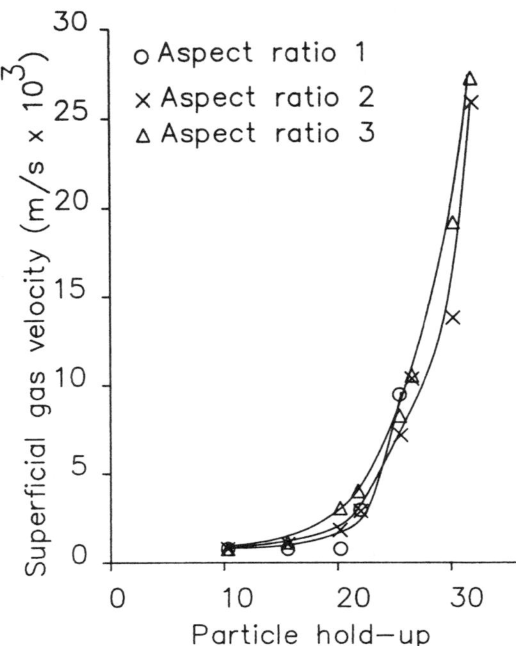

Figure 4: Minimum superficial air velocity for adequate mixing

Sheet Bioreactor

As already mentioned, the most efficient immobilisation is achieved by holding the foam matrix stationary within the bioreactor. A bioreactor was therefore designed as an adaptation of a stirred tank reactor with foam matrices in the form of sheets held stationary as vertical baffles around the central impeller (Fig.5). In this configuration, the foam hold-up can be about 30%; however, it is possible to achieve higher hold-up values by arranging the sheets as horizontal discs.

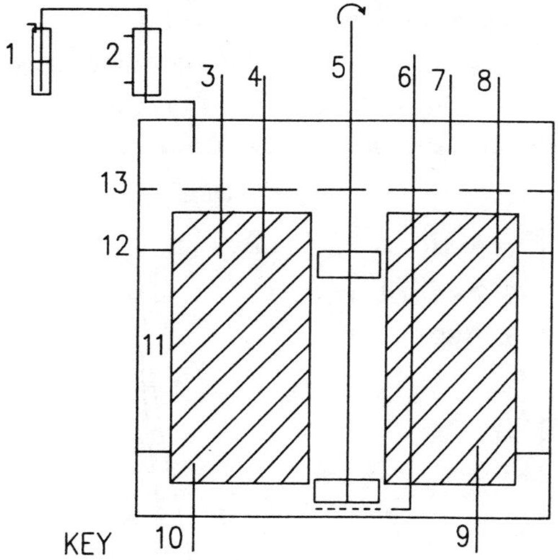

Figure 5: Sheet bioreactor

BIOREACTOR PERFORMANCE

For the growth of plant cells in both bioreactors Schenk and Hildebrandt [8] medium was used. After the immobilisation and growth of the cells were completed, either the operation was continued in the batch mode with depleted growth medium or the spend medium was replaced with a production medium. All production media were nitrogen deficient and hormone free, and in some cases metabolic precursors such as ferulic acid, vanillylamine and valine were included in the media. Sinapic acid was also incorporated in some of the production media as an attempt to divert the metabolic pathway from lignin synthesis [9].

It was also discovered during the operation of the circulating bed bioreactor that the concentration of capsaicin in the medium could be increased substantially by reducing the dissolved oxygen concentration to almost zero [10]. However, on restoration of aeration capsaicin disappeared from the medium (Fig.6). This elusive nature of capsaicin, its feedback inhibition on production and its low solubility in aqueous media suggest the necessity for the removal of capsaicin as it is formed. This has been achieved using a prototype liquid-liquid extraction column integrated with the bioreactor [11]. Food grade sunflower oil was selected as a suitable non-toxic extractant since the capsaicin would be intended for use by the food and pharmaceutical industries.

An example of the performance of the circulating bed bioreactor is given in Fig.6 and of the sheet bioreactor in Fig.7. The average biomass concentration in the circulating bed bioreactor measured after the termination of the run was 10.45 g dry weight/litre (based on total working volume). Referring to Fig.6, the maximum concentration of capsaicin was 17.34 mg/l corresponding to a maximum specific production rate of 0.5 mg capsaicin/g dry weight-day over the period of production.

In the sheet bioreactor, the Schenk and Hildebrandt medium was slightly modified [12] by replacing sucrose with glucose. As shown in Fig.7, a change of growth medium was necessitated by the detection of contamination within the bioreactor. This medium and the subsequent production medium contained 10,000 units of penicillin and 10 mg of streptomycin per litre. The production medium also contained 0.1 mM ferulic acid and was deficient

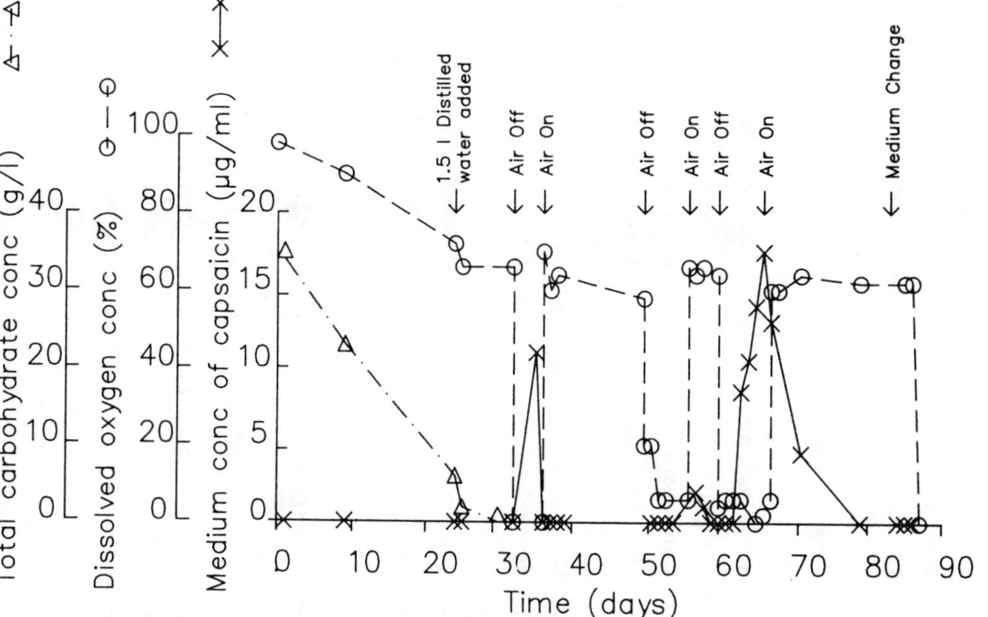

Figure 6: Performance data for the circulating bed bioreactor indicating the effect of dissolved oxygen on capsaicin production

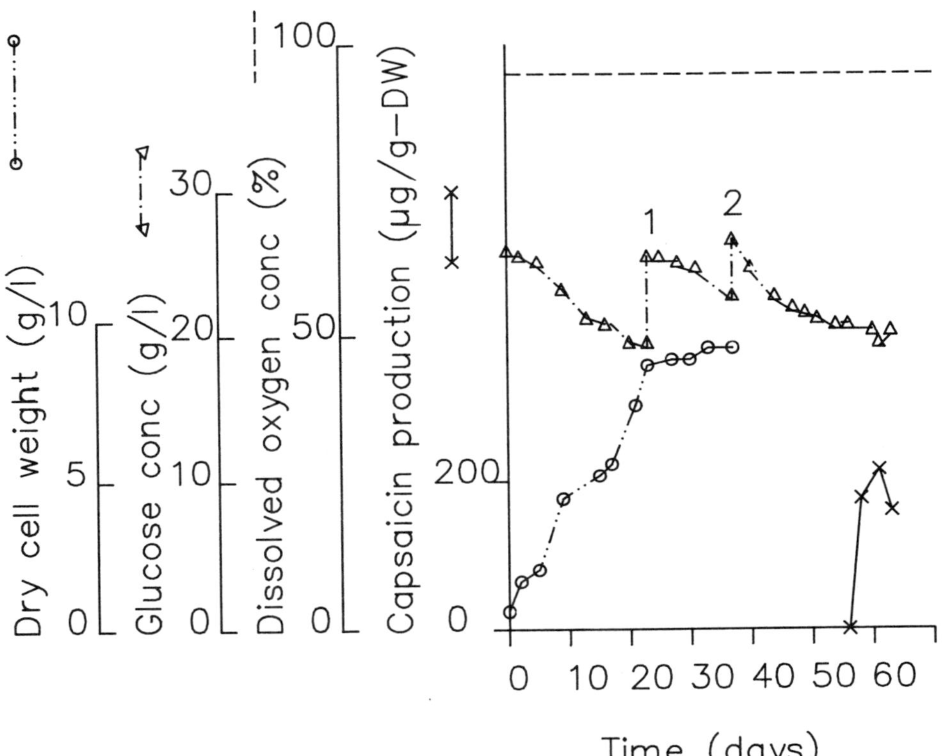

1 New growth medium plus antibiotics
2 New production medium plus antibiotics

Figure 7: Performance data for the sheet bioreactor

in nitrogen. The maximum specific production rate for capsaicin was 0.04 mg/g dry weight-day over the period of production. This lower productivity may be because of the high concentration of dissolved oxygen and the contamination problem.

CONCLUSION

With minor alterations to the conventional stirred tank bioreactor, it is possible to obtain practicable bioreactor systems suitable for use with immobilised plant cell cultures. Furthermore, the bioreactor configurations developed here can be easily scaled-up. The maximum specific production rate of the immobilised cells obtained with simultaneous oil extraction at 0.1 mg capsaicin/g dry weight-day compares well with the rate of 0.5 mg/g dry weight-day during the accumulation of capsaicin in the ripening *Capsicum frutescens* fruit [13].

ACKNOWLEDGEMENTS

The authors would like to acknowledge the Biotechnology Directorate of the SERC and Albright & Wilson Ltd. for financial support.

REFERENCES

1. Fowler, M.W., Commercial application and economic aspects of mass plant cell culture. In *Plant Biotechnology*, ed. S.H. Mantel and H. Smith, Cambridge University Press, Cambridge, 1983, pp3-37.

2. Lindsey, K. and Yeoman, M.M., The synthetic potential of immobilised cells of *Capsicum frutescens*, Mill. C.V. annum. *Planta*, 1984, **162**, 495-501.

3. Hall, T.L., Flowers, T.J. and Roberts, R.M., *Plant-Cell Structure and Metabolism*, 2nd ed. Longman, London, 1981.

4. Street, H.E., *Plant Tissue and Cell Culture*, 2nd ed., Blackwell, London, 1977.

5. Yeoman, M.M., Lindsey, K., Miedzybrozka, M.B. and McLauchlan, W.R., Accumulation of secondary products as a facet of differentiation in plant cell and tissue cultures. In *Differentiation in Vitro*, ed. M.M. Yeoman and D.E.S. Truman. British Soc. for Cell Biol. Symposium 4, Cambridge University press, 1982, pp65-82.

6. Brodelius, P., Immobilised plant cells and protoplasts, in Handbook of Plant Cell Culture, Vol.4, ed. D.A. Evans, W.R. Sharp and P.V. Ammirato, Macmillan Publ. Co., New York, 1986, pp287-315.

7. Lindsey, K., Yeoman, M.M., Black, G.M. and Mavituna, F., A novel method for the immobilisation and culture of plant cells, FEBS Lett., 115, 1983, pp143-149.

8. Schenk, R.U. and Hildebrandt, A.C., Medium and techniques for induction and growth of monocotyledonous and dicotyledonous plant cell cultures, Can. J. Bot., 50, 1972, pp199-204.

9. Mavituna, F., Park, J.M., Wilkinson, A.K. and Williams, P.D., Characteristics of immobilised plant cell reactors. In Plant and Animal Cells: Process Possibilities, ed. C. Webb and F. Mavituna, Ellis Horwood, Chichester, 1987, pp92-115.

10. Wilkinson, A.K., Williams, P.D. and Mavituna, F., The effect of oxygen stress on secondary metabolite production by immobilised plant cells in bioreactors. In, Plant Cell Biotechnology, ed. M.S. Pais, Springer-Verlag, Heidelberg, (in press).

11. Wilkinson, A.K., Williams, P.D. and Mavituna, F., Effect of continuous extraction on production of a secondary metabolite by immobilised plant cells. In Proceedings of the 4th European Congress on Biotechnology: Vol.2, ed. O.M. Neijssel, R.R. van der Meer and K.Ch.A.M. Luyben, Elsevier, Amsterdam, 1987, pp574-576.

12. Mavituna, F. and Park, J.M., Growth of immobilised plant cells in reticulate polyurethane foam matrices. Biotech. Letts., 7(9), 1985, pp537-640.

13. Suzuki, T. and Iwai, K., Constituents of red pepper species: chemistry, biochemistry, pharmacology and food science of the pungent principle of Capsicum species, in The Alkaloids, ed. Brossi, Academic Press, 23, 1984, pp227-299.

INTERNATIONAL CONFERENCE ON BIOREACTORS AND BIOTRANSFORMATIONS
GLENEAGLES, SCOTLAND, UK: 9-12 NOVEMBER 1987

Paper A4

V. good
Presentation a little haltering.

FERMENTATION STUDIES OF TRANSFORMED ROOT CULTURES

Peter D. G. Wilson, Martin G. Hilton, Richard J. Robins
and Michael J. C. Rhodes

Plant Cell Biotechnology Group,
AFRC Institute of Food Research (Norwich Laboratory),
Colney Lane, Norwich NR4 7UA,
UK

ABSTRACT

Organised cultures, consisting of interconnecting root systems, have been established by transformation of a range of plants with the bacterium *Agrobacterium rhizogenes*. These fast-growing cultures have favourable biochemical and biotechnological characteristics, making them a system of high potential for the *in vitro* production of valuable plant-derived metabolites. The non-homogeneous and structured nature of the cultures has necessitated an investigation into novel modes for their exploitation in fermentation systems. This presentation describes some of the features of these roots growing in fermenters up to 1.5 l capacity and experiments designed to investigate their productive capacity over extended periods.

INTRODUCTION

The potential of *in vitro* cultures for the production of valuable plant secondary metabolites for use as medicinal compounds, fragrances or food colours and flavours has long been recognised. Only one process based on a cell suspension culture, namely that to produce the pigmented anti-inflammatory drug, shikonin from cells of *Lithospermum erythrorhizon* has, however, been exploited on the industrial scale [1]. The principal problem with cell suspension cultures is their high degree of genetic and biochemical heterogeneity [2], which frequently leads to low and unstable production of the desired metabolites [3]. Similar problems attend the use of immobilised plant cells since cell suspensions are used to inoculate the support matrix. Immobilised cell cultures, however, may have improved properties over cell suspensions in terms of higher and more stable product formation rates and may offer scope for the design of novel processes [4]. The improved productive capacity of immobilised cultures is attributed to the development of differentiation of cells and of organisation within the immobilised cell particles [5]. We have extended this concept and have investigated plant organ cultures as a stable source of secondary metabolites. It is possible to grow cultures of plant roots or shoots but these generally grow slowly and are difficult to maintain [6]. We have shown, however, that roots (so-called "hairy" roots) formed by transformation of plant material by the soil pathogenic bacterium, Agrobacterium rhizogenes have many of the properties desirable for exploitation of their capacity to synthesise secondary metabolites [7]. This paper will describe the properties of hairy root cultures and their growth in fermenters. The potential of such systems in the de novo synthesis of secondary products and in specific biotransformation reactions will be outlined.

RESULTS AND DISCUSSION

1. The induction and maintenance of hairy root cultures

The method used to establish hairy root cultures is illustrated in Figure 1. As yet, this approach is only applicable to dicotyledons.

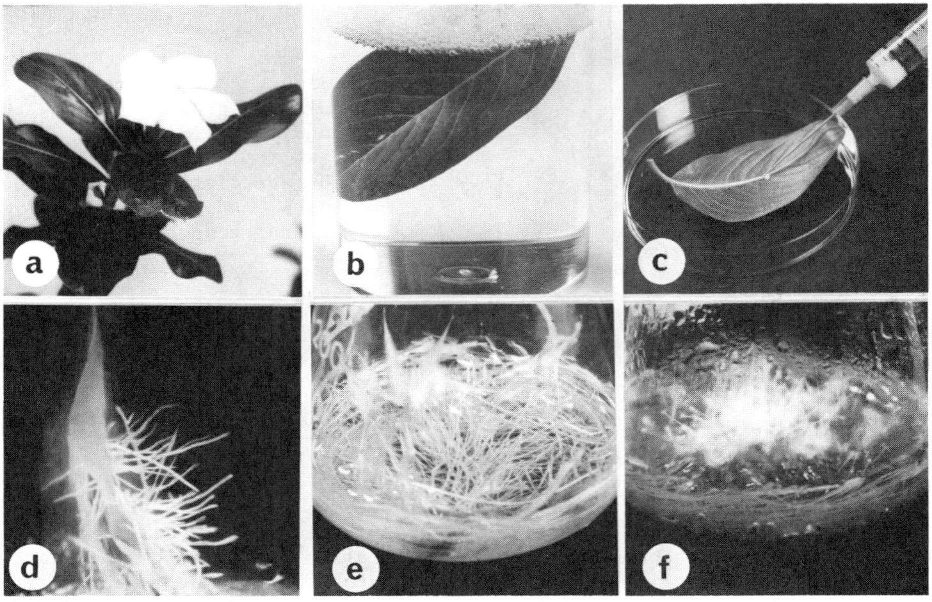

Figure 1. The method of initiating hairy root cultures: a) a plant of Catharanthus roseus; b) the surface sterilisation of a leaf for use as an explant; c) infection of the explant with a suspension of Agrobacterium rhizogenes; d) profuse growth of roots from points of infection 2 - 3 weeks later; e) treatment of excised roots with antibiotics to remove contamination by A. rhizogenes; f) axenic culture of transformed roots.

A suitable section of sterile plant material (eg. in vitro seedling, or surface sterilised leaf, petiole or stem) is inoculated with a dense suspension of A. rhizogenes (10^8 organisms/ml). After incubation at 26 °C for a period of 2-3 weeks, a profuse growth of roots appears at the point of infection. The bacterium transfers two sections of its plasmid DNA into the plant genome [8]. This transferred bacterial DNA includes the information that leads to alteration of hormone metabolism in the transformed tissue to induce root formation [9]. These roots may be excised from the plant material and maintained indefinitely in independent culture on either solid or liquid media (Fig. 2). In contrast to cell cultures, hairy roots do not require either plant hormones or vitamins in the culture media and relatively simple mixtures of salts with sucrose as the carbon source are sufficient to promote rapid growth.

Figure 2. A hairy root culture growing under the conditions described in the text, illustrating its characteristic profuse root hairs and high degree of lateral branching.

2. The properties of transformed root cultures

Table 1 sets out a comparison of the main properties of hairy root cultures with those of cell suspension cultures. In general, there is very little difference in growth rate between the two types of culture. In fact, this variation in growth rate is often less than that observed between different lines of the same species. The relatively high growth rate in hairy root cultures is perhaps surprising since cell division is limited to the tip meristem. Hairy roots, however, are highly branched and generate many such growth points. In addition, of course, cell expansion makes an important contribution to the overall biomass increase. Growth rate in hairy root lines does not as yet match that of the exceptionally fast growing selected cell suspension culture of Nicotiana tabacum [10] with a doubling time of about 15 hours. There have not, however, been comparable selection studies to isolate fast growing root lines.

Figure 3 shows the pattern of growth of a line of Datura stramonium in batch culture in a shake flask. The culture fresh weight increases 55-fold to over 11 g in about 20 days and enters stationary phase after 25 days. The figure illustrates the relatively high degree of reproduciblity in growth between experiments.

TABLE 1
Properties of hairy root and cell suspension cultures

	Hairy root	Suspension
BIOCHEMICAL PROPERTIES		
Growth rate (doubling time)	2 - 7 days	0.7 - 14 days
Medium requirements	Simple: no vitamins or hormones	Complex: requires specific vitamins and hormones
Genetic stability	Stable: euploid, homogeneous	Unstable: poly- and aneuploids, intra-chromosomal re-arrangements, heterogeneous
Metabolite production	Level and spectrum characteristic of parent plant	Unpredictable, often low
Product release	Some released (see Table 2)	Some released
BIOENGINEERING PROPERTIES		
Ultimate biomass density	c. 30 g dry wt/l	c. 20 g dry wt/l
Stability at high density	Easily maintained	Oxygen needs difficult to meet
Inoculum size	Largely independent, minimal lag	Size dependent, 'conditioning' of medium needed
Biomass handling	Not easily pumped, large particles	Pumpable slurry, small particles
Shear sensitivity	Cultures shear sensitive	Sensitive: lines resistant to shear selected
Rheology	Newtonian liquid phase	Non-Newtonian broth
Continuous operation	Easy: cells self-immobilised	Requires support, mixing problems

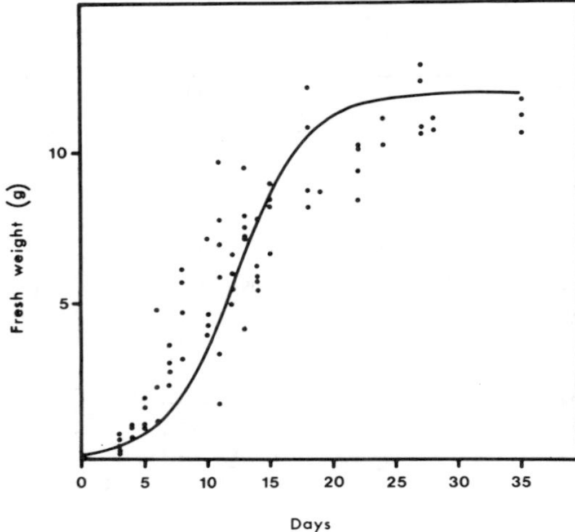

Figure 3. Time course of growth of D. stramonium in batch culture. Root tips (0.2 g fresh weight) were inoculated into 50 ml of B50 medium [14] in 250 ml conical flasks. Flasks were incubated at 26 °C under low light conditions (30 µE/sec/m^2 on a gyratory shaker (90 strokes/min). The contents of individual flasks were harvested at intervals, the roots separated from the medium by filtration under standard conditions and the fresh weight of the roots determined. This graph shows a compilation of data from six separate time-course experiments.

The most significant properties of hairy root cultures relate to their high degree of genetic stability. The initiation of disorganised cultures results in the appearance of cell types which differ in chromosome number (both polyploid and aneuploid types) and in intra-chromosomal organisation from the plant tissue from which they were developed. Hairy root cultures retain the chromosome number of the parent plant and this is maintained over many generations provided the structural integrity of the root system is preserved [11]. Factors such as hormone treatments, which cause the root structure to breakdown with the formation of callus, induce genetic variation. The genetic stability of hairy root lines underlies their stability in growth and in product formation. The levels and spectrum of products formed by a hairy root culture reflect in a general way the capacity for product formation of the roots of the parent plant [12]. This level of product formation is maintained over extended periods. In N. rustica, in which the major product is nicotine, production has remained at its initial high level throughout the two years the culture has existed. This stability of production is dependent on the maintenance of root organisation and is lost if the root structure breaks down [13].

In both hairy root and cell suspension cultures, product formation may closely follow growth, i.e. is growth-related such that the concentration of product per unit weight remains constant throughout the growth cycle. Alternatively, product formation may be divorced from growth (growth un-related) so that typically the bulk of product synthesis occurs at the late part of growth phase or during stationary phase. Thus the concentration of product per unit weight varies over the cycle. Figure 4 illustrates both these relationships between growth and product formation in two species of hairy root culture. In these batch cultures nicotine production in N. rustica is growth-related while in D. stramonium, in which the major product is hyoscyamine, the formation of product is dissociated from growth. In the latter case, there is an initial phase associated with the early part of growth in which the concentration remains constant at about 80 µg/g fresh weight followed by a rapid rise to a level of over 500 µg/g fresh weight after the roots reach stationary phase.

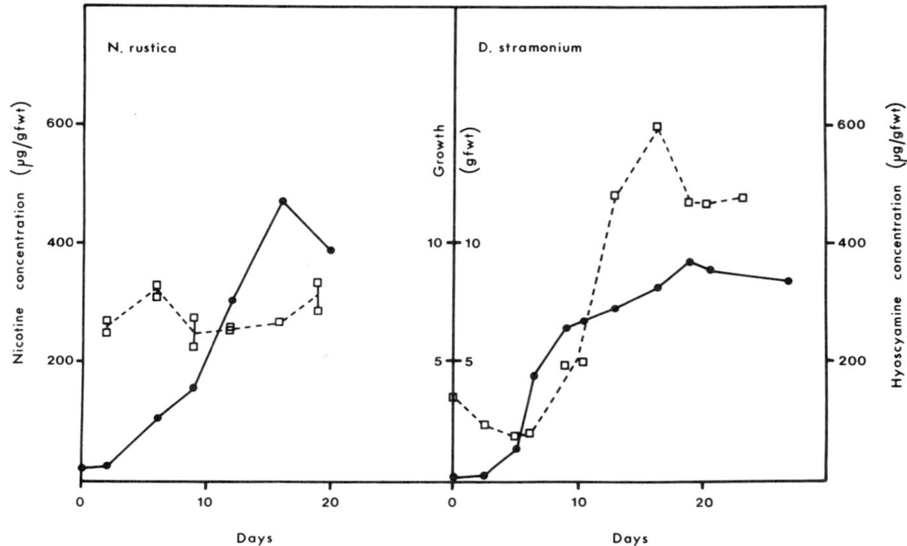

Figure 4.The relationship between growth and product formation in transformed root cultures of N. rustica and D. stramonium. Growth (●) was determined as the fresh weight (see legend Fig. 3); product formation (□) by HPLC following standard extraction procedures [14,24]. Products are nicotine from N. rustica and hyoscyamine from D. stramonium.

In both types of culture, product may be held in the tissue or released into the medium. In hairy root lines of beetroot (Beta vulgaris), the pigments (betacyanin and the vulgaxanthins) are almost entirely retained in the roots, while in other species substantial amounts of product are released into the medium. This is illustrated in Table 2, which gives examples of transformed root cultures which release products. Although in some cases the percentage release is small in batch culture we have shown that, at least in the case of N. rustica, much greater release (up to 80% of the total) is possible in systems in which either the medium is replaced continuously or suitable adsorbents are added to the medium [17]. Figure 5 shows that the time course of release of hyoscyamine into the medium closely follows its appearance in the roots and it is thus unlikely that it represents release by cell necrosis even though the release is low (about 4% of the total in the culture). The level of release varies between different root lines and in some selected lines of D. stramonium, which produce high levels of alkaloid (over 1 mg/g fresh weight), up to 15% is released into the medium (Payne, J. and authors, unpublished).

TABLE 2
Release of product from transformed root cultures

Species	Product	Release (%)	Reference
N. rustica	nicotine	10 - 50	14
N. tabacum	nicotine	5 - 30	12
D. stramonium	hyoscyamine	5 - 15	24, unpubl.
L. erythrorhizon	shikonin	5 - 10	15
Tagetes patula	thiophene	nq	16

nq = not quantified

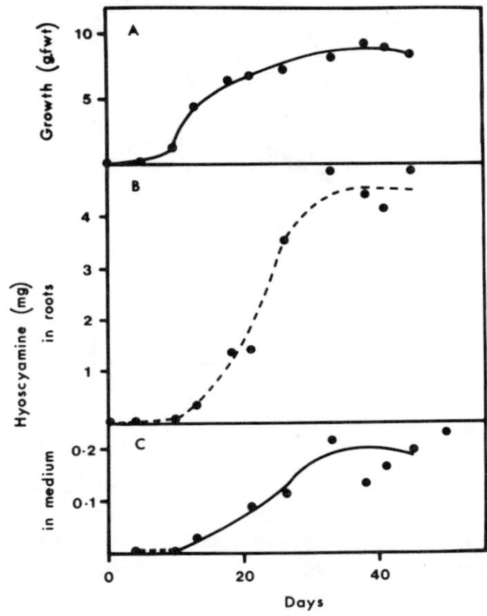

Figure 5. Growth (A) and hyoscyamine production in D. stramonium showing the distribution of product between the cells (B) and the medium (C). Parameters were determined as described in the legends to Figs. 3 and 4.

3. Fermenter studies with transformed root cultures

Table 1 includes a comparison of the properties of hairy roots and cell suspensions in relation to their biotechnological exploitation. Hairy root cultures grow in fermenters from low inocula to high densities without significant conditioning effects and can be maintained in a viable, non-growing state for extended periods. In the examples where product is released, they can readily be used in continuous systems with product recovery from the medium without destruction of the biomass. For continuous culture of cell suspensions without washout of biomass, it is necessary to supply a suitable support and it is often difficult to meet the oxygen demand of such cultures. With transformed root cultures there is very little secretion of glycoproteins and polysaccharides into the medium compared with suspension cultures and the liquid phase remains Newtonian in its rheological properties. Hairy root cultures show sensitivity to shear in the fermenter environment. This parallels the early experience with cell suspensions but with these it has proved possible to select shear-resistant lines by gradual adaption of mixed cultures to increasing shear forces [18]. It is less likely that such resistant types can be found in hairy root cultures since the mechanical stress appears to cause disorganisation of the root and callus formation. There are problems to be solved in the inoculation of fermenters with hairy roots, since this tissue cannot be readily rendered into a pumpable form without inducing damage and loss of root integrity with attendant loss of product synthetic capacity.

Figure 6. The appearance of roots of *N. rustica* grown in (a) a stirred tank and (b) an air-sparged fermenter (b).

Initial experiments to grow hairy roots in stirred tank fermenters (Fig. 6a) were unsucessful due to the problems of shear damage. When roots were grown in air-sparged systems [17], however, root morphology was maintained (Fig. 6b). In this paper we report studies with two species, *D. stramonium* and *N. rustica*. We have investigated the growth and productivity of these cultures in continuous fermenters under a range of conditions. Figure 7a-d shows the growth of transformed roots of *D. stramonium* in a column fermenter about 60 cm tall with an aspect ratio in the main compartment of about 5.7 and a total volume of about 1.5 l. The column was inoculated (Fig. 7a) with 1.7 g fresh weight of roots placed in a stainless steel cage which was lowered aseptically into the fermenter. There was rapid growth of roots from the cage at the base of the fermenter to fill the volume of the column. It is an interesting feature of hairy roots that they lack the geotropism of normal roots and grow upwards as well as downwards. At the end of a 60-day period (Fig. 7d), a final biomass concentration of 1.12 kg (63.8 g dry weight) was achieved, equivalent to about 40 g dry weight/l.

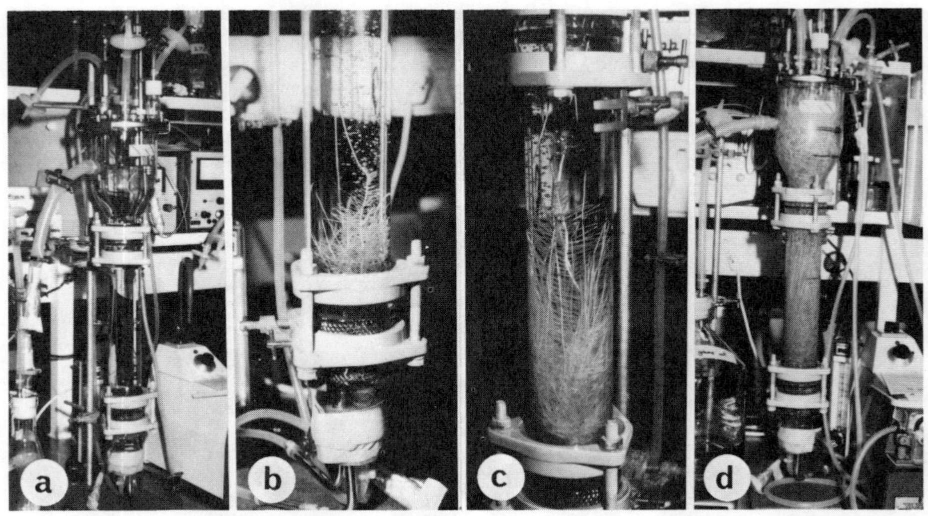

Figure 7. Illustrations of the growth of D. stramonium in a column fermenter at various ages after inoculation (a) 4 days; (b) 11 days; (c) 18 days; (d) 33 days.

A study of the growth and productivity of nicotine in N. rustica was made in these same column fermenters. The production of nicotine was shown in batch culture experiments [14] to be growth-related with a significant proportion of the nicotine released into the medium. We have investigated this system in continuous culture to see if constant removal of nicotine in the liquid outflow would stimulate overall synthesis. Transformed roots of N. rustica (6.0 g fresh weight) were transferred into an inoculation cage and grown in a batch phase with 800 ml of mediun B50 [14] for 21 days (Fig. 8). They were then run continuously in medium of the same composition for 65 days. During the batch phase the medium nicotine concentration rose to a peak (3.9 mg/l) after 10 days and then fell to below 1.0 mg/l. In the continuous phase, the daily rate of release of nicotine was initially low but steadily rose to a peak value of 8.2 mg/day and then during the later stages fell again to a low value by 86 days. At this stage the biomass had grown to 174 g fresh weight (equivalent to 16.8 g dry weight/l). At 86 days, 132 mg of nicotine had been recovered from the medium and less than 0.04 mg was found in the roots at 86 days. If the concentration found in the inoculum (0.18 mg/g fresh weight) was maintained, as occurs in batch operations, then the final yield of alkaloid would have been 32 mg. Thus the continuous operation had led to more than a four-fold stimulation in nicotine production. It is likely that during the period of rising nicotine release, synthesis of nicotine in the roots is active and the internal concentration is at least maintained. The subsequent period of declining nicotine release is associated apparently with a declining rate of synthesis and lowering intracellular alkaloid content. This experiment highlights the difficulty of analysing these non-mixed systems where sampling of the roots from the fermenter presents problems.

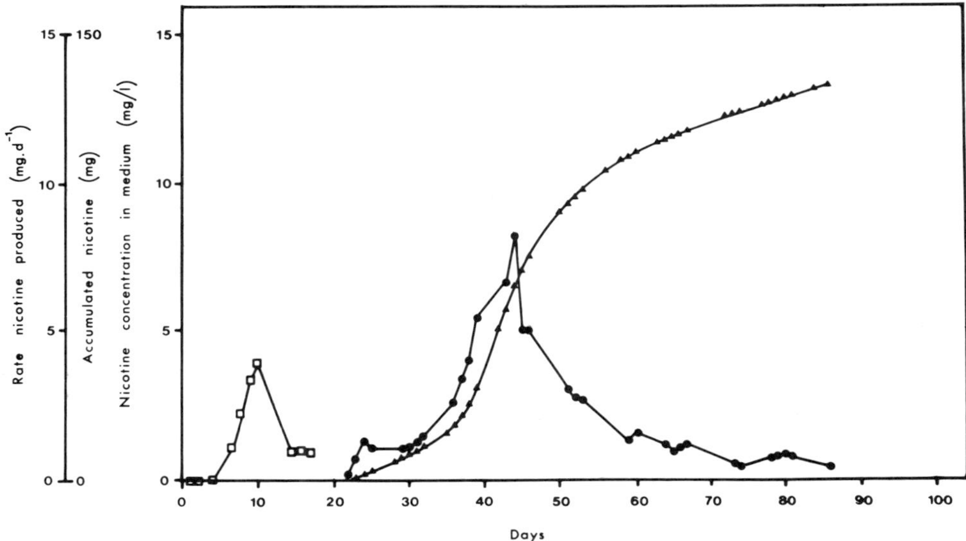

Figure 8. Time-course of nicotine production in root cultures of N. rustica growing in a column fermenter (see Fig. 7) in a full strength B50 medium. The fermenter was operated for 21 days in batch mode followed by 65 days in continuous mode (flow rate 138 ml/day). (□), nicotine accumulation in medium during batch mode; (●), rate of nicotine production in the medium during the continous mode; (▲), cumulative nicotine production in the medium during the continuous mode.

TABLE 3
Growth and nicotine production during continuous culture of N. rustica roots

	Fresh weight (g)	Nicotine production				
		Roots		Medium	System	
		Concn. (mg/g)	Total (mg)	(sum to 64 d) (mg)	(mg/g)	(mg)
Inoculum	6.8	0.064	0.4			0.4
After 64 days	436	0.050	22	72	0.215	94
Total yield if initial concentration maintained:						28

Increased production due to continuous operation: 94/28 = 3.4-fold

Table 3 shows data from another experiment similar to that described in Figure 8 but in which the roots were harvested after 45 days of the continuous phase while the rate of release of nicotine was still high. In this experiment an inoculum of 6.8 g grew to 436 g in this period. The initial nicotine concentration was 0.064 mg/g fresh weight (0.44 mg total) and this concentration was maintained to the end of the experiment (final concentration of approximately 0.050 mg/g fresh weight: 22 mg total in roots). During this experiment 72 mg of nicotine was recovered from the effluent stream, giving an overall production of 94 mg. This value exceeds that which would have been predicted on the basis of growth and maintenance of the initial concentration (28 mg) by a factor of more than three. Thus a clear benefit results from the continuous operation.

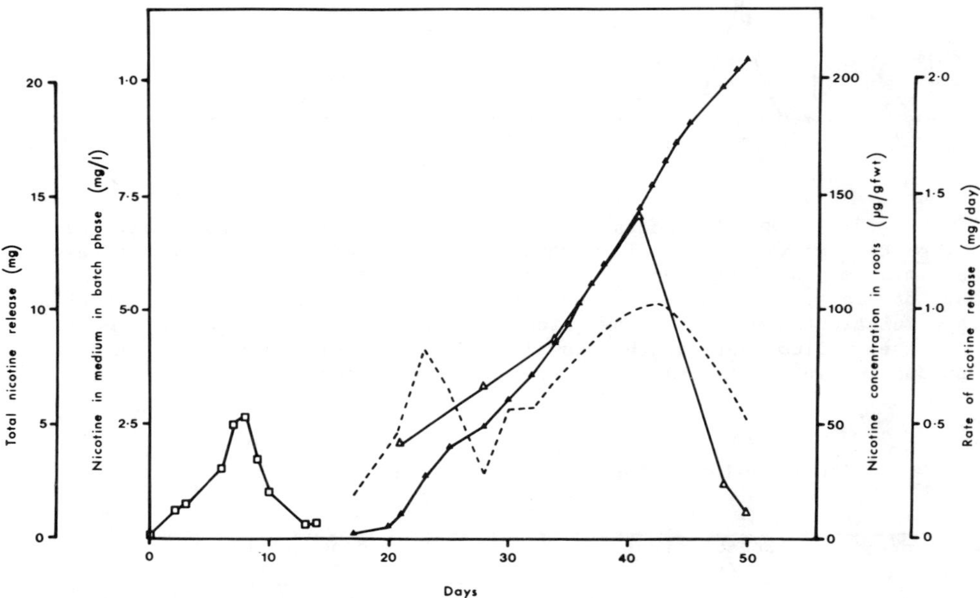

Figure 9. Time-course of nicotine production in a root culture of N. rustica growing in a column fermenter, showing the effect of phosphate limitation during the continuous phase of operation. The fermenter was operated in batch mode for 16 days on full strength B50 medium and then for 34 days in continuous mode in B50 medium lacking phosphate ions. (□), nicotine content of medium during batch mode; (---) (▲), rate and cumulative nicotine production in the medium during continous mode; (△), nicotine concentration in the roots.

In a further experiment (Fig. 9), roots were grown for 16 days in batch culture in B50 medium, after which the culture was fed with B50 medium lacking inorganic phoshate and the fermenter operated in the continuous mode for a further 34 days. The rate of release in the continuous phase rose to a peak (1.0 mg/day) around day 40 and then declined. Measurement of the root concentration of nicotine showed a peak of nicotine content at 41 days and the internal concentration was found closely to follow the rate of release to the medium. As a result of the sequestration of inorganic phosphate by roots during the batch phase, considerable growth occurred during the early stages of the continuous phase using this stored phosphate. Neverthless, over this period the concentration of nicotine in the roots rose three-fold. Beyond 40 days both the rate of release and the internal concentration of nicotine declined so that by day 50 the level was down to 0.020 mg/g fresh weight.

CONCLUSIONS

These experiments demonstrate that it is possible to grow transformed roots to high density and to maintain them for extended periods in continuous culture. They can be maintained either in full media, where ultimately non-nutritional factors limit growth, or in media lacking a specific nutrient. Although the example given relates to a de novo synthesis, the packed bed reactor described here would be equally valuable in biotransformation reactions. The potential of plant cell cultures for biotransformations has been investigated in several laboratories [19]. The most promising system studied is the formation of the cardiac glycoside drug, digoxin, by upgrading of digitoxin (a by-product) via 12-hydroxylation using cells of Digitalis lanata [20]. Here, high rates of hydroxylation were achieved in cell suspensions but, since the product accumulates in the culture medium, immobilised systems using calcium alginate as support have proved equally valuable [21]. The biotransformation capacity of hairy roots has yet to be surveyed in any detail but our knowledge of their complement of enzymes capable of modifying complex subtrates suggests that they may be useful in this context.

In addition to biotransformations based on endogenous enzymes, transformed hairy roots may provide a vehicle for genes from other species and for biotransformations based on heterologous expression of these foreign genes. Examples of this potential is provided by genes for antibiotic resistance which code for proteins which modify and inactivate antibiotics. For instance, the enzyme neomycin phosphotransferase catalyses the phosphorylation of kanamycin and inactivates it, while chloramphenicol acetyltransferase, inactives chloramphenicol by acetylation. These two genes have been incorporated into hairy root systems via a modified Agrobacterium strain bearing the genes for the antibiotic modifying enzymes [22,23]. In the plant system the engineered hairy roots biotransform the antibiotics and are able to grow in the presence of antibiotics at levels toxic to lines not bearing the foreign genes. This system enables other genes coding for useful biotransformations to be incorporated in tandem with these antibiotic resistance genes. Successful transformations may then be selected by challenging the cultures with antibiotics. Thus, useful genes from plant or other sources may be inserted and stably expressed in the hairy root system where the favourable fermenter characteristics of the roots to carry out biotransformation reactions may be exploited.

Acknowledgements: We are most grateful to our colleagues, in particular Drs. J. Payne, A. J. Parr and D. M. Evans for some of the unpublished results used herein. Also to Chris Waspe and Abbi Peerless for technical assistance.

REFERENCES

1. Tabata, M. and Fujita, Y., Production of shikonin by plant cell cultures. In Biotechnology in Plant Science, ed. M. Zaitlin, P. Day and A. Hollaender, Academic Press, New York, 1985, pp. 207-218.

2. Deus, B. and Zenk, M. H., Exploitation of plant cells for the production of natural compounds. Biotech. Bioeng., 1982, **24**, 1965-74.

3. Deus-Neumann, B. and Zenk, M. H., Instability of indole alkaloid production in Catharanthus roseus cell suspension cultures. Planta medica, 1984, **50**, 427-31.

4. Rhodes, M. J. C., Immobilised plant cells. In Topics in Enzyme and Fermentation Technology, ed. A. Wiseman, Ellis Horwood, Chichester, 1985, **10**,pp. 51-87.

5. Lindsey, K. and Yeoman, M. M., Novel experimental systems for studying the production of secondary metabolites by plant tissue cultures. In Plant Biotechnology, ed. S. H. Mantell and H. Smith, Cambridge University Press, Cambridge, 1983, SEB Seminar series, **18**, pp. 39-66.

6. Butcher, D. N. and Street, H. E., Excised root culture. Biol Rev., 1964, **30**, 513-86.

7. Hamill, J. D., Parr, A. J., Robins, R. J., Rhodes, M. J. C. and Walton, N. J., New routes to secondary products. Bio/Technology, 1987, in press.

8. Jouanin, L., Restriction map of an agropine Ri plasmid and its homologies with Ti plasmids. Plasmid, 1984, **12**, 91-102.

9. Tepfer, D., Transformation of several species of higher plants by Agrobacterium rhizogenes: sexual transmission of transformed genotype and phenotype. Cell, 1984, **37**, 959-67.

10. Noguchi, M., Matsumoto, T., Hirata, Y., Yamamoto, K., Katsuyama, A., Kato, A., Azechi, S. and Kato, K., Improvement of growth rates of plant cell cultures. In Plant Tissue Culture and its Bio-technological Application, ed. W. Barz, E. Reinhard and M. H. Zenk, Springer-Verlag, Berlin, 1977, pp. 85-94.

11. Aird, E. L. H., Hamill, J. D. and Rhodes, M. J. C., Cytogenetic analysis of hairy roots from a number of plant species transformed with Agrobacterium rhizogenes. Theor. Appl. Genet., submitted.

12. Parr, A. J. and Hamill, J.D., Relationships between biosynthetic capacities of hairy root cultures of Nicotiana spp. and the plants from which they were derived. Phytochemistry, submitted.

13. Flores, H. E., Hoy, M. W. and Pickard, J. J., Secondary metabolites from root cultures. Trends Biotechnol., 1987, **5**, 64-69.

14. Hamill, J. D., Parr, A. J., Robins, R. J. and Rhodes, M. J. C., Secondary product formation by cultures of Beta vulgaris and Nicotiana rustica transformed with Agrobacterium rhizogenes. Plant Cell Rep., 1986, 5, 111-4.

15. Kamada, H., personal communication.

16. Ketel, D. H., Breteler, H. and de Groot, B., Thiophene-biocides from cell cultures of Tagetes patula (marigolds). Poster presented at meeting Process possibilities for plant and animal cell cultures, Inst. Chem. Engineer., UMIST, Manchester, 1986.

17. Rhodes, M. J. C., Hilton, M. G., Parr., A. J., Hamill, J. D. and Robins, R. J., Nicotine production in "hairy root" cultures of Nicotiana rustica: fermentation and product recovery. Biotech. Lett., 1986, 8, 415-20.

18. Scragg, A. H., Allan, E. J., Bond., P. A. and Smart, N. J., Rheological properties of plant cell suspension cultures. In Secondary Metabolism in Plant Cell Cultures, ed. P. Morris, A. H. Scragg, A. Stafford and Fowler, M. W., Cambridge University Press, Cambridge, 1986, pp. 178-94.

19. Reinhard, E. and Alfermann, A. W., Biotransformation by plant cell cultures. Adv. Biochem. Eng., 1980, 16, 49-83.

20. Alfermann, A. W., Bergmann, W., Fijur, C., Helmbold, U., Schwantag, D., Schuller, P. and Reinhard, E., Biotransformation of B-methyldigitoxin by cell culture of Digitalis lanata. In Plant Biotechnology, ed. S. H.Mantell and H. Smith, Cambridge University Press, Cambridge, 1983, SEB Seminar series, 18, 67-74.

21. Alfermann, A. W., Schuller, I. and Reinhard, E., Biotransformation of cardiac glycosides by immobilised cells of Digitalis lanata. Planta medica, 1980, 40, 218-223.

22. Hamill, J. D., Prescott, A. and Martin, C., Assessment of the efficiency of co-transformation of the T-DNA of disarmed binary vectors derived from Agrobacterium tumefaciens and the T-DNA of A. rhizogenes. Plant Mol. Biol., submitted.

23. Evans, D. M., unpublished.

24. Payne, J., Hamill, J. D., Robins, R. J. and Rhodes, M. J. C., Production of hyoscyamine by hairy root cultures of Datura stramonium. Planta medica, 1987, in press.

INTERNATIONAL CONFERENCE ON BIOREACTORS AND BIOTRANSFORMATIONS
GLENEAGLES, SCOTLAND, UK: 9-12 NOVEMBER 1987

Paper A5

SEED PREPARATION FOR RAPID GERMINATION - ENGINEERING STUDIES

A.W. Nienow
Department of Chemical Engineering
Uinversity of Birmingham
Birmingham B15 2TT

and

P.A. Brocklehurst*
AFRC Institute for Horticultural Research
Wellesbourne
Warwick CV35 9EF

*Present address Shell, Sittingbourne Research Centre
Sittingbourne, Kent

ABSTRACT

When seeds are treated with a solution of suitable osmotic pressure on filter paper for a period of time, they can be brought to the point of germination. However, they do not actually germinate. Upon subsequent sowing (following washing, drying and storage), germination of all viable seeds occurs rapidly (within approximately 48 hours) with many advantages in reducing losses and mechanising harvesting. The technique is known as seed priming. In order to use such a process commercially, a suitable method must be found for treating large quantities of seeds at high concentrations. Preliminary studies with a bubble column device gave poor results compared to filter papers and stirred three-phase reactors have been investigated. As background data, oxygen uptake rates of the seeds have been measured over a 7 day period in a suitable polyethylene glycol priming solution. K_La values have been measured for a variety of aeration rates, agitator types and speeds. Finally, conditions have been found in which priming in such a reactor was able to produce results equivalent to those found from priming on filter papers. Lower aeration or agitation speeds lead to oxygen starvation and higher speeds to seed damage. Thus, the seed priming reactor resembles in many ways the classical fermenter. Some anomalous results probably due to the extreme sensitivity of seeds to trace toxic materials have been found. Seed concentration could usefully be increased. The work is continuing.

NOMENCLATURE

D	impeller diameter, cm
H	vessel unaerated liquid height, cm
k_La	mass transfer coefficient, s^{-1}
N	agitator speed, s^{-1}
OUR	oxygen uptake rate, μlitres O_2/g seed/hr
P	power drawn in liquid, W

Po power number = $P/\rho N^3D^5$, dimensionless

Q_G sparged gas rate, m^3s^{-1}

Re Reynolds number = $\rho ND^2/\mu$, dimensionless

T vessel diameter, cm

t_m time constant of an oxygen electode, s

t_{50} the time taken for half of the viable seeds to germinate, days

vvm (volume of air/minute)/(volume of fluid in reactor), min^{-1}

Greek Symbols

μ_a apparent dynamic viscosity, Nsm^{-2}

μ dynamic viscosity, Nsm^{-2}

ρ liquid density, kg m^{-3}

Subscripts

g gassed

JS seeds just suspended

INTRODUCTION

In many vegetable crops, the marketable yield and timing of maturity are influenced by the time taken for the seedlings to emerge [1]. To improve crop earliness and uniformity and thus to increase the potential for mechanised harvesting, seedling emergence should be as rapid and uniform as possible. A technique of treating seeds to increase the speed and uniformity of germination and seedling emergence is known as seed priming. Seeds are imbibed on filter paper moistened with a solution of suitable osmotic pressure for a period of days, which has the effect of controlling the water uptake of the seeds such that they reach the brink of germination, but visible germination does not occur [2]. Upon subsequent sowing (following washing, or washing and drying) germination occurs rapidly and uniformly [2,3].

In order to use such a process commercially, a suitable method must be found to treat large (10-1000 kg) quantities of seed. Moreover, if the process is to be economically viable, it must be possible to treat seeds at high concentrations, since although some vegetable seeds have a high retail value (currently up to £600 per kg), for other species the retail value may be as little as £2 per kg. The essential requirements for successful treatment are controlled osmotic pressure, temperature and oxygen supply to the seeds. A preferred material for the osmotic solution is polyethylene glycol 6000 (PEG) since it is non-toxic and of sufficiently high molecular weight that it does not penetrate the seeds [4]. However, PEG solution does have certain disadvantages for a bioreactor; compared with water, it is viscous and has low oxygen solubility [5].

Onion seeds (cv. White Lisbon) were used throughout, dusted with fungicide (Benlate-T, Dupont Ltd.) at 3.3g per kg seed. PEG 6000 (BDH Chemicals Ltd., average molecular weight 6000 - 7500) solution was used at a concentration of 350 kg per m^3 distilled water, except where specified otherwise.

The experiments reported here were done to develop a bioreactor for priming large quantities of vegetable seeds in PEG solution. They were conducted at the I.H.R. and at the University of Birmingham. The work is continuing as an AFRC Link Scheme entitled the Process Engineering of Seeds. As such, it also includes seed drying and coating.

PRELIMINARY EXPERIMENT AT I.H.R., WELLESBOURNE ON A BUBBLE COLUMN

This experiment was conducted before interaction between the two groups had started. Five grams of seed were primed in a bubble column essentially as described by Darby and Salter [6] who treated seeds in water or inorganic salt solutions. A Perspex cylinder (47.5mm i.d. x 1000 mm) contained 1 l.PEG solution, in which the seeds were suspended by an air flow of 2 l./min. at the base. Foaming was controlled by adding a few drops of silicone-based antifoam (30% w/w dimethyl polysiloxane, BDH Chemicals Ltd.). The seed was primed for 14 days at 15°C; for comparison, seed was also primed under similar conditions on filter paper as described by Brocklehurst and Dearman [3]. After priming, seeds were rinsed in running tap water for 1 minute, then immediately put to germinate on filter paper moistened with distilled water at 15°C. Untreated seeds served as controls and there were four replicates of fifty seeds for each treatment in the germination test.

Percentage germination was not affected by priming, averaging 96%; germination time (measured as the time taken for half of the viable seeds to germinate (t_{50})) was markedly reduced by priming on filter paper (to 0.5 days compared to 4.3 days for untreated seeds) but only slightly reduced to 3.3 days by priming in the column. It was noticed that the air passed through the bubble column in large bubbles, without being finely dispersed. Thus the relatively poor performance of seeds from the bubble column may have been due to an inadequate rate of oxygen transfer to the seeds.

In order to investigate this further, it was decided to carry out a more thoroughly-engineered series of experiments to assess the reason for the poor performance in the bubble column. Because oxygen transfer was suspected to be the cause of the problem, a stirred reactor was selected for investigation.

In addition, the agitation of onion seeds in PEG solution by air was an example of a three-phase system in which considerable work had been done [7]. It was also decided to treat the system as a bioreactor, with onion seeds probably having similar requirements to a microorganism. Thus, oxygen demands, k_La values, the possibilities of "shear" damage, the correct agitation conditions (air-flow and agitator speed), the physical properties of the solution all had to be determined and the performance of the reactor as a seed primer (equivalent to fermenter "yield") had to be evaluated.

The first part of the experimental section reports the determination of the "engineering" parameters; the second part, the biological evaluation. In practice, the two were conducted to some extent both simultaneously and iteratively.

TABLE 1 Viscosities of PEG 6000 Solution (m Pas)

Temp/Conc	10%	15%	20%	25%	30%	35%
15°C	6.2	7.8	14	19	22	27
20°C	-	-	-	-	-	28
22°C	-	-	-	-	-	24
25°C	-	-	-	-	-	24
30°C	-	-	-	-	-	22

PHYSICAL PROPERTY VALUES

Oxygen Solubility and Diffusivity. The solubility of oxygen in PEG 6000 (Union Carbide Ltd.) has been reported [5]. At 25°C, it falls to 56% of its solubility in water at 35% w/v concentration and to 41% at 45% w/v concentration. Diffusivity values are also reported.. As with most polymer solutions, diffusivity is only weakly affected by the dissolved material. Thus a value of 5.5 x 10^{-9} m^2/s (±20%) is given.

Viscosity. Viscosity and viscoelasticity were determined at Birmingham. At 35% w/v, no effective viscoelasticity could be determined on an R17 Weissenberg Rheogoniometer. Flow curves were measured on a Contraves Rheomat 30 instrument over shear rates from 50 s^{-1} to 700 s^{-1}. All solutions up to 35% concentration and at temperatures from 15 to 30°C were essentially Newtonian in behaviour with flow behaviour indices between 1 and 1.08. Table 1 shows the values obtained. Provided evaporation was eliminated, the viscosity did not change significantly during seed priming.

Density. The density of 35% w/v PEG 6000 is 1030 kg/m^3 at 20°C.

STIRRED BIOREACTOR ENGINEERING DATA

Equipment.
Two water-jacketed Perspex vessels were used. In each, the height of PEG solution used was approximately equal to the vessel diameter. Details are given in Table 2. In each case, an impeller clearance above the base of 1/4 and four 10% baffles were used. Humidified air (to reduce evaporation)was introduced through a point sparger, monitored by calibrated rotameters. Nitrogen was sparged unmonitored. Oxygen concentration was measured by an oxygen electrode (T_{22}, manufactured in house; T_{15}, EIL Ltd.). In the case of T_{22}, the power was measured using an air bearing dynamometer [8].

Selection of Agitation Conditions.
The oxygen uptake of seeds is typically of the order of 100 - 200 μlitres O_2/g seeds/hr. This is much lower than in usual fermenters. On the other

TABLE 2 Vessel and Impeller Details (Dimensions in cm)

	T	H	D	Agitator Type	Temp. Control
T_{15}	14.7	13.6	7.6	3 blade marine propeller	Water jacket; 15°C; manual
T_{22}	21.6	24.5	9.7	6 blade 45° pitch turbine	Water jacket; 20-30°C; automatic
T_{19}	18.5	18.5	10.5	6 blade 45° pitch turbine	Air-conditioned room; 15°C

TABLE 3 Minimum Agitator Speed for Suspension in 35% PEG, N_{JSg}

	Seed Concentration	Aeration rate	N_{JSg}
T_{22}, [22]	0.2%	0.17 vvm	350 rpm
	7%	0.17 vvm	600 rpm
T_{15}, [11]	5%	0.25 vvm	350 rpm
T_{15}, [12]	3.6%	0.25 vvm	250 rpm

hand, the large size of seeds compared to other biological materials suggests that they might be more prone to mechanical damage either due to impeller-seed impacts or seed-seed abrasion. Therefore the initial runs were conducted at sparge rates of about 0.2 vvm (0.17 vvm in T_{22} and 0.25 vvm in T_{15}). For such low sparge rates, Nienow et al. [9] showed that 6-bladed, 45° pitch agitators of impeller-to-tank diameter ratio of 1/2, pumping downwards are very energy efficient. From the agitators available, a 6-bladed 45° pitch agitator of 9.7 cm diameter was selected for use in T_{22} and 3-bladed propeller of 7.6cm was chosen for T_{15}.

The minimum speed for seed suspension (size 2 to 3mm and density 1070 kg/m^3) based on visual observations as a function of seed concentration and aeration rate is reported in Table 3. In general experiments were conducted at or above the minimum agitator speed required for suspension at the aeration rate employed.

Oxygen Up-Take Rates (OUR).

Temperature is an important variable in seed priming. The effectiveness of germination is enhanced by priming at lower temperatures but treatment must be for longer. Therefore, temperatures from 15 to $30^{o}C$ were used to assess the optimum.

OUR's were measured in a Gilson Differential Respirometer at $15^{o}C$. Seeds were primed on filter papers in 35% w/v PEG and removed for determinaton of OUR at various times. During assessment, they were placed in shake flasks and either covered with distilled water or 35% w/v PEG solution. After about 1 to 2 days, the OUR reached a plateau of about 40-45 μlitres O_2/g seeds/hr in water. In PEG, however, a plateau rate of 28 to 32 μlitres O_2/g seeds/hr was obtained.

Runs were also carried out in the stirred vessels using the technique of Bandyopadhay et al. [10] in which the fall in dissolved O_2 concentration with time is monitored after sparging is stopped. In T_{22}, 600g of seed were used at $20^{o}C$ and 1 vvm, with agitation conditions such that all seeds were suspended and 100% saturation was obtained except when the air was turned off.

Measurements were made over a 7 day period and the air was generally switched off for between 10 and 15 minutes. Tests with nitrogen sparging over the top surface in the absence of seeds showed that no surface gassing occurred under unaerated conditions at the agitator speeds utilised. After about 1.5 to 2 days, OUR is a little greater than that obtained in the shake flasks in PEG solutions (Table 4). However, since OUR doubles for roughly each $10^{o}C$ rise in temperature, the values are in quite good agreement. However, the stirred vessel OUR's then fall to a plateau of about 75% of the maximum value. Both the maximum and plateau values are lower at higher speeds. This lowering suggests that a) the OUR is not limited by bulk diffusion as if it were, OUR would increase; and b) there is some possibility of impaired metabolism, due to increased mechanical disturbance.

Later experiments in T_{15} [11] broadly supported the above findings. Carried out again at $20^{o}C$, measurements after two days are shown in Table 5. The maximum value at 450 rpm is a little higher but the low value at 350 rpm suggest some diffusion limitation and the large decrease in OUR at the two higher speeds suggests mechanical damage.

In a third set of experiments in T_{15} [12], anomalous results were found. In the first set, very high OUR's of 84 μlitres O_2/g seed /hr were obtained due to fungal attack. Even very small traces of fungi give enormously increased OUR, typical values for micro-organisms being five

TABLE 4 Oxygen Uptake Rates in 35% PEG at 20°C in T_{22}.

Time after seeds were added to the tank (hrs)	oxygen uptake rate (μlitres O_2/g seed/h)			
13.0	6.4	-	-	0
14.5	3.6	-	-	0
16.5	14.7	-	-	4.6
18.5	-	-	18.4	6.4
20 - 22	34.1	30.4	23.0	14.7
22.5 - 24.5	39.6	35.9	30.4	24.9
39.0 - 39.5	47.9	50.7	40.6	41.5
43.0 - 44.0	42.4	35.9	36.9	34.1
45.5 - 64.5	36.9	35.0	33.2	30.4
63.0 - 64.0	26.8	24.9	29.5	23.0
67.0 - 68.0	25.8	26.7	29.5	23.0
69.5 - 70.5	29.5	28.6	25.8	26.7
86.0 - 87.0	27.6	28.6	25.8	23.0
40.5 - 91.5	25.8	24.9	25.8	23.0
93.0 - 94.0	27.6	26.7	24.0	23.0
speed (rpm)	360	450	550	600

orders of magnitude higher; for <u>Azotobacter vimeladii</u>, for example, it is 1.3×10^6 μlitres O_2/g cells/hr! Subsequently values with seeds that gave a poor seed priming performance gave maximum values after 2 days of only 24 μlitres O_2/g/hr at 15°C at a seed loading of 100 kg/m^3; and 30 μlitres/g/hr at a seed loading of 50 kg/m3. The latter figure is in good agreement with the work in the Gilson Respirometer. The implications of these results will be discussed later.

Why is the OUR lower in PEG than in water when measured on the Gilson equipment? Since bulk diffusion does not appear to be limiting, it seems probable that the PEG molecules are blocking the entry of O_2 to the seeds. The large PEG molecules have an advantage over other osmotica in that they do not penetrate the seed. On the other hand, if the molecules cannot enter, they still may be able to offer a partial blockage, thus lowering O_2 transport rates.

The Measurement of k_La Values.

Using the unsteady state technique in which de-aeration is conducted by nitrogen sparging, there are two potentially large sources of error and both of these are enhanced by high values of k_La. Errors can occur either as a result of assumptions about the gas phase mixing [13] and/or because of the lag associated with a sluggish oxygen electrode response [14]. In this work, k_La values were so low that the simple no "depletion" model could be used. Also, the time constant t_m of the oxygen electrode was

TABLE 5 OUR at 20°C after 2 days in T_{15}

Speed (rpm)	350	450	550	650	750	900
OUR (μlitre O_2/g seed/hr)	27	60	42	44	27	15

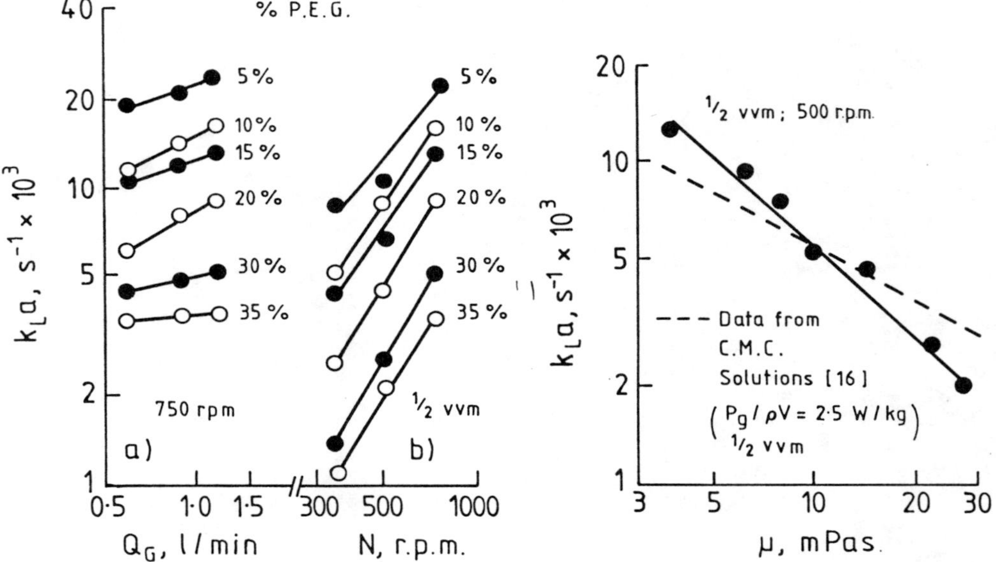

Fig. 1 k_La versus a) aeration rate and b) impeller speed

Fig. 2 k_La versus viscosity (PEG 6000 solutions)

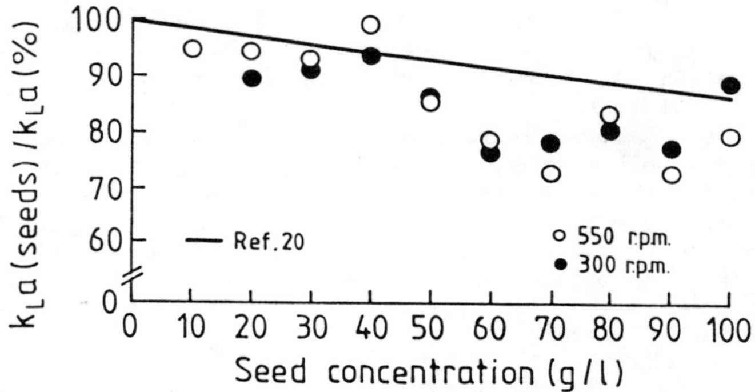

Fig. 3 Effect of seeds on k_La

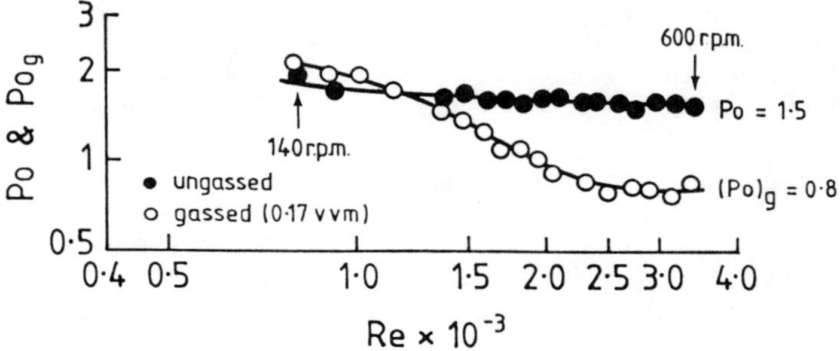

Fig. 4 Power number versus Reynolds number

about 20 sec. Therefore, the lag associated with it is negligible [15] for all k_La values such that $t_m < 1(/(5k_La)$, i.e., for k_La's < 0.01 s^{-1}. This was the case for virtually all experiments in solutions of 10% PEG or greater and only these data are reported

The Effect of Aeration Rate and Agitator Speed. At 10% PEG (viscosity 6.2 m Pas at 15°C), k_La is a function of agitator speed (Fig. 1b) and aeration rate (Fig. 1a). The functional relationships are typical of those commonly reported for low viscosity systems. However, increases in PEG concentration to 35% (viscosity 27 m Pas) make k_La virtually independent of aeration rate. Such an effect has been reported for shear thinning broths such as Xanthan [16]. However, in the shear thinning case, the apparent viscosity is required to be much higher (>300 m Pas) before becoming independent of aeration rate.

The Effect of Viscosity.
To a first approximation, $k_La \, \alpha 1/\mu$ (Fig. 2). Work on shear thinning fluids by a number of workers has suggested that $k_La \; \alpha(\mu_a)$-1/2 [16]. The surprisingly high dependence found here must be ascribed to a greater effect of viscosity on the flow in the agitator region when the fluid is not shear thinning. For example, van't Riet [17] showed that the shear rate in the region of the agitator is approximately 100 times the agitator speed whilst the average shear rate is only of the order of 10 times the speed. However, even in viscous shear thinning fluids, observation of the vortices behind impeller blades shows them to be similar to those found in low viscosity fluids [18]. Typical values of k_La for shear thinning CMC solutions are also shown in Fig. 2 [16].

The Effect of Seed Concentration. Work on the effect of suspended solids on k_La values has shown the extreme sensitivity of k_La to the type and concentration of solids. Both increases and decreases have been reported with increasing solids concentration [19]. The present reduction is quite small and of the same order as that reported [20] when using similarly sized glass Ballotini (Fig. 3).

The Effect of Antifoam and Fungicide on k_La. As usual [21], it is found that k_La falls markedly with small amounts of antifoam, (Table 6). However, it is barely effected by the presence of fungicide.

Power Input.
Power was measured in T_{22} (Fig. 4). Typical of downward pumping agitators, the power number under aerated conditions is greater than that without aeration at low agitator speeds when the impeller is flooded and has to pump against the upward rising gas-liquid plume [9]. However, at the highest speeds the ratio of gassed power to ungassed is surprisingly low (~0.5) because of the larger stable cavities forming behind the blades in these viscous Newtonian fluids [18].

BIOLOGICAL EVALUATION

These results are reported in chronological order. Using tank T_{22} [22], 10-20g seeds were primed in 8.7 l.PEG solution with an aeration rate of 1.5 l./min. at 360 rpm, at 20° and 25°C for 7 days. Seeds were also primed at the same temperatures on filter paper. All primed seeds were rinsed in running tap water for 1 minute, then germinated (along with untreated controls) on filter paper moistened with distilled water held at room temperature. There were four replicates of 100 seeds per treatment in

TABLE 6 Effect of Additives on k_La in 35% w/v PEG 6000 Solution

	$k_La(s^{-1} \times 10^3)$ 300 rpm	550 rpm
PEG	0.80	2.78
PEG + antifoam (0.02%)	0.68	1.96
PEG + fungicide (3.3g/kg seed)	0.80	2.83
PEG + antifoam + fungicide	0.64	1.96

TABLE 7 Effect of Temperature and Agitation Speed on the Germination Rate of Primed Seeds (expressed as t_{50} (days)) in T_{22}

Temp.	Speed	Primed in tank	Primed on filter paper	Untreated
25°C	360 rpm	1.3	1.1	2.7
20°C	360 rpm	1.0	0.7	3.3
20°C	360 rpm	0.9	-	-
20°C	400 rpm	1.6	0.7	3.3
20°C	500 rpm	2.1	-	-

TABLE 8 Germination Rates (t_{50}) Using 5g. Seeds in T_{19} [12]

Duration (days)	Tank	Filter Paper	Untreated
10	0.6	0.6	4.8
14	0.5	0.5	4.3

the germination test. In a second series, seeds were primed at 20°C but with agitation speeds of 360, 400 and 500 rpm.

A typical example of a successful priming is shown in Fig. 5. Germination was not affected by treatment, averaging 90%. Table 7 indicates that the lower the treatment temperature and agitation speed, the faster was seed germination. At the higher speeds, there were clear signs of abrasion of the seed coat. Under the best conditions tested, i.e., 20°C

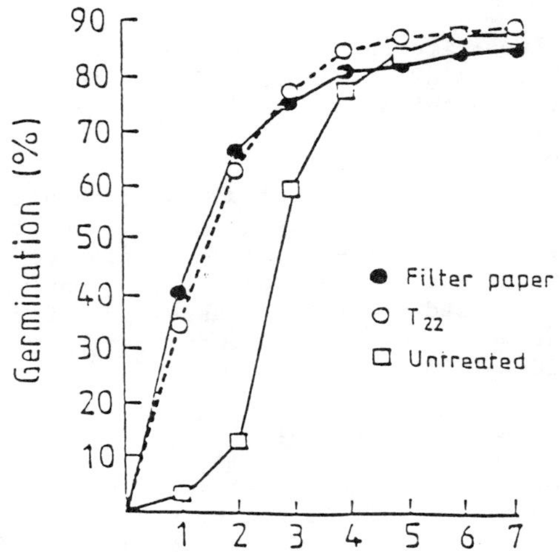

Fig 5 Germination time after 7 days priming at 25°C

TABLE 9 Effects of Concentration and Speed on t_{50} on Seeds Primed for 10 days in T_{19} [12]

Speed (rpm)	Concentration (g/l)	Tank	Filter Paper	Untreated
400	70	3.6	0.6	4.4
400	18	3.1	0.6	4.3
250	36	2.0	0.6	4.5
250	18	1.6	0.7	4.6

and 360 rpm, the germination of seeds from the tank was only slightly slower than those from filter paper.

Following the improvements in germination rate with lower temperatures and slower speeds, a T_{19} tank (see Table 2 for dimensions) was placed in a controlled-temperature room at 15°C, at I.H.R. In a preliminary experiment, 5g seed was primed in 5.5 l.PEG solution with an airflow of 2 l./min. at 250 rpm. After priming for 10 and 14 days, the seeds were rinsed and four replicates of 50 seeds germinated at 15°C. Once again, germination was not affected averaging 94%. Priming both in the tank and on filter paper reduced t_{50} by the same amount (Table 8). Since there was little difference between the 10 day and 14 day treatments, 10 days was used as the standard thereafter.

The final experiments [12] examined the effects of seed concentration and agitation speed on germination. Seeds were primed at 15°C in T_{19} containing 5.2 l.PEG solution at 1.3 l./min. and 250 and 400 rpm. Seed concentrations of 18, 36 and 70 g/l were used and antifoam was incorporated into the solution at a standard rate of 0.02% (v/v). Seeds were also primed on filter paper and after priming, the seeds were rinsed and four replicates of 50 seeds were germinated at 15°C.

Priming did not affect germination which averaged 94%. The germination of seeds primed in the tank was consistently slower than that of seeds primed on filter paper, and germination was slowest at the highest seed concentration and agitation speed (Table 9). As in the experiments with T_{22}, there were signs of abrasion of the seed coat at the highest agitation speed.

To determine whether this relatively poor performance of the seeds primed in the tank was due to the amount of antifoam used, seeds were primed in PEG solution containing either 0, 0.02% or 0.2% (v/v) antifoam. Priming was done on filter paper, and also using 2g seeds in 100ml PEG solution, contained in glass separating funnels (50 mm i.d.) at an aeration rate of 600 ml./min. supplied via a glass capillary tube. After 10 days at 15°C, seeds were rinsed and four replicates of 50 seeds germinated at 15°C. Viability was not affected by antifoam and germination averaged 92%. The rate of germination was markedly slower only for seeds primed in the glass funnels with the highest level of antifoam; seeds on filter paper were not affected (Table 10).

DISCUSSION

From the data collected, the operating conditions for a large seed

TABLE 10 Effects of Antifoam on t_{50}

Concentration (% v/v)	Glass Funnel	Filter Paper	Untreated
0.00	1.2	0.9	-
0.02	1.5	0.8	5.2
0.20	2.0	0.8	-

priming reactor can be determined. However, the last unsatisfactory series of tests (Table 9) throw up the question that existed at the beginning. How sensitive to mechanical (or other sorts of) damage are the seeds? Though oxygen solubility and k_La values are low in 35% w/v PEG 6000, the low oxygen demand should not be difficult to meet with dissolved oxygen concentrations near to saturation. Therefore some further consideration of the results is worthwhile.

Firstly, the OUR in Table 5 show a fall at high speeds indicating that a deleterious effect on metabolism is being produced. High speeds at high concentrations (400 rpm at 70 kg/m^3 seeds in T_{19}) revealed signs of abrasion, the outer black skin of the onion seed being worn away to reveal the white endosperm. Abrasion rates at high solids concentrations increase with N^4 and (solids concentration)^2 [23], so sensitivity to these parameters is acute.

Secondly, Birch [22] did not use anti-foam whilst it was used by Baker [12] and in the preliminary test at I.H.R. It is possible that in addition to lowering k_La values by producing a barrier to mass transfer at the air-liquid interface of bubbles, antifoam inhibits oxygen uptake to seeds in a similar way. Certainly seeds in PEG 6000 had lower OUR's compared to water and on filter paper.

Finally, chemical damage cannot be ruled out. Shortly before the results in Table 9 were obtained [12], the Perspex vessel T_{19} was repaired using Perspex cement (Tensol cement no.3, I.C.I. Ltd.). Unpolymerised monomers may have been leached out of this cement into the PEG solution, with an adverse effect on germination. Certainly plant tissue is extremely sensitive to certain plasticisers. For example, cabbage seedlings may be killed by a di-butyl phthalate vapour at concentrations as low as 360 pg per l. of air [24].

All of these possibilities are now under investigation. The engineering parameters are reasonably well established.

ACKNOWLEDGEMENTS

We thank Elsoms Seeds Ltd. for onion seed; Messrs. M.R. Baker, S.W. Birch and J. Schofield for the data presented here; Mr. D.F. Oldham for constructing the stirred tank used at I.H.R.; and Messrs S. Chatwin and R.S. Badham for technical assistance at Birmingham.

REFERENCES

1. Currah, I.E.: "Plant uniformity at harvest related to variation between emerged seedlings". Acta Horticulturae, 1978, 72, 57-68.
2. Heydecker, W. and Coolbear, P.: "Seed treatments to improve performance - survey and attempted prognosis". Seed Sci. & Tech., 1977, 5, 353-425.
3. Brocklehurst, P.A. and Dearman, J.: "Interactions between seed priming treatments and nine seed lots of carrot, celery and onion, Part I: Laboratory germination", Ann. of Appl. Biol., 1983, 102, 577-584.
4. Brocklehurst, P.A. and Dearman, J.: "A comparison of different chemicals for osmotic treatment of vegetable seeds", Ann. of Appl. Biol., 1984, 105, 391-398.
5. Mexal, J., Fisher, J.T., Osteryoung, J. and Reid, P.C.P.: "Oxygen availability in polyethylene glycol solutions and its implications in plant-water relations", Plant Physiol., 1975, 55, 20-24.
6. Darby, R.J. and Salter, P.J.: "A technique for osmotically pre-treating and germinating quantities of small seeds", Ann. of Appl. Biol. 1976, 83, 313-315.

7. Chapman, C.M., Nienow, A.W., Cooke, M. and Middleton, J.C.: "Particle-Gas-Liquid-Mixing in Stirred Vessels: Part 1-IV", Chem.Eng.Res.Des., 1983, 61, 71-81. 82-95, 167-181, 182-185.
8. Nienow, A.W. and Miles, D.: "A dynamometer for the accurate measurement of mixing torque", J.Sci.Inst., 1969, 2, 994-5.
9. Nienow, A.W., Kuboi, R., Chapman, C.M. and Allsford, K.: "The dispersion of gases into liquids by mixed flow agitators", Proc.Int.Conf. on Physical Modelling of Multi-Phase Flows, BHRA, Cranfield, 1983, p.417-438.
10. Bandyopadhyay, B., Humphrey, A.E. and Taguchi, H.: Biotech. Bioeng., 1967, 9, 533-544.
11. Schofield, J.: "The measurement of mass transfer properties in a seed priming reactor for scale-up purposes", B.Sc. Project Report, University of Birmingham, 1984.
12. Baker, M.: "The Evaluation of a Novel Bioreactor", M.Sc. Project Report, University of Birmingham, 1984.
13. Chapman, C.M., Gibilaro, L.G. and Nienow, A.W.: "A dynamic response technique for the estimation of gas-liquid mass transfer coefficients in a stirred vessel", Chem.Eng.Sci., 1982, 37, 891-896.
14. Nienow, A.W. and Wisdom, D.J.: "The effect of scale on specific absorption rates in aerated vessels with a constant gas residence time", Proc.Int.Symp. on Mixing in the Chemical Industry, Polytechnique de Mons. 1978, C12-1 to C12-12.
15. van't Riet, K.: "Review of measuring methods and results in non-viscous gas-liquid mass transfer in stirred vessels", Ind.Eng.Chem. (Proc.Des.Dev.), 1979, 18, 357.
16. Nienow, A.W.: "Mixing studies on high viscosity fermentation processes - Xanthan gum", in Biotechnology Europe, 1984, Online, London, 1984, 293-304.
17. van't Riet, K.: "Turbine agitator hydrodynamics and dispersion performance", Ph.D. Thesis, Delft Technical University, 1975.
18. Allsford, K.V.: "Gas-liquid dispersion and mixing in mechanically agitated vessels with a range of fluids", Ph.D. Thesis, University of Birmingham, 1985.
19. Nienow, A.W., Konno, M. and Bujalski, W.: "Studies on three-phase mixing, a review and recent results", Proc. 5th European Mixing Conf. BHRA, Cranfield, 1985, pp.1-13.
20. Frijlink, J.J., de Jong, P.G.T. and Smith, J.M.: "Properties of gas-liquid dispersions with suspended solids", CHISA Conference, Prague, 1984, Paper V.3.53.
21. Andrew, S.P.S.: "Gas liquid mass transfer in microbiological reactors", Trans.I.Chem.E., 1982, 60, 3.
22. Birch, S.W.: "PEG pre-treatment of seeds in a stirred aerated tank: A possible commercial process", M.Sc. Project Report, University of Birmingham, 1982.
23. Nienow, A.W. and Conti, R.: "Particle abrasion at high solids concentrations in stirred vessels", Chem.Eng.Sci., 1978, 1077-86.
24. Hardwick, R.C., Cole, R.A. and Fyfield, T.P.: "Injury to and death of cabbage (Brassica deracea) seedlings by vapours of di butyl phthalate emitted from certain plastics", Ann. of Appl. Biol., 1984, 107, 97-105.

INTERNATIONAL CONFERENCE ON BIOREACTORS AND BIOTRANSFORMATIONS
GLENEAGLES, SCOTLAND, UK: 9-12 NOVEMBER 1987

Presentation — language problem

Paper B1

BIOREACTOR CONTROL AND MODELING: A SIMULATION PROGRAM BASED ON A STRUCTURED POPULATION MODEL OF BUDDING YEAST

L. Cazzador[1], L. Alberghina[2], E. Martegani[2] and L. Mariani[1,3]

[1] *LADSEB-CNR, Padova, Italy*

[2] *Dipartimento di Fisiologia e Biochimica Generali,
Sezione di Biochimica Comparata,
Università di Milano, Italy*

[3] *Dipartimento di Elettronica e Informatica,
Università di Padova, Italy*

ABSTRACT

The yeast Saccharomyces cerevisiae is an eukaryotic microorganism used for biotechnological processes both for the production of biomasses and for the production of specific chemicals including recombinant DNA products. The knowledge of biomass properties is of fundamental importance for the monitoring and controlling of the fermentation processes. In previous papers we have shown that the analysis of segregated parameters, and in particular of the protein distributions obtained by flow cytometry, may be a powerful tool to monitor the growth conditions of the microbial biomass, providing information about changes in growth rate, population heterogeneity and age distribution. We used a structured model of yeast population to develop a bioreactor model that integrates the dynamics of the process with the dynamics of the microbial population, allowing an approach to a control based upon the properties of the individual cells that compound the biomass. The model describes the behaviour of the cellular system only in terms of growth rate and critical cell mass (or protein content) and is used in conjunction with a simple bioreactor model based on a limiting concentration of nutrient. To compare the experimental results with the prediction of the model, a simulation procedure has been developed, that describes in a quantitative manner steady and transient states that originate during growth perturbations in continuous cultures. The program is based on general population balance equations and considers discretized protein distributions and variable time intervals. The results are compared with experimental protein distributions obtained during perturbed continuous cultures and show a satisfactory fit.

INTRODUCTION

Biotechnological processes, which are based on the utilization of living organisms, are dependent on the intracellular properties of biomass, thus the knowledge of the kinetics of cell growth is an important aspect for the design and operation of a microbial reactor. Flow cytometry is a powerful tool for the analysis of cell population dynamics, since it gives data on relevant cellular parameters (such as DNA, RNA, protein content and enzymatic activities) on a cell basis (segregated data) [1,2]. In particular, protein distributions are easy to obtain experimentally and contain much information, as they are strictly related to the age distributions and directly depend on the law of growth of the single cell, i.e., the law of protein accumulation during a cell cycle [2]. Their use could also become increasingly important for the monitoring and control of processes, as soon as defined relations are established between these products and biochemical and physiological activities of the cell population in the reactor [3]. To achieve this aim, it is often necessary to use mathematical models and computer programs.

In previous papers [4,5] we have presented a growth-controlled structured model of cell populations of the budding yeast Saccharomyces cerevisiae, one of the eukaryotic microorganism most widely used for the production both of biomasses and of specific chemicals, and we have shown its use for the analysis of experimental histograms of protein distributions obtained by flow cytometry in exponentially growing cell cultures [4]. The growth cycle of Saccharomyces cerevisiae is more complex than that of other eukaryotic or prokaryotic microorganisms due to the occurrence of unequal division, which is dependent on the growth rate and on the genealogical age of the cell, and this fact has to be taken into account for the analysis of the distributions of cellular parameters [6].

In subsequent papers [7,8] more general versions of this structured model have been presented and their predictions have been compared with experimental distributions obtained in continuous cultures.

The protein distributions have a predictive value on the process dynamics, being very sensitive to changes in synthesis and division rates, as experiments in batch reactors and in chemostats have shown [4,7], therefore they offer a very interesting possibility to develop new control procedures in continuous and fed-batch reactors. For this purpose, a difficult task is the identification of correlations between the cell environment and the bioreactor dynamics with the population structure, in such a way to be able to make predictions about the population changes in response to environmental changes. The approach just mentioned in [7,8], to integrate the population model with the bioreactor model, has been further exploited and a new computer simulation program, based on it, has been developed, able to quantitatively describe the steady states and the transient states that follow growth perturbation in cultures.

In the following Sections the assumptions, which are the basis of the single-cell mathematical model on which the behaviour of the population strictly depend, are brieafly summarized and the distinctive features of the simulation program are presented. The potentiality and the usefulness of the latter is illustrated afterwards by some examples of perturbed conditions and by the comparison of simulated results with experimental findings.

CELL CYCLE MODEL

We refer to the model used in [8,9], by assuming that the internal state of the cell is described by three variables a, m and m_p, that respectively indicate: a the cell genealogical age which is related to the number of scars left on the cell surface by the buds after their abscission; m the cell mass (or, more generally, the level of any very expressive chemical component of the cell) and m_p the value of the same component at the time of budding. The following assumptions are at the basis of the model [8,9].

[A1] The environment of the cell population is an ideal bioreactor with perfect mixing and with a limiting substrate whose concentration s is the only external physical variable influencing the cells.

[A2] The mass of the cell increases during the time increment dt according to $dm = \mu m dt$, with μ being a given function of the current substrate concentration s.

[A3] A new bud emerges on the surface of an unbudded cell, when its mass m reaches a critical mass value m_b, which is a specified function of the genealogical age a and of the substrate concentration s.

[A4] The state variable m_p represents the mass of the cell at budding time and does not vary after budding, therefore $m_p = 0$ is assumed for unbudded cells.

[A5] The bud grows until the mass m of the cell reaches the critical mass value hm_b ($h > 1$), at which time the cell divides into a daughter cell of mass $m - m_p$ and a parent cell of mass m_p.

[A6] The length of the budded phase can never become shorter than a minimum time T_m required at least for DNA replication to occur.

The state of the population, described as the mass distribution of the cells with the same genealogical age, has been analytically computed in steady state balanced growth [8]. For example, assuming that $\mu(s)$ and $m_b(a,s)$ vary with s and a as shown in Figure 1, the steady state solution for the total mass density function $f(m)$ (such that $\int_{m_1}^{m_2} f(m)\,dm$ gives the number of cells of the population with mass between m_1 and m_2) is shown in Figure 2.

$$\mu(s) = \mu_{max}\frac{s}{K_s + s}, \quad \mu_{max} = 0.48\ hr^{-1}, \quad K_s = 25\ mg\ l^{-1}$$

$$m_b(1,s) = m_b(0,s)[1 + Q]$$

$$m_b(2,s) = m_b(0,s)[1 + Q(1 + A)]$$

$$m_b(a,s) = m_b(0,s)[1 + Q\frac{1-A^a}{1-A}]$$

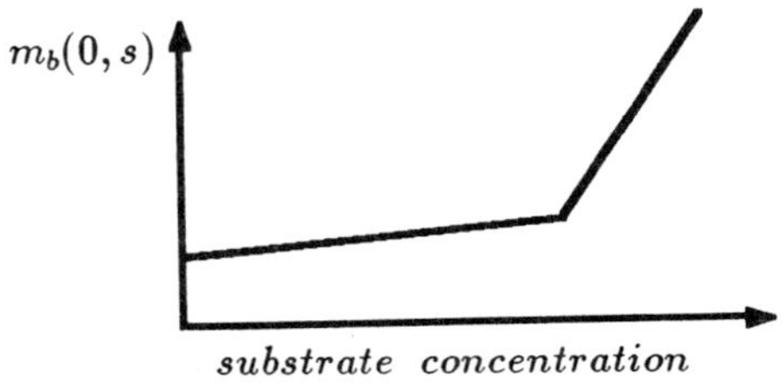

$$A = 0.5, \quad Q = 0.2, \quad h = 1.65$$

$m_b(0,s)$

substrate concentration

Figure 1 Specific growth rate and critical threshold of budding

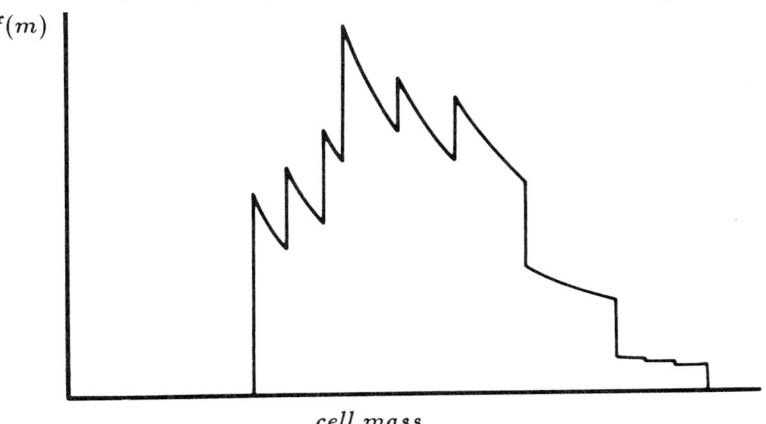

Figure 2 Mass distribution in steady state

Transient and perturbated states of growth have been analyzed only with complex approaches, as different subpopulations and different classes of cells between each subpopulation are to be taken into account [10]. To do it in a easier way, a simulation program has been developed, which will be briefly described in the following Section.

SIMULATION PROGRAM

To develop a computer simulation program from the previous model, it is necessary to take into account only a finite number of sub-populations, for example the first n_p. Thus we will assume that cells with genealogical age $a \geq n_p - 1$ all present the same cell cycle and therefore the same critical mass level. This assumption is easily accepted as cells with a high number of scars are a small fraction of the population: for instance, in steady state, cells with five or more bud scars are in fact only about 3 % [8]. Moreover, the set of all the possible cell mass values (m_{min}, m_{max}) has to be properly discretized. Thus we will consider a partition of this set into n_c steps, not necessarily of equal length, that will be referred to as "channels"

$$(m_j, m_{j+1}) \qquad j = 1, 2, ..., n_c; \qquad 0 < m_{min} = m_1 < ... < m_{n_c} < m_{n_c+1} = m_{max} \tag{1}$$

each characterized by its mean value, defined as

$$M_j = \phi(m_j, m_{j+1}) \qquad j = 1, 2, ..., n_c \tag{2}$$

where ϕ is a suitable given function (see later). Since for an unbudded cell the mass at the budding time m_p is initially set to zero, the "channel zero", with $m_0 = 0$ and $M_0 = 0$, must be also considered.

The simulation program approximates all the cell mass (and therefore the current mass m, the mass at the budding time m_p and the critical mass levels for budding m_b) by using the mean values M_j $(j = 0, 1, .., n_c)$ of the channels to which they respectively belong. In the following, the discrete density function $N(a, i, j, k)$ gives the number of cells, per unit of culture volume, at the sampling time t_k, with genealogical age a and mass M_i, which budded with mass M_j. The indexes vary according to $a = 0, 1, .., n_p - 1$, $i = 1, 2, .., n_c$, $j = 0, 1, .., n_c$ and $k = 0, 1, ...$ The dynamics of the population is described by updating this density function after a finite time interval (t_k, t_{k+1}). The sampling times t_k, the channel partition (1) and the mean mass function (2) have been chosen so as to facilitate the computation of the density function at time t_{k+1}, once given that at time t_k. As, from assumption A2, in the time interval (t_k, t_{k+1}) the mass of the single cell increases according to

$$m(t_{k+1}) = m(t_k)\alpha_k, \qquad \alpha_k = exp[\int_{t_k}^{t_{k+1}} \mu \, dt] \tag{3}$$

and then by a factor $\alpha_k > 1$ independent of the mass, the following conditions have been imposed:

$$\alpha_k = const. = \alpha; \qquad m_{j+1} = \alpha m_j; \qquad M_j = m_j\sqrt{\alpha}. \tag{4}$$

In this way, on a logarithmic scale, in each time interval (t_k, t_{k+1}), all the cells increase their mass by a constant amount $ln(\alpha)$, equal to the channel length, and the mean mass of the channel corresponds to the geometric mean.

The state of the population is then updated by using the following procedure, comprising three main steps:

[S1] Estimate the next sampling time t_{k+1} such that $\alpha_k = \alpha$.

[S2] Update $N(a,i,j,k)$ into the intermediate function $\bar{N}(a,i,j,k+1)$ as a result of growth and dilution.

[S3] Update the intermediate function $\bar{N}(a,i,j,k+1)$ into $N(a,i,j,k+1)$ as a result of budding and division.

The first step requires integrating, with the aid of a suitable numerical algorithm, starting from the initial conditions computed at t_k and going on until the time t_{k+1} for which in the equation (3) $\alpha_k = \alpha$ holds, the following equations, which represents the dynamical model of the bioreactor:

$$V\frac{dx}{dt} = \mu x V - F_{in} x \tag{5}$$

$$V\frac{ds}{dt} = F_{in}(s_{in} - s) - \frac{1}{Y}\mu x V \tag{6}$$

$$\frac{dV}{dt} = F_{in} - F_{out} \tag{7}$$

In such equations V represents the working volume of the bioreactor, x is the biomass concentration (which is related to the cell mass by $x(t_k) = \sum_{a,i,j} M_i N(a,i,j,k)$), s is the substrate concentration, F_{in} and F_{out} are the input and output flow rates, s_{in} is the substrate concentration in the input flow, μ is the specific growth rate and Y is the yield coefficient. In batch culture $F_{in} = F_{out} = 0$, in continuous culture $F_{in} = F_{out} = F$ and in fed-batch culture $F_{in} = F$ and $F_{out} = 0$.

The second step requires the computation of the following coefficient:

$$\beta_k = exp[\int_{t_k}^{t_{k+1}} D\, dt] \tag{8}$$

which takes into account the fraction of lost cells, in the time interval (t_k, t_{k+1}), by the dilution rate $D = F_{in}/V$. In a batch culture $\beta_k = 1$ as there is not dilution $(F_{in} = 0)$. The growth process shifts the cells exactly from a channel to the following and therefore the discrete function $N(a,i,j,k)$ can be easily updated in the form:

$$\bar{N}(a,i,j,k+1) = \beta_k N(a,i-1,j,k). \tag{9}$$

The third step accomplishes the processes of budding and division. For the cells that neither bud nor divide in the interval it is:

$$N(a,i,j,k+1) = \bar{N}(a,i,j,k+1). \tag{10}$$

For the unbudded cells (i.e. $j = 0$) with mass $M_i \geq M_b(a, s(t_{k+1}))$, for which bud emerges in the interval, the density function becomes:

$$N(a,i,i,k+1) = N(a,i,i,k+1) + \bar{N}(a,i,0,k+1). \tag{11}$$

For the budded cells (i.e. $j > 0$) with mass $M_i \geq hM_b(a, s(t_{k+1}))$, and for which the minimal time T_m has elapsed from budding time, division occurs in the interval, so that the density function $N(a,i,j,k+1)$ must be modified to take into account the newborn parent and daughter cells, given respectively by:

$$N(a_p,j,0,k+1) = N(a_p,j,0,k+1) + \bar{N}(a,i,j,k+1) \tag{12}$$

where a_p, the genealogical age of the parent cell, is the minimum between $a+1$ and $n_p - 1$, since cells with genealogical age greater than $n_p - 1$ are not considered;

$$N(0,n,0,k+1) = N(0,n,0,k+1) + \bar{N}(a,i,j,k+1) \tag{13}$$

where n is the integer for which M_n approximates the mass of the daughter cell $M_i - M_j$.

The discrete function $N(a,i,j,k)$ at time t_k consists of $(n_p)(n_c)(n_c+1)$ data. It is neither necessary nor useful to work with all of them because almost all the cells with equal mass and genealogical age have also equal budding mass (this is only true in steady state) and moreover because some channels (most probably the first and last ones) can be empty of cells. Thus a suitable data structure has been introduced to make less time consuming the computation and to manage only the non zero values of $N(a,i,j,k)$. In this way the minimum number of data necessary to be updated can be made less than $(n_p)(n_c)$.

The program has been developed on a DIGITAL VAX 11/750 computer in FORTRAN language. Many parameters of the cell cycle model and of the bioreactor model can be introduced in an interactive way, so as to facilitate simulations of different growth conditions with different numerical approximations. In particular, it is possible to set the number of subpopulations n_p and the number of channels n_c to be considered. Various graphic outputs can be selected to display in real time the evolution of the process variables and the time course of the population distribution. The calculations involved in the algorithms used are not too long and allow the simulation of one hour of a standard real process with less than one minute of CPU time.

RESULTS

To compare the simulated distributions with the experimental ones, the program must also account both for the various sources of variability in the physiological activity of the cells and for the measurement errors [7]. For this a further sub-program can be inserted which assumes that cells in the channel i-th be statistically spread over other channels according to a truncated gaussian function with mean M_i and given coefficient of variation CV.

The program has been tested in many different conditions and its predictions have been compared with experimental results, both in steady state and in perturbed conditions in order to also validate the cell cycle model and its underlying assumptions. Until now only qualitative results are available while quantitative analyses are under study. Two different situations have been mainly investigated: a) variation of the dilution rate in a continuous process (dilution rate step); b) variation of the substrate concentration in a continuous process (glucose pulse).

A dilution rate step, of a larger enough width, can produce, during the transient period connecting the previous and the new steady state, distributions of different shapes and in particular with a bimodal structure. For example, in Figures 3, 4 and 5 the consequences of a dilution rate step up on the mass distribution of the population are shown. As the model predicts, the dilution rate step up increases the substrate concentration and therefore the critical mass level that the cell must achieve to bud or divide (see Figure 3). During the first part of the transient period, the number of cells per unit of culture volume therefore decreases because of both the greater dilution rate of the culture and the greater critical mass levels which prevent cell division. In Figure 4 some samples of the time course of the theoretical distribution (without statistical spread) evidence the main occurrences. Pictures with $T = 0, 1$ show that the distribution moves to the right and consequently the mean mass of the cells increases. Moreover the peaks, which arise on some channels, account for those cells that budded with a mass M_i equal to the old budding level, divided with a mass M_j greater than the old division level hM_i, because of the threshold increase. In this way, if the step width is great enough, the newborn daughter cells have a mass value above the current budding level and therefore it buds immediately. Pictures with $T = 2, 3, 4$ show that the daughter cells, budded with a mass greater than the current budding level, at the next division generate daughter cells with a mass value smaller than usual, and therefore they form a separate group on the left side of the distribution. One has to recall that the simulation program assures a minimal time period T_m for the DNA replication, so, also in this case, the time elapsed from budding to division is anyway greater or equal to T_m. In Figure 5 the distributions obtained by applying an appropriate statistical spread, to take into account the noise introduced by the measurement process, show that the only features that can be really observed, after this filtering, are the shift of the population to the right and the emergence, during the transient period, of a bimodal distribution. The peak on the left side corresponds to the group of newborn daughter cells which assume a mass much smaller than the mean mass of the population.

As an another example, a glucose pulse in the medium is considered. As the model predicts, the pulse gives rise to an immediate and significant effect on the distribution of cell mass. During a first period, in fact, cells can no more bud or divide, because of the critical threshold increase induced by that of the substrate concentration. Therefore the distribution moves continuously to the right as long as the cells become bigger and the number of cells per unit of culture volume decreases until division occurs for the cells which eventually reach the new critical level for dividing. In Figure 6 the time courses of the substrate and biomass concentrations and of the growth rate and of the number of cells are shown, while in Figure 7 the theoretical distributions are represented. Pictures with $T = 2, 3, 4$ in particular evidence that, when the excess of substrate is exhausted, a group of cells with small mass emerges on the left side of the distribution. This set represents the newborn daughter cells generated from parent cells which budded before that the critical levels came down to their steady state values and which divided at the new steady state critical level. Also in this case the simulation program assures that a cell divides only if it has spent at least the minimum time T_m in the budded phase. By applying an appropriate statistical spread to the resulting distributions, the time course shown in Figure 8 is obtained, which evidences the emergence, during a transient period, of a bimodal distribution.

The simulate distributions, in both the cases before illustrated, are quite similar to those experimentally observed [5].

ACKNOWLEDGEMENTS

This work was supported by a grant of the Italian National Research Council (CNR), Progetto Strategico Biotecnologie, to LADSEB.

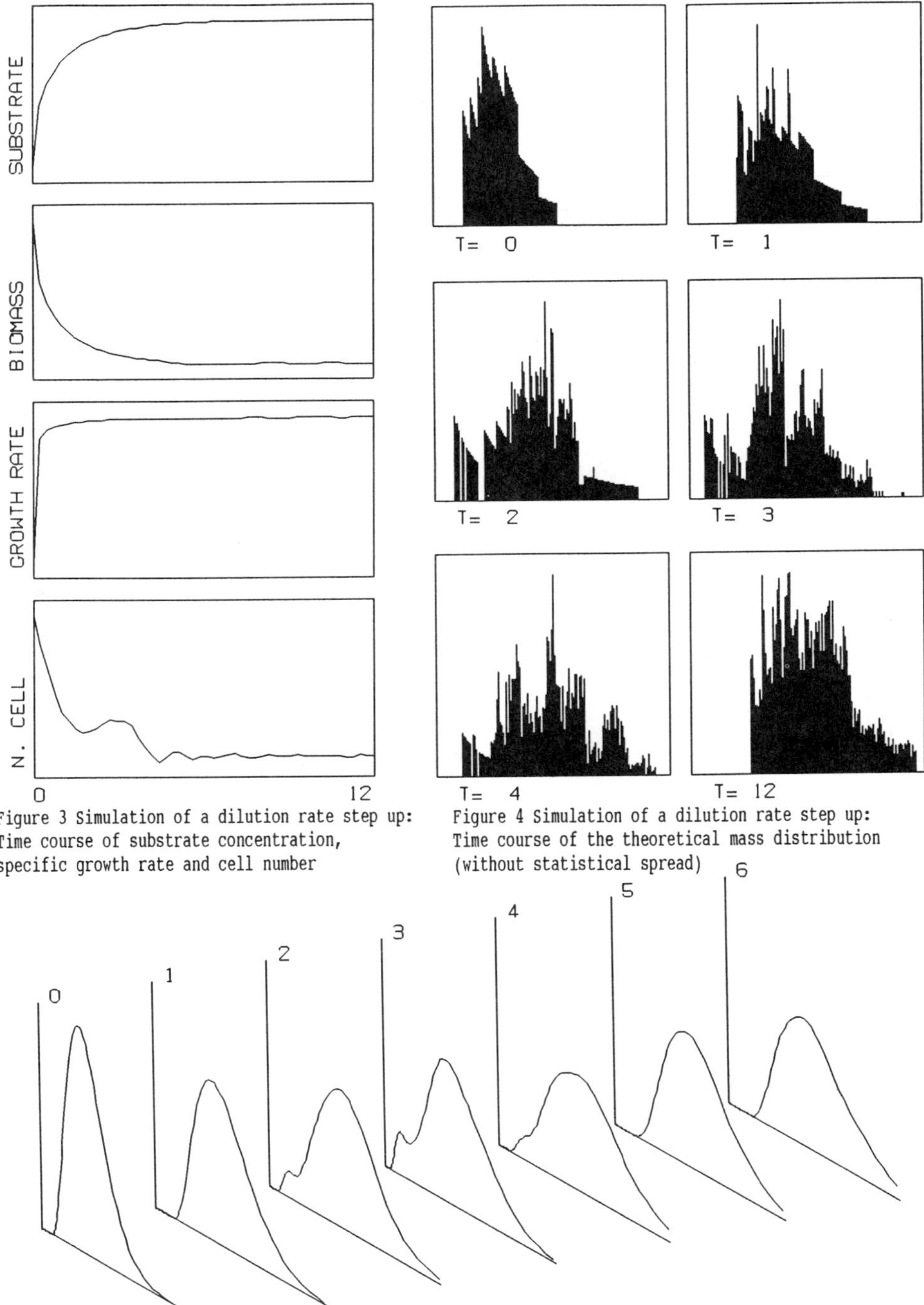

Figure 3 Simulation of a dilution rate step up: Time course of substrate concentration, specific growth rate and cell number

Figure 4 Simulation of a dilution rate step up: Time course of the theoretical mass distribution (without statistical spread)

Figure 5 Simulation of a dilution rate step up: Time course of the mass distribution applying a Gaussian spread

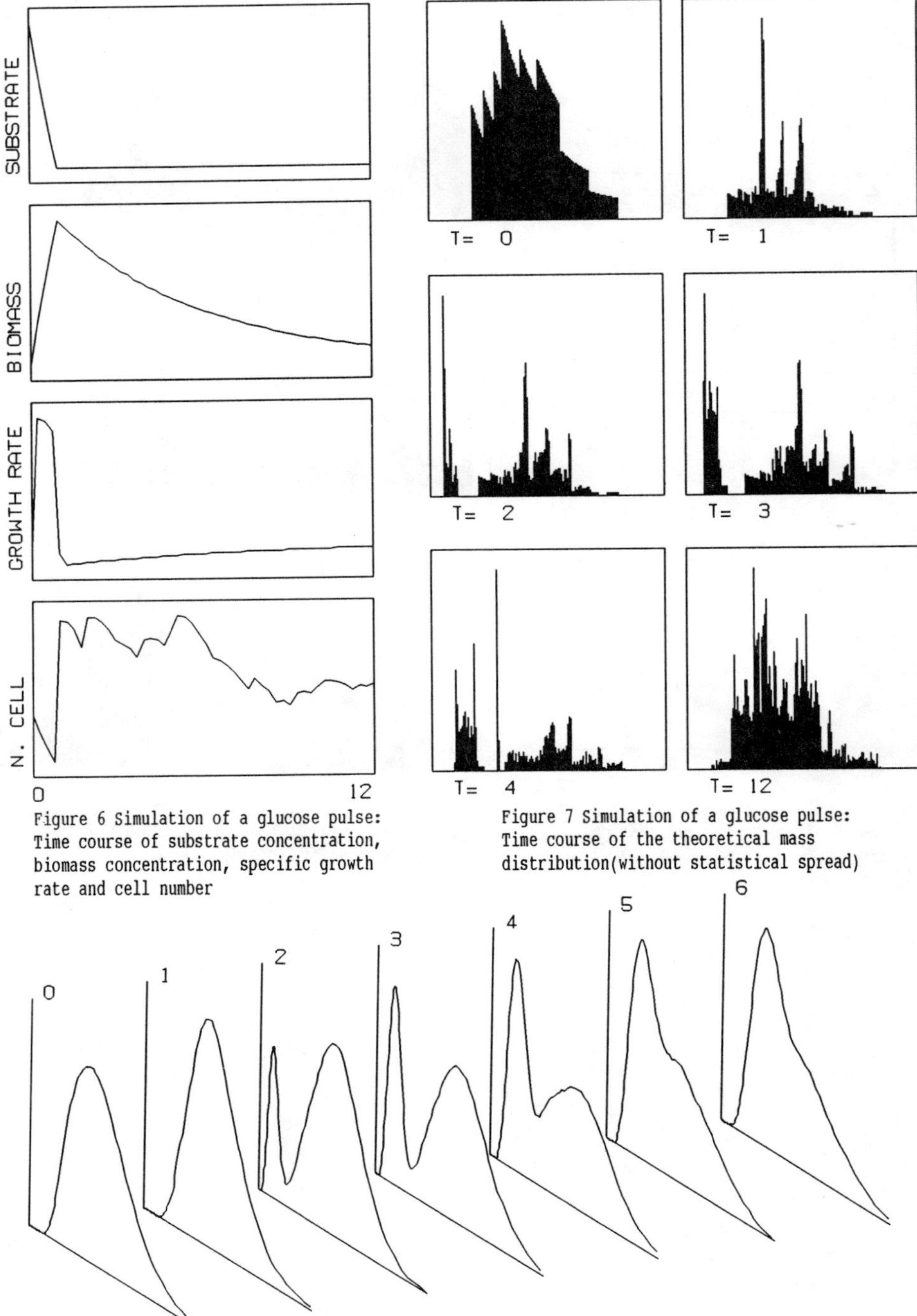

Figure 6 Simulation of a glucose pulse: Time course of substrate concentration, biomass concentration, specific growth rate and cell number

Figure 7 Simulation of a glucose pulse: Time course of the theoretical mass distribution(without statistical spread)

Figure 8 Simulation of a glucose pulse: Time course of the mass distribution applying a Gaussian spread

REFERENCES

1. Bailey, J.E., Fazel-Madjlessi, J., McQuitty, D.N. and Gilbert, M.F., Measuring Microbial Population Dynamics, Ann. N.Y. Acad. Sci., 1979, 326, 7-16.

2. Agar, D.W. and Bailey, J.E., Continuous Cultivation of Fission Yeast: Analysis of Single-Cell Protein Synthesis Kinetics, Biotechnol. Bioeng., 1981, 23, 2315-2331.

3. Bailey, J.E., Agar, D.W. and Hjortso, M.A., Acquisition and Interpretation of Flow Microfluorimetry Data in Microbial Population, Comp. Appl. in Ferm. Tech., 1982, Soc. Chem. Ind., 157-165.

4. Alberghina, L., Mariani, L., Martegani, E. and Vanoni, M., Analysis of Protein Distribution in Budding Yeast, Biotechnol. Bioeng., 1983, 25, 1295-1310.

5. Ranzi, B.M., Compagno, C. and Martegani, E., Analysis of Protein and Cell Volume Distribution in Glucose-Limited Continuous Cultures of Budding Yeast, Biotechnol. Bioeng., 1986, 28, 185-190.

6. Vanoni, M., Vai, M., Popolo, L. and Alberghina, L., Structural Heterogeneity in populations of the Budding Yeast Saccharomyces cerevisiae, J. Bacteriol., 1983, 156, 1282-1291.

7. Martegani, E., Mariani, L. and Alberghina, L., Yeast Biotechnological Processes Monitored by Analysis of Segregated Data with Structured Models, Modeling and Control of Biotech. Processes, ed. A. Johnson , Pergamon Press, 1985, 237-243.

8. Mariani, L., Martegani, E. and Alberghina, L., Yeast Population Models for Monitoring and Control of Biotechnical Processes, IEE Proc., 1986, 133, Pt.D., 210-216.

9. Cazzador, L. and Mariani, L., A Simulation Program for Yeast Cell Population Models, Int. Rep. LADSEB-CNR 02/87, April 1987

10. Hjortso, M.A. and Bailey, J.E., Transient Responses of Budding Yeast Populations, Math. Biosc., 1983, 83, 121-148.

INTERNATIONAL CONFERENCE ON BIOREACTORS AND BIOTRANSFORMATIONS
GLENEAGLES, SCOTLAND, UK: 9-12 NOVEMBER 1987

Not presented

Paper B2

ON-LINE MEASURING METHODS FOR THE CONTROL OF THE FLUID DYNAMIC BEHAVIOUR OF BIOREACTORS

A. Lübbert, B. Larson and T. Korte
Institut für Technische Chemie
Universität Hannover, D-3000 Hannover 1, German Federal Republic

ABSTRACT

Improvements in the development of two modern techniques for the characterisation of bioreactors are discussed. The first is the 'Heat pulse technique' for measuring the motions in the continuous liquid phase of submerged bioreactors and the second the 'Ultrasound-pulsed-Doppler method' for measuring the properties of the dispersed gas phase. Examples are given for both to furnish evidence of their on-line applicability in real fermentation processes, at any scale.

INTRODUCTION

Very little information is available on the fluid dynamic behaviour of large bioreactors. The main reason is that there are no reliable measuring devices, that can be applied to actual cultivation broths under aseptic conditions. In this paper, improvements in two modern techniques which allow local measurements of the important fluid dynamic properties in bioreactors during the cultivation process are discussed. These two techniques comprise the 'Heat pulse method' to determine the continuous liquid phase behaviour and the 'Ultrasound-pulsed-Doppler method' to determine information on the dispersed gas phase.

HEAT PULSE TECHNIQUE

The 'Heat pulse method' is a correlation method, which is used primarily to determine the time of flow distribution of fluid elements flowing from one specified point in the dispersion to a second. Usually correlation measurements are conducted by measuring the fluctuations of an appropriate variable at two adjacent points in the flow. The resulting time-of-flow distribution can be obtained from the two signals by calculating the cross- correlation function.

The usual correlation technique has severe constraints when applied to turbulent flow within chemical and biochemical reactors, because these correlations are possible without real transport of matter from one measuring point to the other. Thus any synchronizing effects in the flow, eg waves or turbulent eddies can result in misleading correlations and it becomes impossible to measure times-of-flow or velocities. Unambiguous results however can be obtained by replacing one fluctuating measuring signal by a predefined signal, introduced into the flow. This can be accomplished by means of a tracer input. The tracer concentration, which is measured at the other point, can then only originate from the input, if one uses a rapidly decaying tracer. The transfer of information is then coupled with a transport of matter from the source point to the detector.

In such tracer experiments mean times-of-flow are in the order of 100 ms. Consequently, one must use extremely short tracer pulses to fulfil the requirements of system theory. In a real implementation of the technique there are some problems with material tracers, since it is not possible to add quickly even a small amount of material to the flow without transferring considerable momentum and disturb the time-of-flow measurement.

Heat as a non-material tracer matches all requirements and therefore provides a solution to this problem. Any additional buoyancy is negligible in our case since this raises the temperature in the fluid by not more than 1 K. This can be done electrically by heating the fluid between two electrodes utilizing the resistance of the conductive biosuspension. Extremely simple probes such as the improved device shown in Fig. 1 must be introduced into the flow. Insteady of using d.c. we supplied the probe with high frequency current to act as a heating source for the medium. 700 kHz is an appropriate frequency to minimize electrochemical effects at the electrodes and to suppress cross-talk effects to the receiver channel by means of simple cut-off filters. Sensitive resistance thermometers can be used to detect the temperature fluctuations at the second point in the flow.

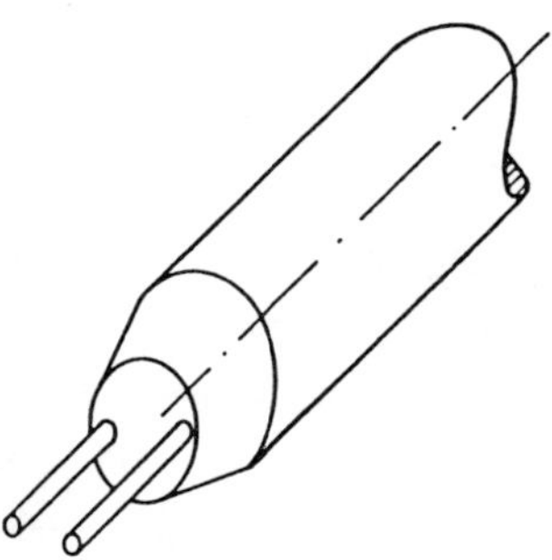

Figure 1. Heating probe which can be used to mark fluid elements locally. Heat is introduced into the fluid phase by current flow through the conducting cultivation broth.

To obtain a sufficiently high signal-to-noise ratio it is essential to use more sophisticated measuring techniques than simple Dirac pulse methods. We used a pseudo statistical tracer input, which emulates temperature fluctuations sufficiently well. Thus the cross-correlation of the input signal with the system's response delivers the time-of-flow distribution as indicated in Fig. 2.

Good signal-to-noise ratios are obtainable even in strongly disturbed systems. This opens the possibility of reducing the input signal intensity to very low levels. Thus the method can be applied even in very sensitive biological systems. As already mentioned, the temperature enhancement may be kept lower than 1 K for real applications. Nevertheless, it is possible to increase the probe distance over which one can detect correlations between input signal and system's response in the riser of an airlift loop reactor. Compared to earlier experiments, it is now possible to operate with about 20 cm probe distances. Because it was not possible to move the detector over such large distances during a fermentation run, we demonstrated this property with simple air-in-water dispersions. Using the method of analysis of the time-of-flow curves described in [1] the mean flow times, shown in Fig. 3, could be determined.

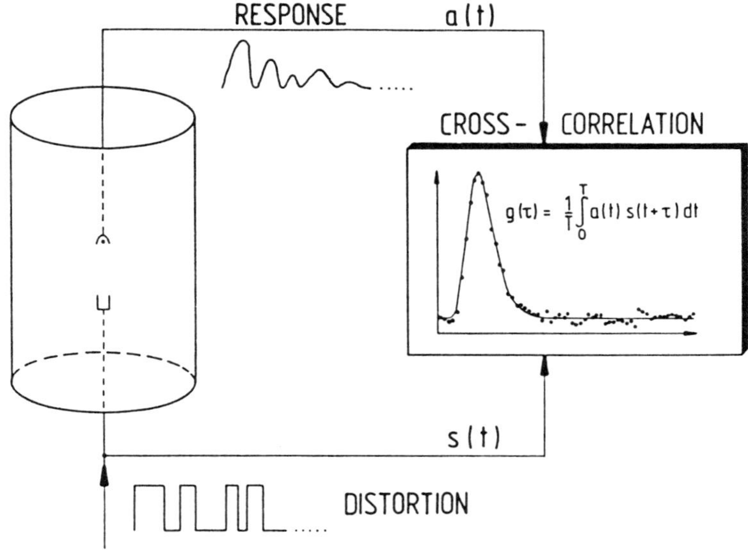

Figure 2. Principle of the recovery of time-of-flow curves from the pseudo statistical test signal and the system's response by calculating cross-correlation functions.

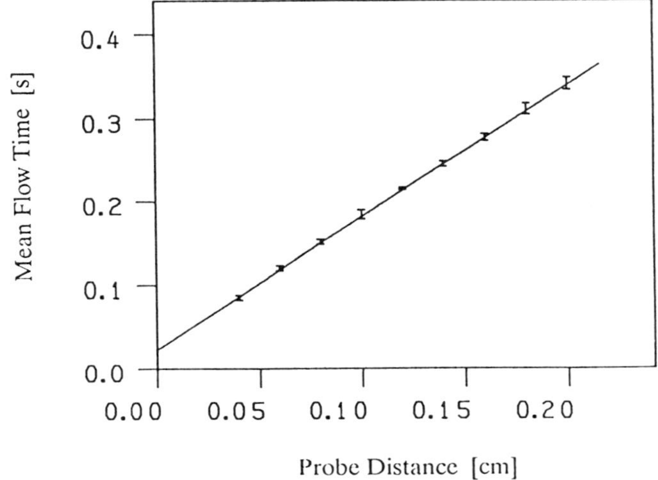

Figure 3. Mean flow times versus probe distance measured in the draught tube of a laboratory scale airlift tower loop reactor.

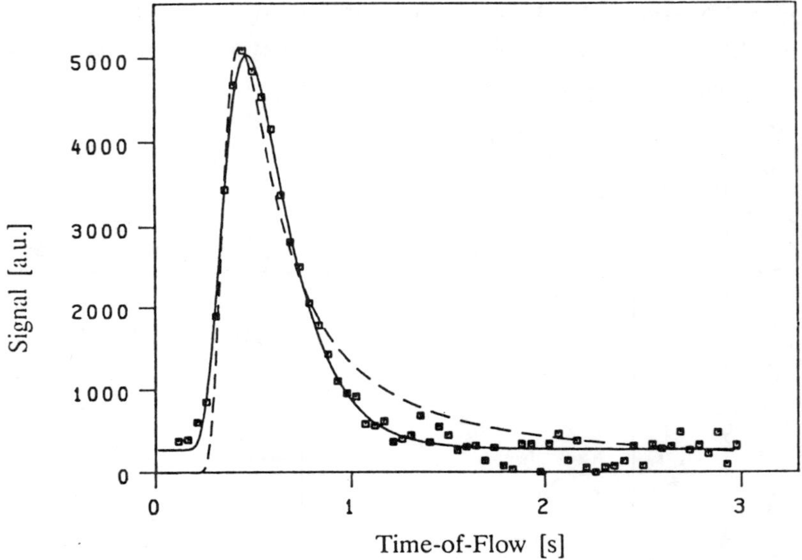

Figure 4. Time-of-flow distribution density measured at the axis in the draught tube of an airlift tower loop reactor during a cephalosporine production. The symbols are the measuring data the lines are results from model calculations.

Similar distances may be obtained during real fermentation runs. Fig. 4 indicates a result obtained, during a cultivation run in an extremely complicated medium for a cephalosporine production on a viscous medium containing about 20 volume per cent peanut meal. The sensor probe was at a distance of 11 cm from the heat source.

In Fig. 4 the symbols represent the data and the curves are results of a least square fit of models to the data. Principally such models contain two types of information concerning the flow: first, the time of transport from the point where the fluid has been labelled to the detection point and secondly, information on the dispersion process in the fluid during this flow time.

The broken curve is based on the assumption that the mixing process is by isotropic turbulence and it is clear that turbulence is not the dominating process here. The full curve results from a model in which the dispersion is less intensive. This assumes that convective and diffusive effects govern the mixing process. From both models it is easy to extract the characteristic numbers which describe the dispersion process, ie the diffusivity.

ULTRASOUND-PULSED-DOPPLER TECHNIQUE

In many aerobic cultivation processes the oxygen-supply to the microorganisms is limited. In investigations on oxygen transfer, one is interested in the transport cross section, especially in the specific interfacial area as well as in the mass-transfer coefficient, both being dependent on the fluid dynamic motion in the bioreactor. Unfortunately, there was a lack of reliable measuring devices which could be used during biochemical cultivations in sterile environments until very recently. The 'Ultrasound-pulsed-Doppler technique' is a newly developed method which matches the requirements necessary in cultivation processes.

The general set up of the Ultrasound-pulsed-Doppler technique for measuring local bubble velocity distributions and local specific interfacial areas is shown schematically in Fig. 5. Ultrasound pulses are transmitted into the disperson. Because bubbles are excellent reflectors, part of the ultrasound power is reflected to a detector. In the case of moving bubbles, the detected signal is shifted in frequency according to Doppler's principle. This frequency shift is proportional to the bubble's velocity component in the direction of the mid-angle between the input and reflected beams.

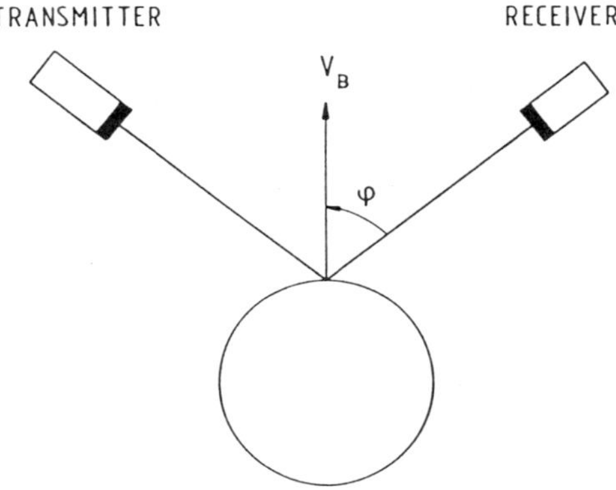

Figure 5. Schematic representation of the set up of an ultrasound-pulsed Doppler anemometer.

Typical probes used in our investigations are shown in Fig. 6. The probes are steam-sterilizable and can be introduced into reactors of all sizes. The same probes can be used as both transmitters and receivers. If one uses ultrasound pulses instead of a continuous ultrasound wave, each probe may be used alternately as transmitter or receiver. Using electronic gating techniques, the measuring volume can be adjusted appropriately within several centimetres from the probe.

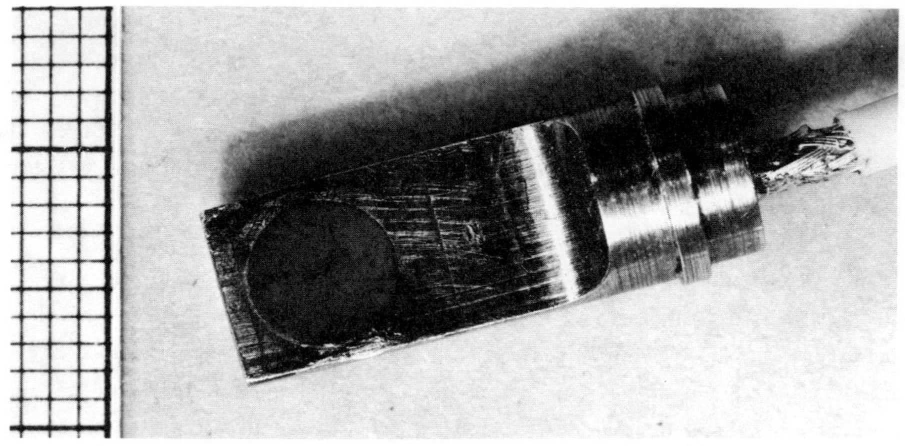

Figure 6. Ultrasound probes used during the experiments.

An ultrasound frequency of 4 MHz proved to be appropriate for bioreactor applications. A Doppler signal was derived electronically which oscillates with the difference frequency between the transmitted and the received ultrasound frequencies. The Doppler shifts were typically in the frequency band between 0 and 5 kHz. Such signals can be analysed digitally and in real-time by means of modern microelectronic components.

A simple calculation of power spectral density function is not sufficient for the analysis of the data, because this function does not contain the phase information, which must be used to distinguish between positive and negative bubble velocity values. This, however, can be accomplished with a quadrature detection, which requires the analysis of the original Doppler signal and an additional Doppler signal formed with the 90° phase-shifted transmitter frequency. Both signals were sampled and digitized at a rate of 10 kHz each.

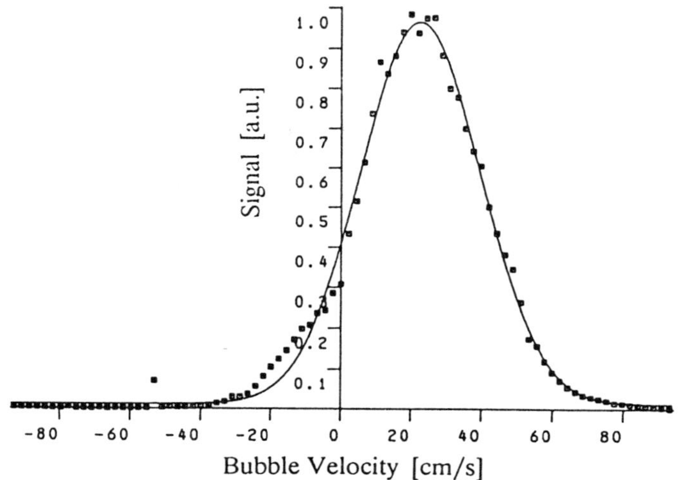

Figure 7. Bubble velocity distribution in a stirred tank reactor.

We developed an analysing unit based on a Texas Instruments' signal processor TMS320, which is able to calculate the bubble velocity distribution on-line during the data sampling. Thus, the result is available directly at the end of the data acquisition period.

A typical result obtained in a stirred tank reactor is depicted in Fig. 7. In such flows we observe positive as well as negative velocity values in most flow directions. Because it is not known a priori, the detector must be turned around the measuring volume to probe the flow direction. In Fig. 8 the mean velocities obtained at several directions are summarised. In this discussion, these data are shown to demonstrate, that the distortion of the flow by the probe is minimal. Otherwise we would not observe such a clear cosine function representing the mean bubble velocity components along the different directions.

A typical result for the mean bubble velocity measured during a yeast cultivation in an airlift loop reactor with concentric draught tube is shown in Fig. 9. The plot contains the values measured on the axis of the draught tube and in the annulus.

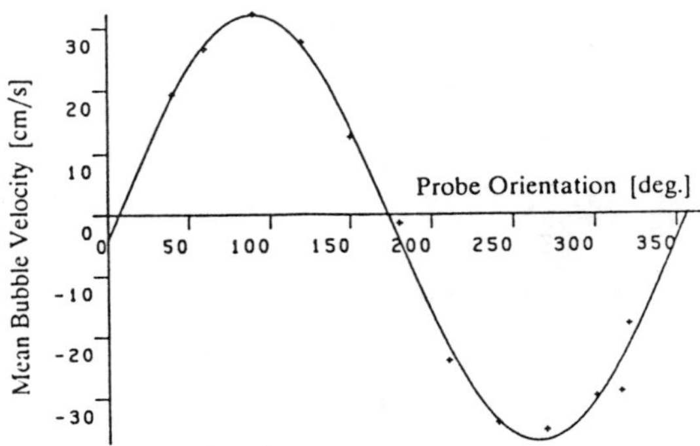

Figure 8. Components of the mean bubble velocity along several directions within a horizontal plane.

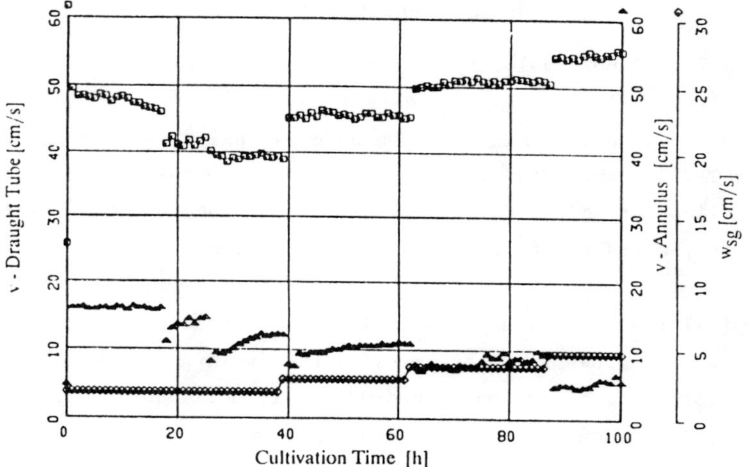

Figure 9. Typical results of a measurement of the mean bubble velocity along the axis of the riser and in the downcomer section of an airlift tower loop reactor during the yeast cultivation.

The scatter in the data appears to be minimal. It is interesting to note that the velocity sensitively responds to changes in the superficial velocity which is also displayed in the plot. At low fermentation times there are some additional jumps in the velocity signals. These are responses to additions of antifoam reagents leading to significant changes in the hydrodynamics as considerable changes in the energy input.

There are several possibilities of measuring specific interfacial areas by transmission techniques. Ultrasound is one type of radiation which is well suited to biotechnological applications because it is simple, non-destructive, and can be transmitted through optically opaque media. The transmission path, however, is also limited. This is no real disadvantage as far as one is interested in local measurements. As already shown, one can select a suitable path length for the ultrasound wave in the single probe technique by time gating. Thus, relatively small path lengths are possible, and they can be positioned anywhere in the bioreactor. Locations are only limited by their accessibility to probes. Hence, the method can be used in arbitrary large vessels, as opposed to the usual transmission techniques.

The detected ultrasound power density depends on two competing effects. It increases with the number density and the size of the bubbles in the measuring volume. On the other hand, it decreases according to an extinction law on the path through the dispersion from the transmitter via the measuring volume to the detector. The increase is proportional to the specific interfacial area, 'a', whereas the decreasing effect is proportional to exp(-a*d/4), d being the length of the ultrasound path. The overall signal power is therefore proportional to the product a*exp(-a*d/4). The proportionality constant can be obtained experimentally.

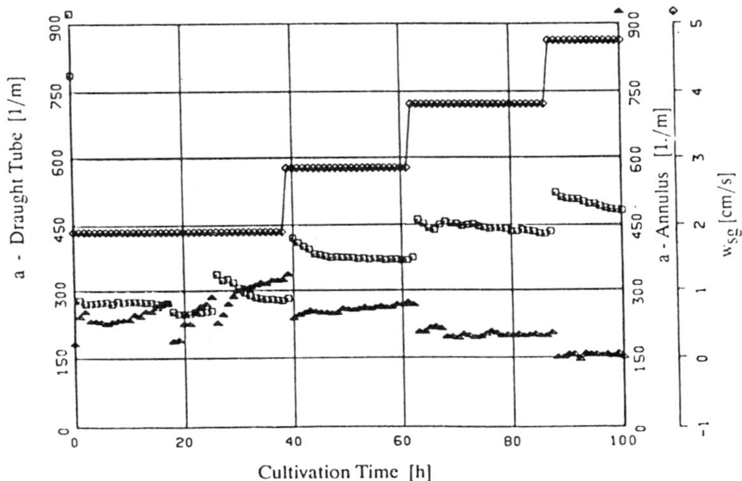

Figure 10. Specific interfacial area measured during a yeast fermentation as a function of the fermentation time.

This can then be used to determine values for the interfacial area from the measured ultrasound power, provided the specific interfacial area, 'a', is greater than a minimal value which can be adjusted by the path length. This condition can be easily fulfilled in practical cultivations. Fig. 10 contains results obtained during the fermentation from which the velocity values displayed in Fig. 9 were obtained.

CONCLUSIONS

We have described two methods, which can be used to measure the fluid dynamic properties of the flow in bioreactors during the cultivation processes. Computer controlled, both methods work on-line and are fully automized. Thus, they can be used to control the bioreactor, eg control of gas feed rate to maintain specific interfacial area, energy input for mixing etc.

REFERENCES

1. Lübbert, A., Korte, T. and Larson, B., Simple measuring techniques for the determination of bubble- and bulk-phase velocities in bioreactors, Proc. Int. Conf. on Bioreactor Fluid Dynamics, Cambridge, UK, BHRA, The Fluid Engineering Centre, Cranfield, Bedford, 1986, pp. 33-44.

2. Lübbert, A. and Larson, B., A new method for measuring local velocities of the continuous liquid phase in strongly aerated gas-liquid multiphase reactors, Chem. Eng. Technol., 1987, 10, 27-32.

3. Lübbert, A., Korte, T. and Schügerl, K., Ultrasonic Doppler measurements of bubble velocities in bubble columns, In Measuring Techniques in Gas-Liquid Two-phase Flows, IUTAM Symposium Nancy, France, 1983, eds, J.M. Delhaye, G. Cognet, Springer Verlag, Berlin, 1984, pp. 479-494.

4. Stravs, A. A. and von Stockar, U., Measurements of interfacial areas in gas-liquid dispersions by ultrasonic pulse transmission, Chem. Eng. Sci., 1985, 40, 11169-1175.

ACKNOWLEDGEMENT

The financial support of the DFG is gratefully acknowledged.

**INTERNATIONAL CONFERENCE ON BIOREACTORS AND BIOTRANSFORMATIONS
GLENEAGLES, SCOTLAND, UK: 9-12 NOVEMBER 1987**

Paper B3

A CALORIMETRIC STUDY OF MICROBIAL GROWTH ENERGETICS

U. von Stockar, I.W. Marison, and B. Birou
Institute of Chemical Engineering
Swiss Federal Institute of Technology
CH-1015 Lausanne, Switzerland

ABSTRACT

Isothermal reaction calorimetry has been employed on the bench scale to measure the enthalpy efficiency and oxygen efficiency of growth of several microbial strains cultured on a range of substrates with varying degrees of reduction. The observed values of these energetic efficiency factors agreed well with simple theoretical predictions based solely on the degree of reduction of the substrate as independent variable. Calorimetry is thus shown to be a useful tool for investigations of microbial growth energetics. Measuring heat dissipation in large scale bioreactors provides on-line information on the energetics of growth and should therefore be useful for process control purposes.

INTRODUCTION

The study of microbial energetics is frequently limited by the availability of sensitive and precise measurement techniques.

Calorimetry, the study of heat evolution during microbial growth, and non-growth related conditions, offers a simple technique for investigating microbial energetics. Heat is liberated by all micro-organisms as the result of the metabolism of the carbon and energy source for production of the energy required for biosynthetic reactions, transport, maintenance etc. The amount of heat dissipated is a measure of the free enthalpy dissipation necessary to drive the growth process at a non-zero rate. Consequently it is not correct to regard this heat as 'wasted' energy although it clearly depends on the efficiency of energy coupling. That part of the enthalpy present in the substrates which is retained in the biomass can however be used as a measurement of the growth efficiency of an organism. Such data may be obtained from the determination of the enthalpy content of the biomass, using combustion calorimetric techniques, which combined with heat evolution measurements (calorimetry) can be used to develop energy balances to give an insight into the energetics of a fermentation process.

Heat dissipation measurements on an industrial size reactor would then offer continuous information on the energetics and state of a culture thereby facilitating process control and design.

The aim of the present paper is to explore the application of novel calorimetric techniques, on a laboratory scale, for the study of microbial energetics.

THEORY

Definition and Significance of Energetic Growth Efficiency Factors

Microbial aerobic growth efficiency is often characterized in terms of two efficiency factors, termed η_O and η_H, which are defined as follows [1-4]:

$$\eta_O = \frac{\gamma_X}{\gamma_S} Y'_{X/S} \qquad (1)$$

$$\eta_H = \frac{\Delta H'_X}{\Delta H'_S / Y'_{X/S} + e_{X_3} \Delta H'_N} \qquad (2)$$

In these expressions, $Y'_{X/S}$ is the C-molar biomass yield on substrate (see eq 3), γ_i indicate the degrees of reduction defined by eq (5), $\Delta H'_i$ represent the C-molar heats of combustion, whereas e_{X_3} indicates the number of nitrogen atoms per carbon atom in the microbial biomass. The indices i assume values of X, S, P, and N for, respectively, dried biomass, substrate, and nitrogen source.

The meaning of these two efficiency coefficients can best be understood by an analysis of combined elemental and enthalpic balances of the growth process.

The overall stoichiometry of a general microbial growth process giving rise to a maximum of one major fermentation product may be described as follows:

$$\frac{1}{Y'_{X/S}} \, C H_{e_{S_1}} O_{e_{S_2}} N_{e_{S_3}} + Y'_{O/X} O_2 + Y'_{N/X} NH_3 \longrightarrow$$

$$C H_{e_{X_1}} O_{e_{X_2}} N_{e_{X_3}} + Y'_{P/X} \, C H_{e_{P1}} O_{e_{P2}} N_{e_{P3}} + Y'_{C/X} CO_2 + Y'_{W/X} H_2O \qquad (3)$$

The equation has been written in terms of C-moles, which means that each chemical formula for substrate, biomass, and for product has been reduced to the basis of one carbon atom. The stoichiometric coefficients appearing in this equation must therefore be regarded as instantaneous C-molar yields defined as the ratio of two conversion rates expressed in C-mol /(L•s).

In order to define the stoichiometry proposed by eq (3), one needs to determine the six yield coefficients appearing in this equation. Fortunately, it is not necessary to determine them all experimentally, as they are not independent from each other. Since four elemental balances can be written for the elements C, H, O and N, which involve the six $Y'_{i/j}$-coefficients, only two of them are truly independent. Hence, it is possible to compute all Y-values if two of them are known.

Thus if the C-molar yield of dry biomass on substrate, $Y'_{X/S}$ and the C-molar yield of product per C-mol of biomass formed, $Y'_{P/X}$, are known, it is possible to compute the oxygen required per biomass formed by solving the elemental balances. The result is:

$$Y'_{O/X} = \frac{\gamma_S}{4\, Y'_{X/S}} - \frac{1}{4}\gamma_X - \frac{1}{4}\gamma_P\, Y'_{P/X} \tag{4}$$

Solutions for the other stoechiometric coefficients may be found from several publications [5,6].

The functions γ_S, γ_X, and γ_P appearing in eq (4) are the degrees of reduction of, respectively, substrate, biomass, and product. For any chemical compound of the formula

$$C\, H_{e_{i1}}\, O_{e_{i2}}\, N_{e_{i3}}$$

the degree of reduction is defined by:

$$\gamma_i = 4 + e_{i1} - 2\, e_{i2} - 3\, e_{i3} \tag{5}$$

The degree of reduction indicates the number of moles of oxygen multiplied by 4 it would take to combust one C-mol of the compound to CO_2, H_2O and NH_3, and is often also called the number of "electrons available" in the compound.

Equation (4) is the basis for defining η_0. It can be rearranged as follows:

$$1 = \frac{4}{\gamma_S} Y'_{O/X} \cdot Y'_{X/S} + \frac{\gamma_X}{\gamma_S} Y'_{X/S} + \frac{\gamma_P}{\gamma_S} Y'_{P/X} \cdot Y'_{X/S} \tag{6}$$

The individual right hand terms of eq (6) show the fractions of oxygen up-take potential or of electrons available in the substrate that are, respectively, used up by the oxygen, conserved in the biomass, and conserved by the product. The fraction conserved in the biomass is equal to η_0 according to eq (1) and has been called "energetic yield coefficient" [3], "generalized substrate yield" [4], or "oxygen efficiency of growth" [1].

Besides being a measure of growth efficiency in terms of reducing power, η_0 is also of practical importance as it determines the oxygen requirement. Substituting eq (1) into eq (6), it follows for a strongly aerobic process without product formation ($Y'_{P/S} = 0$):

$$Y'_{O/X} = \frac{\gamma_X}{4} \left(\frac{1 - \eta_0}{\eta_0} \right) \tag{7}$$

The efficiency η_H defined by eq (2) is based on an enthalpy balance for the growth process according to eq (3):

$$\frac{1}{Y'_{X/S}} \Delta H'_S + e_{X_3} \cdot \Delta H'_N = Y'_{Q/X} + \Delta H'_X + Y'_{P/X} \cdot \Delta H'_P \qquad (8)$$

This equation has been derived by using combined elemental balances in order to express $Y'_{N/X}$ as a function of $Y'_{X/S}$ and $Y'_{P/X}$ and by assuming that neither the substrate nor the product contains nitrogen ($e_{S_3} = e_{P_3} = 0$) (for details, see Marison and von Stockar [5]). $Y'_{Q/X}$ in eq (8) is the so-called "heat yield". It denotes the amount of heat released per C-mol of biomass formed and can be measured calorimetrically.

Equation (8) may be rearranged as follows:

$$1 = \frac{Y'_{Q/X}}{\Delta H'_S/Y'_{X/S} + e_{X_3}\Delta H'_N} + \frac{\Delta H'_X}{\Delta H'_S/Y'_{X/S} + e_{X_3}\Delta H'_N} + \frac{Y'_{P/X} \cdot \Delta H'_P}{\Delta H'_S/Y'_{X/S} + e_{X_3}\Delta H'_N} \qquad (9)$$

The individual right hand terms of eq (9) show the fractions of the chemical energy available in the substrate that are, respectively, dissipated as heat, conserved in the biomass, and conserved in the product. The fraction conserved in the biomass is equal to η_H according to eq (2) and has been called "enthalpy efficiency of growth" [1].

Besides being a measure of growth efficiency in terms of energy, η_H is also of practical importance as it determines the amount of heat released during growth. Substituting eq (2) into eq (9) it follows for a process with $Y'_{P/X} = 0$:

$$Y'_{Q/X} = \Delta H'_X \left(\frac{1 - \eta_H}{\eta_H} \right) \qquad (10)$$

Usefulness of calorimetry

As $Y'_{Q/X}$ can be measured by calorimetry, the enthalpy efficiency of growth can be readily determined using eq (10):

$$\eta_H = \frac{\Delta H'_X}{Y'_{Q/X} + \Delta H'_X} \qquad (11)$$

But calorimetry can also be useful to determine η_O. The determination of $Y'_{O/X}$, from which η_O can be calculated, is in practice often difficult in laboratory fermentors due to the small oxygen depletion of the air stream and the difficulties in setting up oxygen balances this entails.

For strongly aerobic processes, it has been shown [1] many times that oxygen up-take and microbial heat generation are strictly coupled in a linear fashion:

$$Y'_{Q/X} = 4 \cdot Q_O \cdot Y'_{O/X} \qquad (12)$$

Q_0 is more or less constant for all strongly aerobic processes and amounts to about 115 kJ/C-mol [1].

Owing to the linear relationship (12) between oxygen up-take and heat release, it is possible to substitute calorimetry for oxygen up-take measurements. η_0 can then be determined by combining eqs (12) and (7):

$$\eta_0 = \frac{\gamma_X}{Y'_{Q/X}/Q_0 + \gamma_X} \qquad (13)$$

Prediction of efficiency factors

According to eq (1), η_0 may be predicted provided the C-molar biomass yield coefficient is known. As has been demonstrated by several authors [1,4,7-9], $Y'_{X/S}$ can be correlated as a first approximation with the degree of reduction of the substrate, γ_S. Typical correlations are of the following form:

$$Y'_{X/S} = a \cdot \gamma_S \qquad \text{for } \gamma_S < \gamma_S^C \qquad (14a)$$

$$Y'_{X/S} = C \qquad \text{for } \gamma_S \geqslant \gamma_S^C \qquad (14b)$$

Roels [1] assigns the following values to the model parameters:

$a = 0{,}13 \; ; \; C = 0{,}6 \; ; \; \gamma_S^C = 4{,}67.$

Substituting eq (14) into eq (1), one obtains the following prediction of the energetic yield coefficient η_0:

$$\eta_0 = a\,\gamma_X \qquad \text{for } \gamma_S < \gamma_S^C \qquad (15a)$$

$$\eta_0 = C\,\frac{\gamma_X}{\gamma_S} \qquad \text{for } \gamma_S \geqslant \gamma_S^C \qquad (15b)$$

A similar prediction may be made for η_H. It is well known that the C-molar enthalpies of combustion of $\Delta H'_i$ of most organic compounds and hence also of biomass may be estimated quite accurately by assuming that they are proportional to the degree of reduction defined with respect to CO_2, H_2O, and N_2 as combustion products [1,10]:

$$\Delta H'_i = Q_0\,\gamma_i^0 \qquad (16)$$

where

$$\gamma_i^0 = 4 + e_{i1} - 2\,e_{i2} \qquad (17)$$

and where Q_0 may be assumed constant and amounts to about 115 kJ/C-mol [10].

By substituting eqs (16), (17) and (14) into eq (2) and assuming that the nitrogen source is NH_3 for which $\gamma_N^0 = 3$, one obtains:

$$\eta_H = \frac{a\ \gamma_X^0}{1 + 3\ a\ e_{X3}} \qquad \text{for } \gamma_S < \gamma_S^C \tag{18a}$$

$$\eta_H = \frac{C\ \gamma_X^0}{\gamma_S^0 + 3\ C\ e_{X3}} \qquad \text{for } \gamma_S \geqslant \gamma_S^C \tag{18b}$$

MATERIALS AND METHODS

Cultures and media

The yeast strains employed in the present study, *Candida lipolytica* NRRL 1094, *Kluyveromyces fragilis* NRRL 1109, *Candida utilis* NRRL 1084 and *Candida boidinii* NRRL 2332, were obtained from the Northern Regional Research Laboratory (Peoria, Il. USA). The strain of *Methylophilus methylotrophus* was obtained from Imperial Chemicals Industries (Billingham, UK).

Media for the maintenance and cultivation of the strains has been reported elsewhere [5,6,13].

Cultivation conditions and analytical procedures

All yeast strains, with the exception of *C. boidinii*, were grown at 30 °C and pH 5.5. *C. boidinii* was grown at 25 °C, pH 5.5 and *M. methylotrophus* at 37 °C and pH 7.0. pH was maintained at the desired set-point by the automatic addition of NaOH (4N) and HCl (4N).

Agitation was maintained at 700 rpm with an aeration rate of 3 l/min controlled using a mass-flowmeter (Model 5850 TR, Brooks Instruments, Veenedaal, NL). These conditions were chosen in order to maintain a dissolved oxygen concentration above 50 % saturation thus suppressing fermentation by-product formation.

Growth experiments were carried out in a modified BSC 81 heat-flux calorimeter with a working volume of 1.6 l such that the heat dissipated by the cultures could be measured continuously *in-situ.*

Exit gases from the BSC 81 were passed through paramagnetic and infra-red analyzers for the determination of O_2 and CO_2 respectively.

Samples (2 ml) from the batch cultures were removed at intervals for the determination of absorbance (at 600 nm) and carbon substrate concentration (g/l). Cell dry weight (g/l) was determined by relating the measured absorbance to a previously determined calibration curve of cell dry weight versus absorbance at 600 nm.

Glucose, galactose and lactose were determined using standard enzymatic techniques [14].

Methanol and ethanaol were analyzed using gas chromatography (Shimadzu, Kyoto, Japan, Model Mini 1) as previously described [5,6,13].

Glycerol, succinate, citrate and acetate were determined using HPLC (Knauer, West Berlin, FRG) techniques described elsewhere [6,13].

Calorimetric measurement system

A detailed description of the BSC 81 (Ciba-Geigy, Basel, CH) [15,16] heat flux calorimeter, modified for the cultivation of microbial cells, together with the principles of the measurement technique have been described elsewhere [11,12].

Basically the BSC 81 consists of a 2-litre jacketted glass reactor connected to a silicone oil thermostatting system. The reactor is constructed in a similar way to a standard laboratory scale fermentor with a system for monitoring and controlling agitation, aeration, pH and other environmental parameters. The flow rate, temperature and humidity of the air entering the reactor are carefully controlled using a mass-flowmeter, connected to a thermostatted bubble column.

The silicone oil thermostatting system is composed of two interconnecting systems. In the first system the oil is heated to a temperature approximately 3 °C higher than the desired reactor temperature, T_R. In the second system oil is cooled to a temperature of at least 5 °C below T_R. A computer-controlled valve connects the two systems such that the 'hot' and 'cold' oils are mixed rapidly together. The resulting oil, with a temperature T_J, is pumped at a rate of 2 l/s through the jacket of the reactor in order to maintain the reactor temperature, T_R, constant.

Control of the reactor temperature T_R is based on its continuous, precise measurement. A control algorithm counteracts any deviations from the set-point by adjusting T_J by means of the computer controlled mixing valve.

The temperatures T_R and T_J can be used to measure the power, q_F (W) exchange between the reactor and the oil thermostatting system:

$$q_F = U \cdot A \cdot (T_R - T_J) \tag{19}$$

where U is the global heat transfer coefficient (W/m^2-°K) and A is the heat transfer area (m^2).

In the absence of any fermentation the value of q_F will be the sum total of all the heat loss and gain terms from the reactor, through agitation, evaporation, heat lost to the airsteam and to the environment. On the power-time curve (thermogram) this signal represents the base-line which should remain constant if all of the parameters are carefully controlled.

During a fermentation process or other reaction in which heat is dissipated or absorbed, the value of T_J will change in order to maintain a constant T_R. Consequently the rate of heat evolved q, at any time, t, will be given by:

$$q(t) = U \cdot A \cdot [T_{Jo} - T_J(t)] \tag{20}$$

where T_{Jo} is the jacket oil temperature in the absence of a reaction. Equation (20) can consequently be used to determine the rate of heat dissipated by a microbial culture at any time. Integration of equation (20) yields the total amount of heat dissipated by a culture. Changes in A, due to removal of culture samples or addition of pH controlling agent, can be corrected for on-line. Similarly changes in U, as a result of the growth of the culture, may be corrected for using an in-built electrical calibration heater, the power output of which is precisely known.

RESULTS AND DISCUSSION

Relationship between heat generation and growth

Heat flux measurements in the modified BSC 81 as a function of time during aerobic batch cultures have already been reported by Marison and von Stockar in several publications [5,6,11,12]. It was demonstrated by them as well as by other authors that the heat generation rate increases strictly in parallel with the growth rate during the exponential phase. Indeed, when the measured heat generation rate is integrated in order to compute the total amount of heat released up to a certain point in time, and when this is plotted against the total amount of biomass formed, linear correlations are usually obtained such as shown in figure 1.

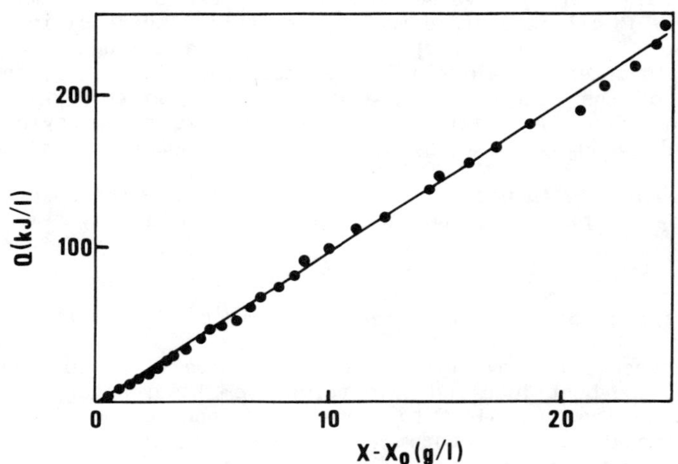

FIGURE 1: The relationship between total heat evolved, Q(kJ/l) and cell dry weight, $X - X_0$ (g/l) for K. fragilis grown on deproteinized whey permeate. Slope = $Y_{Q/X}$ = 9.43 kJ/g cell dry weight Reproduced with permission from [5].

The slope of the correlation is the heat yield $Y_{Q/X}$ expressed in kJ per g cell dry weight. The procedure employed to convert this data into the C-molar heat yield $Y'_{Q/X}$ (kJ/C-mol) may be found in [5].

Heat yields were determined in this manner for batch cultures of various micro-organisms grown aerobically on a range of substrates with varying degrees of reduction. Typical results are compiled in table 1. From the calorimetric data, the oxygen requirement per C-mol of biomass formed, $Y'_{O/X}$, was computed using eq (12) and also included in table 1.

TABLE 1. Measured Heat Yields and Efficiency Factors

In compiling the table, experimental results of several ref. [5,6,10] were used.

STRAIN	SUBSTRATE	γ_S	$Y'_{Q/X}$	η_H	Y'_{OX}	η_O
C. lipolytica	Citrate	3	388.2	0.587	0,844	0.554
	Succinate	3.5	434.0	0.560	0.943	0.527
K. fragilis	Glucose	4	324.5	0.618	0.705	0.602
	Galactose	4	362.9	0.596	0.789	0.570
	Lactose	4	395.9	0.578	0.861	0.547
C. utilis	Acetate	4	460.4	0.545	1.001	0.512
	Glycerol	4.67	279.4	0.664	0.607	0.634
C. boidinii	Ethanol	6	539.0	0.506	1.172	0.473
methylotrophus	Methanol	6	595.6	0.485	1.295	0.446
C. lipolytica	Hexadecane	6.13	663.7	0.454	1.143	0.421

Units: $Y'_{Q/X}$, kJ/C-mol ; $Y'_{O/X}$, mol/C-mol

Energetic Growth Efficiency Factors

The enthalpy efficiency of growth η_H and the oxygen efficiency of growth η_O were calculated using eqs (11) and (13), respectively. The required data on dried biomass, compiled in table 2, was either measured in the laboratory of the authors [10] or taken from the literature [9].

TABLE 2: Heats of combustion and degree of reduction of dry biomass

ORGANISM	SUBSTRATE	$\Delta H'_X$ kJ/C-mol	γ_X^0	γ_X	Ref.
K. fragilis	glucose	523.3	4.71	4.26	[10]
"	galactose	534.9	4.69	4.18	[10]
"	lactose	542.6	4.63	4.16	[10]
Other yeasts	various	552.0	4.80	4.20	[9]
M. methylotrophus	methanol	561.6	4.92	4.17	[10]

As can be seen from table 1, the two growth efficiency factors are very similar for the same culture experiment. This is not surprising since they both characterize growth efficiency from an energetic point of view. As Roels [1] has shown, the two efficiencies are however theoretically equal only if N_2 is used as a nitrogen source. With ammonia, η_H should theoretically be 6 % higher than η_0 at $\eta_H = 0,5$ [1]. A close analysis of the data in table 1 reveals indeed systematically higher η_H values. The ratio η_H/η_0, averaged over the 10 points presented in table 1, turns out to be 1.060 ± 0.017.

Plots of the efficiency factors as a function of γ_S (figs. 2 and 3) reveal two distinct regions. For low degrees of reductions of the substrates, the efficiencies of growth appear to be more or less constant, whereas growth clearly loses efficiency when highly reduced substrates are used. This behaviour is predicted quite accurately by the theoretical models, eqs (15) and (18). The existence of these two regions has been explained by many researchers by the fact that rather oxidized substrates are limited with respect to their content of chemical energy, while growth becomes carbon limited on reduced substrates. Since oxidized substrates are poor in energy, micro-organisms will try to harness a maximum of the available energy. It seems that a maximum of about 60 % can be retained in the biomass, whereas 40 % has to be dissipated as heat to drive biosynthesis at a non-zero rate, and partially also because of inefficiencies of the growth process. Since the energy of the substrate is proportional to γ_S, retaining a constant portion of this energy in the substrate also means that the biomass yield $Y'_{X/S}$ must increase in proportion to γ_S (see Eq (14)). But it seems that, for unknown reasons, not more than about 60 % of the carbon atoms present in the substrate can be incorporated in the biomass, thereby limiting $Y'_{X/S}$ to 0.6 at high values of γ_S. As γ_S and the energy content of the substrate continue to increase, $Y'_{X/S}$ cannot keep pace and therefore the fraction of the energy retained in the biomass must fall, more energy being dissipated as heat as the $Y'_{Q/X}$ values clearly indicate.

There are obviously deviations from this general trend due to the characteristics of individual strains. A point in case is *C. utilis*, which was able to grow on glycerol with an unusually high biomass yield ($Y'_{X/S} = 0.666$). This reflects itself in higher than average energetic efficiency factors and results in a low heat dissipation, in spite of the relatively high degree of reduction of glycerol.

CONCLUSIONS

Isothermal reaction calorimetry is a useful tool for investigating microbial growth energetics. The enthalpic growth efficiency can be measured directly in cases without product formation, and the energetic yield coefficient also can be measured indirectly. Both energetic growth efficiencies can be predicted by simple models based solely on a knowledge of the degree of reduction of the substrate. The models are confirmed by the experimental data presented in this study.

Measuring heat dissipation rates in large scale fermentors offers online information on, and insight into, the energetics of the culture and should therefore be useful for process control. Abnormally low energetic growth efficiencies with respect to figs 2 and 3 might suggest inefficient growth due to a non-optimal medium, high values might be indicative of product formation. Of course the more on-line yield information is available, the better the diagnosis will be.

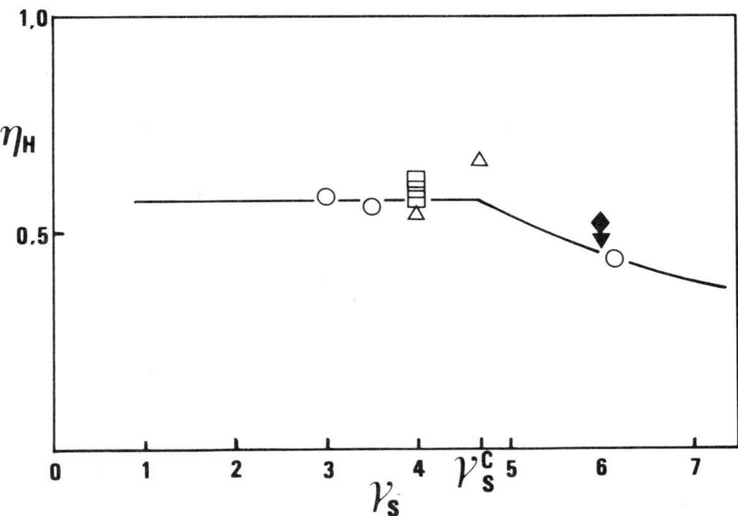

FIGURE 2: Energy efficiency, η_H, of growth on a range of carbon and energy sources having different reductance degrees, γ_S. Symbols: ○ C. lipolytica ; □ K. fragilis ; △ C. utilis ; ◆ , C. boidinii ; ▼ M. methylotrophus. The symbols represent experimentally determined values, the solid line represents the values predicted by the model (eq 18)

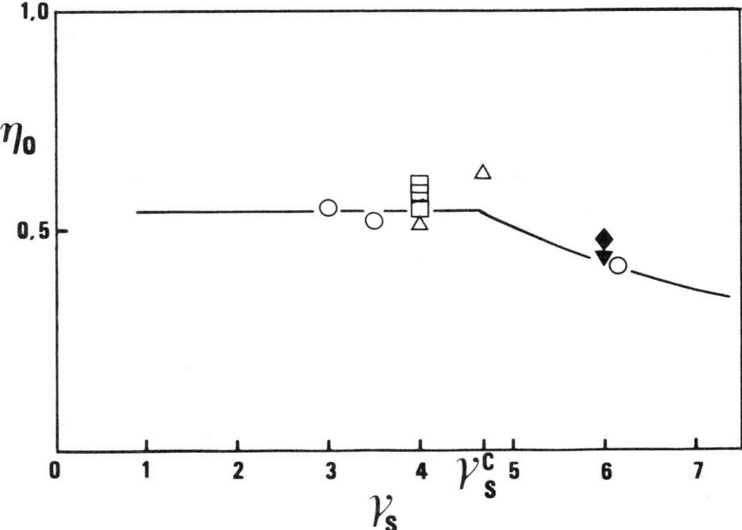

FIGURE 3: Oxygen efficiency, η_O, of growth on a range of carbon and energy sources having different reductance degrees, γ_S. Symbols and explanation as for Figure 2.

For bioprocess control purposes, the concept of the energetic growth efficiencies may not be absolutely necessary. Measured heat yields could be used to compute other coefficients characterizing the instantaneous stoichiometry by using balances such as eq (8). Also, the heat yield $Y'_{Q/X}$, as well as the oxygen requirement, $Y'_{O/X}$, could be predicted directly as a simple function of the degree of reduction of the substrate by substituting eq (18) into eq (10), and, respectively, by substituting eq (15) into eq (7). This is demonstrated for the heat yield in a forthcoming publication [6].

REFERENCES

[1] Roels, J.A.: Energetics and Kinetics in Biotechnology. Elsevier Biomedical Press, Amsterdam (1983)

[2] Minkevich, I.G. and Eroshin, V.K.: Folia Microbiol., 18 (1973) 376

[3] Erickson, L.E., Minkevich, I.G. and Eroshin, U.K.: Biotechnol. Bioeng.,20 (1978) 1595-1621

[4] Heijnen, J.J. and Roels, J.A.: Biotechnol. Bioeng., 23 (1981) 739-763

[5] Marison, I.W. and von Stockar, U.: Enz. Microb. Technol., 9 (1987) 33-43

[6] Birou, B., Marison, I.W. and von Stockar, U.: "Calorimetric Investigation of Aerobic Fermentations", Biotechnol. Bioeng., (accepted for publication, 1987)

[7] Bell, G.H.: Proc. Biochem., 7 (1972) 21-34

[8] Linton, J.D. and Stephanson, R.J.: FEMS Microbiol. Letts., 3 (1978) 95-98

[9] Roels, J.A.: Biotechnol. Bioeng., 22 (1980) 2457-2514

[10] Cordier, J.-L., Butsch, B.M., Birou, B. and von Stockar, U.: Appl. Microbial Biotechnol., 25 (1987) 305-312

[11] Marison, I.W. and von Stockar, U.: Thermochim. Acta, 85 (1985) 493-496

[12] Marison, I.W. and von Stockar, U.: Biotechnol. Bioeng., 28 (1986) 1780-1793

[13] Birou, B.: PhD Thesis No. 612 (1986), EPF-Lausanne, Switzerland

[14] Boehringer Mannheim GmbH, Biochemica, (1977/78) "Methoden der Lebensmittelanalytik"

[15] Giger, G., Aichert, A. and Regenass, W.: Swiss Chem., 4 (1982) 33-36

[16] Regenass, W.: Chimia, 37 (1983) 430-437

ACKNOWLEDGEMENTS

This work was supported by a grant from the "Fonds National Suisse pour la Recherche Scientifique".

INTERNATIONAL CONFERENCE ON BIOREACTORS AND BIOTRANSFORMATIONS
GLENEAGLES, SCOTLAND, UK: 9-12 NOVEMBER 1987

Paper C1

IMMOBILISATION AND GROWTH OF HYBRIDOMAS IN PACKED BEDS

A.D. Murdin[1], J.S. Thorpe[1], N. Kirkby[2], D.J. Groves[3] and R.E. Spier[1]

[1]Dept. of Microbiology,
University of Surrey,
Guildford, Surrey, GU2 5XH,
UK

[2]Dept. of Chemical and Process Engineering,
University of Surrey,
Guildford, Surrey, GU2 5XH,
UK

[3]Dept. of Biochemistry,
University of Surrey,
Guildford, Surrey, GU2 5XH,
UK

ABSTRACT

Packed bed bioreactors have been used for the culture of adherent monolayers of animal cells for a number of years but have only recently been considered for hybridomas. This paper describes the immobilisation of hybridomas in packed beds of polyester foam particles and of other materials. Hybridomas were filtered from growth medium recirculated through a packed bed. The dominant capture mechanism was apparently capture due to gravitational settling. The cells immobilised within the packed bed grew and produced similar quantities of antibody to those produced in static flask or airlift cultures.

INTRODUCTION

Packed bed bioreactors for the cultivation of anchorage-dependant animal cells have been in use for over twenty five years [1]. The commonest support material has been a matrix of 3mm glass beads [2,3,4]. Reactors using a packed bed of such beads have been succesfully operated at the 100 litre scale for the growth of BHK 21 C13 monolayer cells and the subsequent production of foot-and-mouth disease virus vaccine antigen [5]. The growth and productivity of anchorage dependant cells in such systems has been widely studied [2,3,4,5,6], but only recently has much attention been given to the use of packed bed bioreactors for the cultivation of anchorage-independant animal cells, including hybridomas [7,8].

An important feature of the glass bead packed bed system is the ability to inoculate cells into a pre-formed bed, to which they adhere and which they subsequently colonise by growth in situ. For this reason packed beds formed from agarose, alginate or similar beads already containing hybridomas are less desirable than beds formed from some form of matrix which can passively entrap hybridomas introduced into the bed at a later stage. The use of polyester and polyurethane foam matrices for the immobilisation of plant and yeast cells in this way has been described by a number of workers [9,10,11].

The use of a packed bed formed from such materials would appear to have many other advantages for the culturing of hybridomas. Since the cells would be passively trapped within an inert matrix the process should be gentle towards the cells yet have all the benefits of other immobilised cell reactors - the ability to perfuse the cells with fresh medium, provision of a conditioned and protected environment for the cells, increased process intensity, simplified downstream processing and so on. Furthermore, packed bed reactors of the form described [3] are robust, mechanically simple and proven systems.

This paper describes the immobilisation and growth of hybridomas in packed beds formed from various materials, including polyester foam particles.

MATERIALS AND METHODS

Bioreactor

The general configuration of the reactor is shown in Fig. 1, and follows the design of Spier and Whiteside [3]. A vessel holding a bed (3.5cm in diameter, 3.5cm deep) of the support material was connected to a medium reservoir (c. 100ml) which acted as an airlift. Ports for inoculation, sampling, antifoam addition and gas outflow were let into the reservoir. The reactor, including the packed bed, was sterilized by autoclaving before use.

The reservoir was sparged with 5% CO_2 in air at a flow rate of 30ml min^{-1}, which was sufficient to drive medium around the reactor at a flow rate of c.50ml min^{-1}.

Cultures were initiated by inoculating c. $2x10^5$ cells ml^{-1} into the medium reservoir of the reactor.

Packed Bed Materials

Six bed materials were examined. These were stainless steel wool (Radleys, Sawbridgeworth, UK), nylon wool, nylon fibre (Travenol Labs.,Thetford, UK), Pyrex sintered glass discs (J.Bibby Science Products Ltd, Stone, Staffs., UK) and polyether or polyester foam (Flexiplas Ltd, Wisbech, UK).

The sintered glass discs were mixed with 3mm glass beads in the bed, to maintain open channels around the discs. The foam was cut into 5mm cubes before use. All materials were soaked overnight in 7X detergent (Flow Labs. Ltd) then thoroughly rinsed in distilled, deionised water prior to use. With the exception of the glass discs all materials were compressed slightly when the bed was first packed. This reduced further settling of the bed under flow.

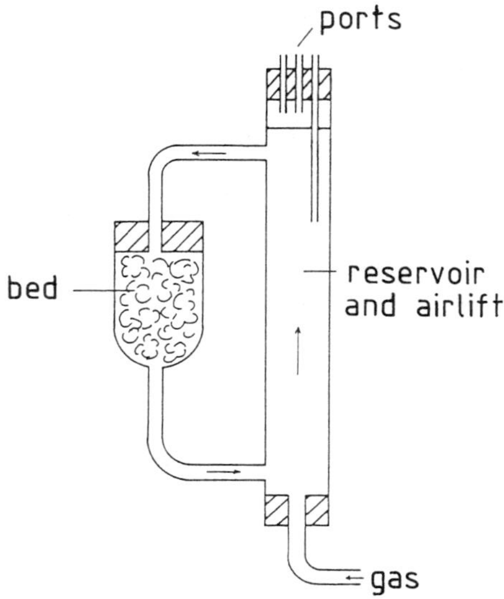

Figure 1. The packed bed bioreactor design of Spier and Whiteside [3]

Hybridomas

Two hybridoma cell lines were used. An NS-1 derived mouse-mouse hybridoma, 321, which secreted IgG to paraquat [12], was cultured in RPMI 1640 (Flow Labs. Ltd) plus 2mM glutamine, 0.2% sodium hydrogen carbonate and 5% adult bovine serum (Advanced Protein Products Ltd, Brierley Hill, UK). A bovine-murine heterohybridoma, B/MT.4A/17.H5/A5.5H8.10, which secreted bovine IgG to testosterone [13], was cultured in DMEM (Gibco Ltd, Uxbridge, UK) plus 2mM glutamine, 0.1125% sodium hydrogen carbonate and 10% myoclone foetal calf serum (Gibco Ltd).

Assays

Where possible, cell numbers and viability were assessed by trypan blue dye exclusion.

Lactate dehydrogenase (LDH) was measured by monitoring spectrophotometrically (λ= 340nm) the rate of oxidation of NADH during the conversion of pyruvate to lactate in the presence of LDH [14].

Antibody production by the 321 cell line was measured using a sandwich ELISA specific for murine IgG. Antibody production by the B/MT cell line was measured using a direct ELISA specific for bovine antibodies to testosterone.

RESULTS AND DISCUSSION

Entrapment of Hybridomas in Packed Beds

The ability of the various support materials to entrap 321 hybridomas was assessed (Fig. 2). Cells were introduced into the medium reservoir of the reactor and counts of cells remaining in suspension were made over the next few hours.

Polyester foam trapped cells most efficiently. Stainless steel wool and sintered glass were also useful, trapping cells more slowly but having a high capacity. After twenty four hours few cells could be observed in suspension in reactors in which these materials formed the packed bed (data not shown). The

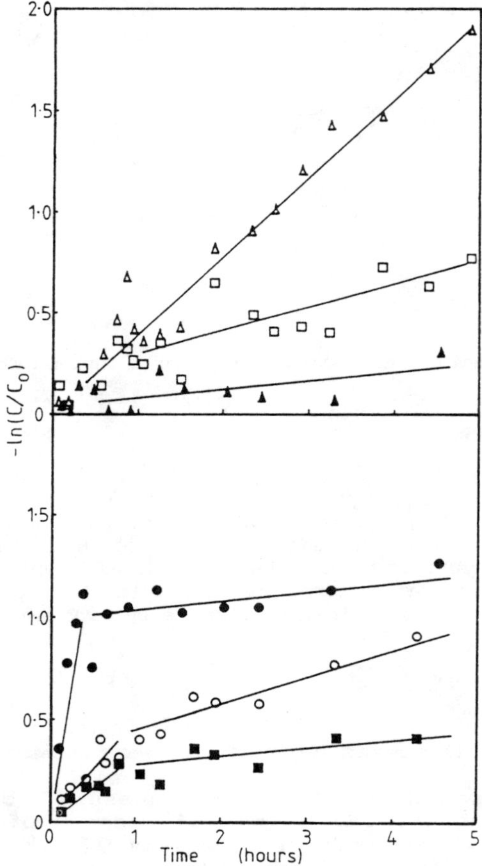

Figure 2. The entrapment of cells by packed beds formed from polyester foam (△), sintered glass (○), nylon fibre (●), nylon wool (▲), polyether foam (■), or stainless steel sponge (□). Cells introduced into the medium reservoir of the reactor were recirculated through the packed bed. C_0 is the initial concentration of cells in the reactor, C is the concentration after time t.

nylon fibre had a high affinity for cells but a limited total capacity, becoming saturated only thirty minutes after the start of the experiment. In reactors utilising nylon fibre packed beds the cells were effectively growing in suspension. Polyether foam and nylon wool were very poor support materials, trapping few cells.

Antibody Production by Hybridomas in the Packed Beds

The amount of antibody produced by the 321 cell line in various culture systems was assessed (Fig. 3). There was no difference in antibody production per unit volume of culture medium by cells cultured in the packed bed reactors, in 175cm^2 static flasks or in 1500ml airlift reactors, with the exception of

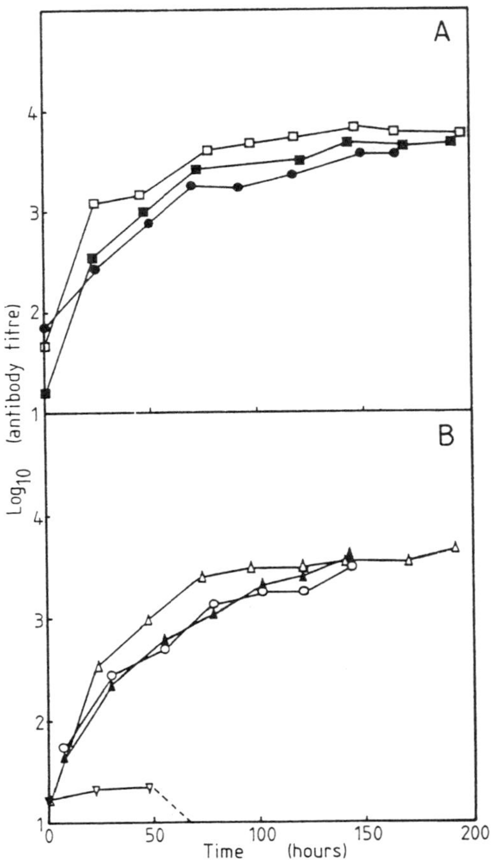

Figure 3. Antibody production by 321 hybridomas in various bioreactors. Panel A shows antibody production in a packed bed of polyester foam (□), in a 175cm^2 static flask (■) and in a 1500ml airlift reactor(●). Panel B shows antibody production in packed beds of stainless steel sponge (Δ), polyether foam (∇), sintered glass (o) or nylon fibre (▲).

the reactor in which the packed bed was composed of polyether foam. Polyether foam was toxic to the 321 cell line. Cell death could be observed in circulating cells within ten hours of inoculation into reactors using polyether foam packed beds, and antibody was not produced (Fig. 3).

Growth of Hybridomas in Packed Beds of Polyester Foam

Further studies were carried out to assess the viability and growth of hybridomas immobilised in packed beds of polyester foam.

The release of LDH into the circulating medium was used to estimate the viability of cells immobilised in a packed bed [14]. Figure 4 shows the relationship between cell growth, viability and the release of LDH in static

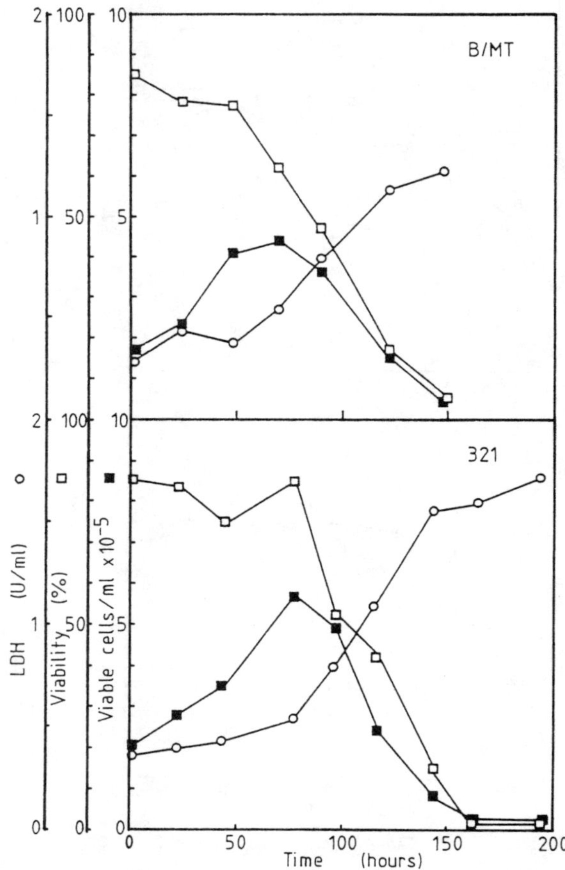

Figure 4. The relationship between growth, viability and release of LDH into the culture medium by 321 and B/MT hybridomas in static flask cultures.

flask cultures of 321 and B/MT cells. The rise in LDH activity corresponded precisely to the decline in cell viability. Figure 5 shows the relationship between cell growth, as indicated by LDH activity, and antibody production by 321 and B/MT hybridomas, respectively, in static flasks and polyester foam packed bed reactors. In all cases LDH activity rose after about four days in culture, indicating the death of a previously viable population of cells. The pattern of antibody productivity differed between the two cell lines. However, in both cases productivity was greatest during the first two to three days in culture, preceding the rise in LDH activity. This indicates that antibody was secreted by viable cells rather than released by dying cells as the cell membrane loses its integrity.

This conclusion is supported by the absence of antibody in the culture medium from the polyether foam packed bed reactor (Fig. 3). $2x10^7$ 321 cells from a three day old culture inoculated into this reactor died within twenty four hours but did not release any antibody. Cells from similar cultures

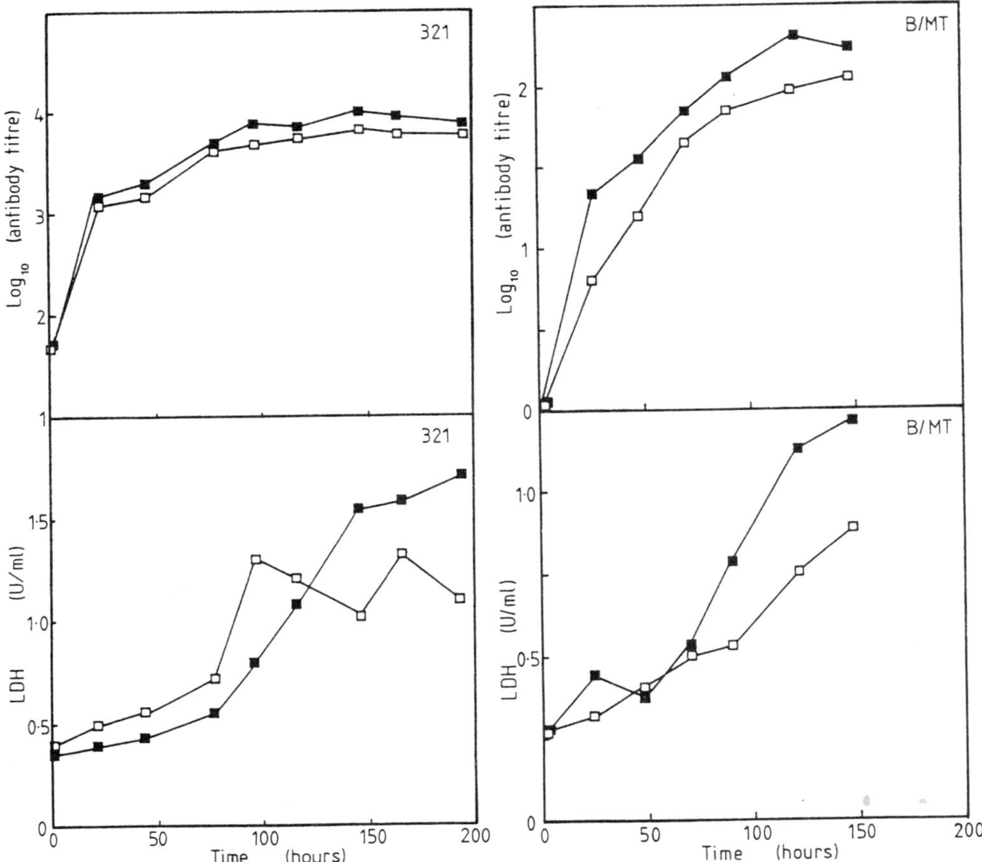

Figure 5. Production of antibody and release of LDH by 321 and B/MT hybridomas in a static flask (■) and in a polyester foam packed bed (□).

inoculated into other reactors remained viable and antibody titres in the culture medium rose rapidly. Thus the initial rise in antibody titre observed was due to the post-inoculation synthesis and secretion of antibody by viable cells, not to the release by dying cells of antibody formed pre-inoculation.

The Mechanism of Entrapment of Hybridomas in a Packed Bed

The way in which cells become entrapped within the packed bed is of some importance. Hybridomas are not robust cells and are susceptible to damage caused by the physical stresses associated with aeration and mixing in reactors [15,16]. Since one object of immobilising the cells is to provide them with a physically protected environment, it follows that the cells should not be subjected to avoidable stress during immobilisation. The results presented above indicate that 321 and B/MT hybridomas survive immobilisation, but provide little information about the actual process of immobilisation. The working assumption throughout the early stages of this investigation was that immobilisation was a result of cells being washed into the support matrix by the flow of medium through the bed, and that a relatively high flow rate would aid this process at the same time as reducing clogging of the bed.

In order to test this assumption culture medium containing various concentrations of 321 cells was pumped downwards at various flow rates through a packed bed of polyester foam cubes in an open, rather than recirculating, system. Cells in the effluent were counted. Figure 6 shows the results from a

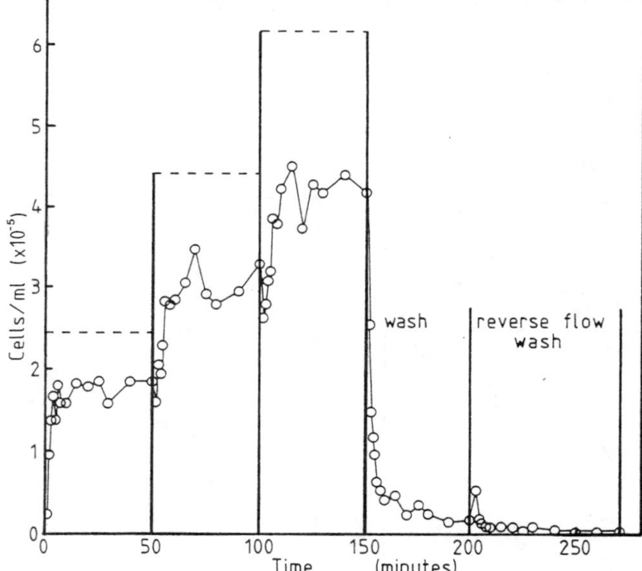

Figure 6. Entrapment of 321 hybridomas by a packed bed of polyester foam. Medium containing succesively higher concentrations of cells (-----) was pumped downwards through the bed, then the bed was washed with cell-free medium pumped downwards (wash) or upwards (reverse flow wash) through the bed. The flow rate was 6ml min^{-1} throughout. Cell concentrations in the effluent medium (O) were determined at frequent intervals.

typical experiment. The packed bed acted as a filter in the way in which it removed cells from the medium. Following an initial period of coating the bed stabilized and thereafter trapped a constant proportion of the input cells. The proportion of cells passing through the bed, f, was independant of the concentration of cells applied to the bed but varied depending on the flow rate through the bed (Table 2).

Coury et al [17] have recently described the capture behaviour of fixed-bed filters. They considered the four conventional mechanisms by which a filter element (in the case of this study a single foam cube) can capture particles passing through a bed. Diffusional, gravitational and inertial capture result from effects which cause the trajectory of a particle to deviate from the medium streamlines around a filter element. Capture due to interception occurs when a particle following a streamline is carried directly into a filter element. Equations modelling each of these mechanisms were presented. That given for inertial capture is applicable only to gaseous systems, and is not considered further here. Inertial capture typically has a theoretical efficiency several orders of magnitude smaller than the other mechanisms, so this is not a major omission. Using the other relevant equations presented by Coury et al [17], summarised in Table 1, it is possible to calculate the theoretical capture efficiences due to the other three mechanisms and to determine the actual capture efficiency of the packed bed.

The results obtained are shown in Table 2 and it is clear that, according to the equations, capture due to gravitational settling should be the dominant capture mechanism acting in this system. The predicted capture efficiencies due to gravitational settling are some two orders of magnitude greater than the predicted capture efficiences due to either inertial or diffusional capture.

The experimentally determined capture efficiencies agree with this finding, greatly exceeding the predicted efficiences due to inertial and diffusional capture. Of the capture mechanisms considered, only gravitational settling has a predicted capture efficiency which is of sufficient magnitude to account for the experimental efficiencies. Direct interception of cells by the foam, which was initially assumed to be the most important capture mechanism, does not appear to contribute significantly to immobilisation. Immobilisation is most efficient at low flow rates, a condition which should reduce the potentially harmful shear effects encountered by the cells.

Note that this does not address the way in which cells finally attach to the support material, merely the way in which they reach a location from which they do not return to the circulating medium. There is evidence that some hybridomas can adsorb to polyester [18], and the B/MT line attaches weakly to plastic surfaces. However, we found that many cells could be recovered from polyester foam particles by gentle squeezing, suggesting that they had simply lodged within the particles without binding.

Two factors which have not been considered in this present study, but which require investigation, are electrostatic and hydrodynamic effects. The electrostatic charge on a surface is an important factor affecting the attachment of anchorage dependant cells [see e.g. 19]. The contribution of electrostatic effects to both the capture and the immobilisation of hybridomas in the system described here remains to be determined, though because of the short range of such forces the contribution to capture is likely to be small. Important hydrodynamic effects may arise because of the porous nature of the foam, which may allow significant flow through the particles. Further investigations into the effects of these factors upon the immobilisation of hybridomas in foam particles are planned.

TABLE 1.
Actual and theoretical capture efficiencies of filters, after Coury _et al_ [17]

Capture mechanism	Capture efficiency, E	
Diffusion	$(4{\cdot}36/\epsilon)(D/Ud_p)^{2/3}$	where U=F/A
Interception	$6{\cdot}3\ \epsilon^{-2{\cdot}4}\ (d_c/d_p)^2$	
Gravitational settling (downflow)	$0{\cdot}0375(u/U)^{0{\cdot}5} + 0{\cdot}21(u/U)^{0{\cdot}78}$	
Experimentally determined	$[-2d_p(\ln f)] / [3H(1-\epsilon)]$	where $f=C_2/C_1$

A is the cross-sectional area of the bed (m^2)
C_1 is the concentration of cells in the medium entering the bed
C_2 is the concentration of cells in the medium leaving the bed
d_c is the diameter of a cell ($12{\cdot}5 \times 10^{-6}$ m)
d_p is the diameter of a foam particle (5×10^{-3} m)
D is the diffusivity of a cell, calculated from the Stokes-Einstein equation. A value of $5{\cdot}4 \times 10^{-14}\ m^2\ s^{-1}$ was used.
F is the volumetric flowrate ($m^3\ s^{-1}$)
f is the proportion of cells passing through the bed. A value for f may also be determined from Fig.2. The slope of a plot of $-\ln(C/C_0)$ against time is $(F/V)(1-f)$
H is the depth of the bed ($3{\cdot}5 \times 10^{-2}$ m)
U is the superficial velocity of medium through the bed ($m\ s^{-1}$)
u is the terminal velocity of a cell settling freely through the medium. This has been determined to be about $14 \times 10^{-6}\ m\ s^{-1}$ from previously published results [16]
V is the reactor reservoir volume (100 ml)
ϵ is the bed voidage (a value of 0·4 was used).

TABLE 2.
Actual and theoretical capture efficiencies of a polyester foam packed bed.

F (ml/min)	n	f	Ea	Eg	Ei	Ed
3	1	0·60	0·081	0·0949	$3{\cdot}55\times10^{-4}$	$3{\cdot}82\times10^{-2}$
6	6	0·69 ± 0·02	0·059	0·0577	$3{\cdot}55\times10^{-4}$	$2{\cdot}54\times10^{-4}$
9	1	0·91	0·015	0·0433	$3{\cdot}55\times10^{-4}$	$1{\cdot}84\times10^{-4}$
50	*	0·987	0·0021	0·0132	$3{\cdot}55\times10^{-4}$	$5{\cdot}86\times10^{-5}$

n is the number of determinations of f.
* calculated from Fig.3 according to the definition in Table 1
Ea is the experimental capture efficiency
Eg is the theoretical gravitational capture efficiency
Ei is the theoretical inertial capture efficiency
Ed is the theoretical diffusional capture efficiency

CONCLUSIONS

The results presented here indicate that a variety of support materials may be used to form packed beds suitable for immobilising hybridomas, and that such materials may be chosen to meet other design criteria (e.g. mechanical strength, price, reuseability, product quality) than compatibility with cell growth and antibody productivity.

The immobilisation of hybridomas in packed beds of this type is apparently due to gravitional settling of the cells into the support material. Immobilisation is more efficient at slower flow rates, in terms of the overall capture efficiency, allowing the use of an immobilisation protocol likely to reduce shear effects experienced by the hybridomas.

Packed bed bioreactors of the design described are able to support the productive growth of immobilised cells of two hybridoma lines. There does not appear to be a major difference between the growth, viability and productivity of hybridomas immobilised in polyester foam packed beds or cultured in conventional static or suspension systems.

Similar reactors have been scaled up to 100 litres capacity for the production of anchorage dependant BHK cells [5], so it is reasonable to expect that it will be possible to scale up the culture of hybridomas. The reactor design can easily be modified to allow perfusion with fresh medium, offering the potential for long-term culture. Studies are in progress of factors affecting the growth, metabolism and productivity of hybridomas in the high cell concentration environment characteristic of long term immobilised cultures.

ACKNOWLEDGEMENTS

The authors wish to acknowledge Dr. Andrew Wright of ICI plc for the kind gift of the 321 cell line, and Dr. J.P. Whiteside, Mrs D. Simpson, Dr. S.C. Musgrave, Mr. Paul Hayter, Dr. A. Handa, Prof. R. Clift and Dr. M. Al-Rubeai for their various contributions to this work. ADM and JST were funded by SERC grant GR/D 04892. The development of the B/MT cell line was supported by AFRC grant 90/9 to Mr. B.A. Morris.

REFERENCES

1. McCoy, T.A., Whittle, W. and Conway, E., A glass helix perfusion chamber for massive growth of cells in vitro. Proc. Soc. Exp. Biol. (NY), 1962, 109, 235-237

2. Burbidge, C., The mass culture of human diploid fibroblasts in packed beds of glass beads. Develop. biol. Standard., 1979, 46, 169-172

3. Spier, R.E. and Whiteside, J.P., The production of foot-and-mouth disease virus from BHK 21 C13 cells grown on the surface of glass spheres. Biotechnol. Bioeng., 1976, 18, 649-657

4. Wohler, W., Rudiger, H.W. and Passarge, E., Large scale culturing of normal diploid cells on glass beads using a novel type of culture vessel. Exp. Cell. Res., 1972, 74, 571-577

5. Whiteside, J.P. and Spier, R.E., The scale up from 0.1 to 100 litres of a unit process system for the production of four strains of FMDV from BHK monolayer cells. Biotechnol. Bioeng., 1981, 23, 551-565

6. Whiteside, J.P. and Spier, R.E., Factors affecting the productivity of glass sphere propagators. Develop. biol. Standard., 1984, 60, 305-311

7. Lazar, A., Reuveny, S., Mizrahi, A., Avtalion, M., Whiteside, J.P. and Spier, R.E., Production of biologicals by animal cells immobilised on a polyurethane foam matrix. Paper presented at the ESACT-OHOLO joint meeting on 'Modern approaches to animal cell technology', Tiberias, Israel, 1987

8. Murdin, A.D., Thorpe, J.S. and Spier, R.E., Immobilisation of hybridomas in packed-bed bioreactors. Paper presented at the ESACT-OHOLO joint meeting on 'Modern approaches to animal cell technology', Tiberias, Israel, 1987

9. Felix, H.R. and Mosbach, K., Enhanced stability of enzymes in permeabilized and immobilised cells. Biotechnol. Lett., 1982, 4, 181-186

10. Black, G.M., Webb, C., Matthews, T.M. and Atkinson, B., Practical reactor systems for yeast cell immobilisation using biomass support particles. Biotechnol. Bioeng., 1984, 26, 134-141

11. Mavituna, F., and Park, J.M., Growth of immobilised plant cells in reticulate polyurethane foam matrices. Biotechnol. Lett., 1985, 7, 637-640

12. Wright, A.F., Green, T.P. and Smith, L.L., The development of mouse monoclonal antibodies that bind and neutralize paraquat. Develop. biol. Standard., 1987, 66, 495-501

13. Groves, D.J., Morris, B.A. and Clayton, J., Preparation of a bovine monoclonal antibody to testosterone by interspecies fusion. Res. in Vet. Sci., accepted for publication.

14. Spier, R.E., Determination of the time to harvest foot-and-mouth disease virus cultures by measurements of the supernatant concentration of lactic dehydrogenase. Biotechnol. Bioeng., 1977, 19, 929-932

15. Smith, C.G., Greenfield, P.F. and Randerson, D.H., Shear sensitivity of three hybridoma cell lines in suspension culture. Paper presented at the ESACT-OHOLO joint meeting on 'Modern approaches to animal cell technology', Tiberias, Israel, 1987

16. Handa, A., Gas-liquid interfacial effects on the growth of hybridomas and other suspended mammalian cells, Vol. II, Ph.D Thesis, 1986, Department of Chemical Engineering, University of Birmingham, U.K.

17. Coury, J.R., Thambimuthu, K.V. and Clift, R., Capture and rebound of dust in granular bed gas filters. Powder Technology, 1987, in press.

18. Katinger, H., Principles of animal cell fermentation. Develop. biol. Standard., 1987, 66, 195-209

19. Butler, M., Hassell, T. and Rowley, A., The use of microcarriers in animal cell cultures. In Plant and Animal Cells: process possibilities, ed. C. Webb and F. Mavituna, Ellis Horwood Ltd, Chichester, UK, 1987, pp64-74

INTERNATIONAL CONFERENCE ON BIOREACTORS AND BIOTRANSFORMATIONS
GLENEAGLES, SCOTLAND, UK: 9-12 NOVEMBER 1987

Paper C2

THE DEVELOPMENT OF A LARGE SCALE PRODUCTION PROCESS FOR TISSUE CULTURE PRODUCTS

P J WILKINSON
CELLTECH LTD
228 BATH ROAD
SLOUGH
BERKSHIRE

ABSTRACT

Analysis of the process economics and operability of a prototype 1000L airlift fermenter system for the production of monoclonal antibodies from hydridoma cells led to a series of design changes when a second system was installed in a production facility. These included increased automation of the process control, installation of an in-place cleaning system and enhanced product protection. The nature of these changes and the benefits experienced are discussed.

INTRODUCTION

In 1984 Celltech commissioned the world's first 1000L airlift fermenter to produce monoclonal antibodies from hybridoma cells in suspension. This was in response to a growing market for antibodies in which demands for multi-kilogram amounts per annum were forseen [1]. It was anticipated that the new system would yield upwards of 100g/antibody per fermentation run lasting two weeks. For comparison, the traditional in- vitro method using mouse ascites tumours yields only 50mg/mouse, i.e. 2000 mice would have to be raised to produce as much antibody as one 1000L fermenter batch. The design of the airlift fermenter vessel at the heart of the process was the result of extensive theoretical engineering work and development. The effects of fluid mixing, shear stress, heat transfer and mass transfer in the fermenter were considered and have been discussed elsewhere[2].

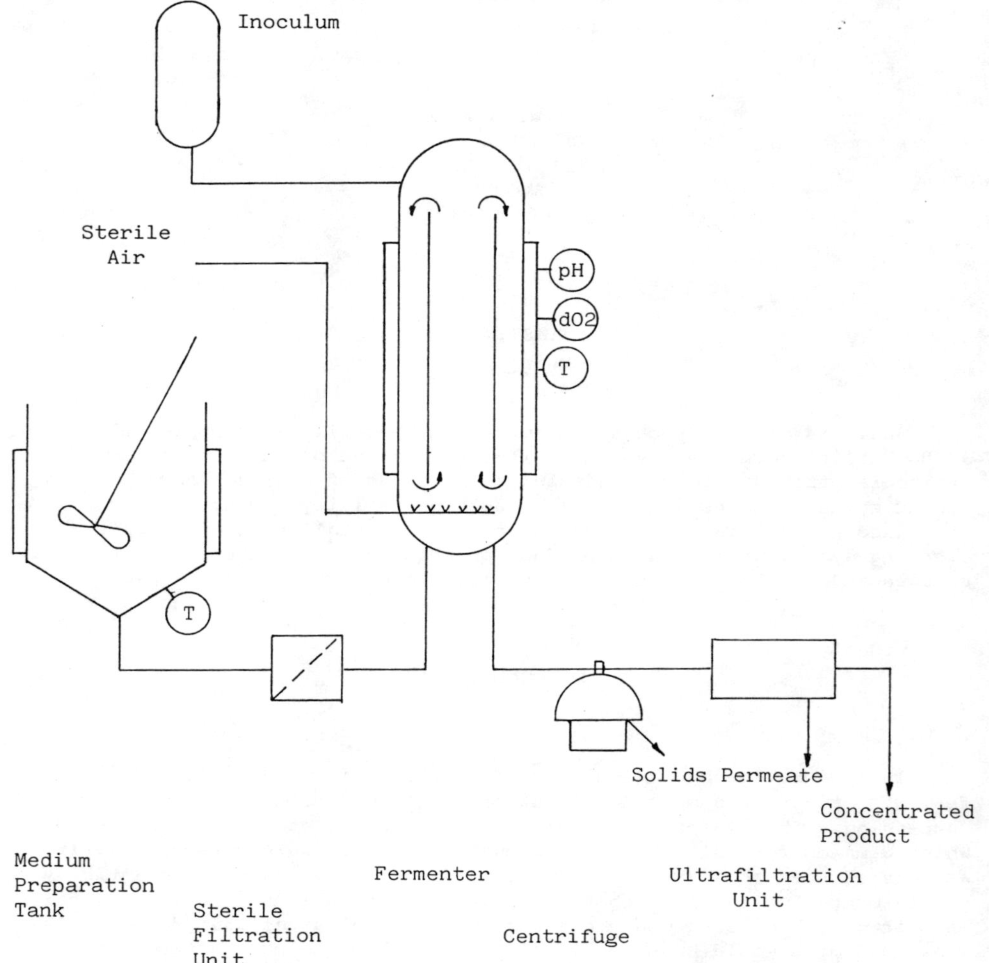

Figure 1 Monoclonal antibody production in an airlift fermenter.

The prototype system met the following requirements:

1) It was installed within tight time, cost and space constraints to enable Celltech to start large scale antibody production as soon as possible.

2) It proved that an airlift fermenter vessel could be used for the large scale production of monoclonal antibodies.

3) The predictions of the physical properties of the fermenter were verified.

4) The suitability of the other unit operations in the process were confirmed.

5) Our operators, who were used to working on laboratory and pilot-scale equipment, gained experience of industrial scale processes.

THE PROCESS

In 1985, a second 1000L fermenter system was ordered to be installed in Celltech's new purpose built manufacturing facility. This provided an ideal opportunity to make improvements to the process in the light of our operating experience.

In order to explain the context in which these design changes were made, a brief description of the process must be given. An outline flowsheet of the process is shown in Figure 1; no fundamental changes have been made to this since the installation of the prototype.

900L of medium is formulated and pre-heated in a jacketed make up vessel. The medium is then pumped through sterilizing filters into the fermenter, which, along with its ancilliary pipework has been steam sterilized in situ beforehand. The pH and dissolved oxygen probes are calibrated when the medium is in the fermenter. The control system brings those parameters and the temperature to the required set points prior to inoculation. When the cells in a 100L inoculum fermenter reach an optimum cell density, the contents of that fermenter are used to inoculate the main fermenter.

The kinetics of animal cell growth are very different to microbial or yeast cells, and it is 10 to 14 days before the maximum product concentration in the main fermenter is reached. During this time the fermenter has to remain sterile and tight control of physical parameters must be maintained.

At the end of the fermentation the contents of the fermenter are harvested through a continuous disc stack centrifuge to remove cell debris. The clarified broth, which contains the product, is stored in a chilled harvest vessel where the protein contents are concentrated by ultrafiltration. The concentrate is then removed for further purification of the product.

To save space and capital cost, a single vessel was used in the original system for medium make up, to receive product after centrifugation and to prepare cleaning solutions.

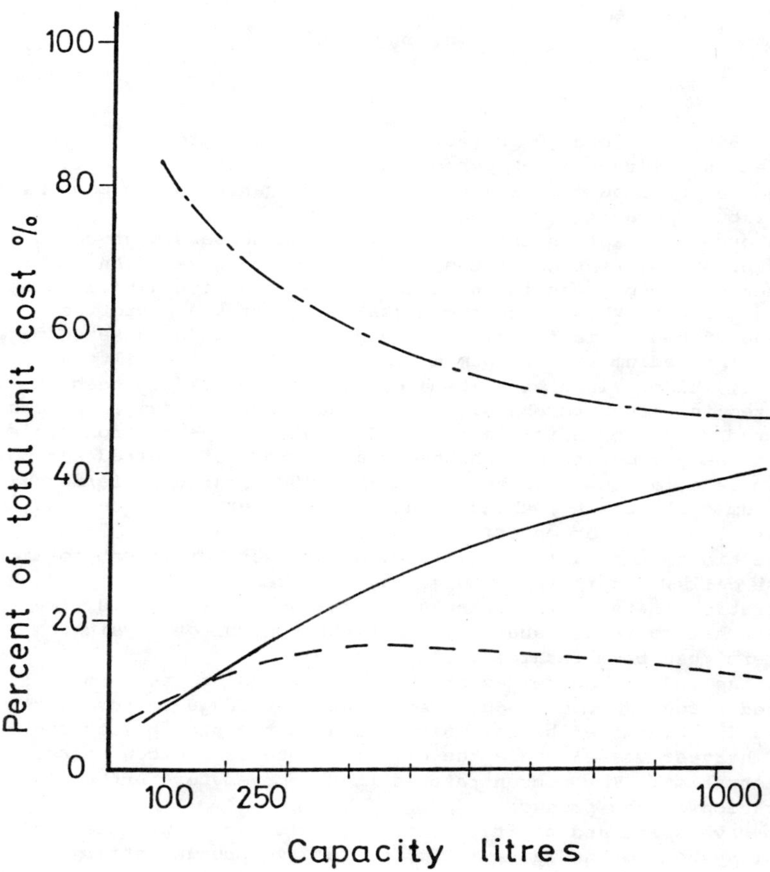

Figure 2. Effect of scale on components of unit cost.

THE MOTIVATION FOR PROCESS IMPROVEMENT

Figure 2 shows the proportions of production cost attributed to labour, materials and depreciation at various scales of operation, based on processes of the prototype design [2]. Capital cost depreciation is less than 20% of the total at all scales. Medium costs increase linearly with scale; at the 1000L scale they account for 40% of the unit costs. Celltech is maintaining a development programme to eliminate costly ingredients such as serum from its medium formulation. The largest cost component, more than 50% at all scales, is labour, indicating that the most effective way to lower production costs would be to reduce the labour component.

When the fermentation is running the only labour requirement is for monitoring and sampling. There then follows a period of intense activity when the batch is harvested, the plant is cleaned and sterilized and the next run started. By reducing the manual input to a minimum in this period the work load can be evened out, enabling a team of operators to run several fermenters simultaneously. Not least of the considerations is the saving in effort made by eliminating the need for operators to walk or climb to valves on a large process plant or carry out mundane repetitive activities.

At Celltech we have been able to ensure that all activities requiring manual intervention can take place during a normal working day by speeding up the activities with the help of automation or by carring them out automatically out of hours. In addition, critical activities are carried out repeatably and reliably by the control system, which can provide written confirmation that they have been completed successfully.

By reducing the turnround time it is possible to produce more batches per year thereby increasing the revenue from the plant.

Some components of the first system had their origins in laboratory or pilot plant equipment. The opportunity was taken to replace these by industrial standard equipment giving greater robustness and thus reducing maintenance costs.

It is of paramount importance to ensure that the plant is safe to operate at all times and complies with the requirements of G.M.P.

INCREASED AUTOMATION

The original process had a limited computerized control system which was operated semi-automatically. Fermentation parameters were controlled by the computer and individual operations could be initiated as part of a pre-programmed sequence or as part of a screen menu selected through a keyboard. Some key valves were automatically driven, others were opened or closed by operators in response to prompts on the screen.

The programme could not progress until the operator had confirmed on the keyboard that he had performed the necessary actions.

The new system has 110 automatic valves installed compared with 40 on the original and the amount of software to control the plant has been approximately tripled.

Sterilizations of lines and vessels were improved by automating all valves involved, for example bleed valves on filters and key valves were regularly pulsed to clear condensate. Manual prompts are retained to ensure filter cartridges have been fitted when appropriate.

The medium preparation protocol requires the medium to be heated and held at four distinct temperatures during the formulation. The temperature control of the medium preparation tank was automated and prompts the operator when the contents are at the correct temperature to proceed.

Calibration of the dissolved oxygen sensor in the fermenter used to be a lengthy operation. Nitrogen would be sparged through the medium for up to three hours until no further decrease in dissolved oxygen was noted, at which point the operator would manually zero the instrument, alter the control point to 100% and open the air supply. After a further three hours, when the level of dissolved oxygen stopped increasing, the full span of the instrument was set. This system was replaced by field mounted transmitters which could be coarsely adjusted. The control system then automatically sparges the fermenter with nitrogen and air and makes the final calibration adjustment of zero and full span in the software.

In the prototype system the inoculum was grown in a proprietary laboratory fermenter and the transfer was an entirely manual operation. There is a time span of about four hours during which the cells are at their optimum density for transfer. This time cannot be determined by direct on-line measurement as yet but can be determined by plotting a growth curve from regular samples. The new control sequence checks that conditions in the main fermenter are correct, and at a pre-set time sterilizes the inoculum line, transfers the inoculum by overpressurising with sterile filtered air, and on completion initiates an automatic clean of the inoculum fermenter. The inoculum fermenter itself is now a scaled-down version of the main fermenter, sharing the same control system, cleaning system and operating in a similar manner.

The cleaning of the plant is automated. Cleaning solutions are made up in separate storage vessels. The control system instigates a clean when it is safe to do so, after checking that the part of the plant to be cleaned is empty of product and all interlocks are in the correct position. The route to be cleaned is automatically opened up and a series of detergent washes and rinses are pumped round at preset temperatures. These operations used to be entirely manual and could only be carried out when the medium make up tank was empty. Cleans can now be carried out immediately the equipment requires it and with minimal operator intervention. For example, an eight-hour cleaning regime on the ultrafiltration rig can now be carried out unattended overnight.

CHOICE OF CONTROL SYSTEM

We chose the APV Automation Ltd Accos 2S control system for both plants. It offered the following facilities:

1) A proven track record in process industries, especially dairies and breweries.

2) Extra items, such as valves or pumps, can be easily incorporated at a later date.

3) The software can be easily edited or expanded in house, subject to safeguards to prevent unauthorised changes.

4) Every item with on/off or analogue control has a manual override (essential during commissioning).

5) Because the system requires positive feedbacks from the items it controls, it can readily diagnose fault conditions and be programmed to take appropriate emergency actions.

6) The start and end of every operating sequence is logged.

These features have been specified on other process control systems installed by Celltech.

CLEANING-IN-PLACE

The techniques of cleaning-in-place have become firmly established in the process industries in recent years. Pre-prepared detergent solutions or rinse water can be heated if required and pumped either through a continuous pipe circuit or into spray nozzles to create jets in tanks. The liquid can be recirculated through a break tank or run to drain. The combination of the physical scouring action and the chemical effects of the detergents removes fouling material from internal surfaces which come into contact with the product. Careful engineering is required to ensure that:

1) There are no dead legs, crevices or "shadowed" areas where dirt can collect and not be cleaned.

2) Distinct cleaning routes are identified which can be cleaned whilst other operations are being performed elsewhere on the plant.

3) The cleaning operations do not endanger personnel and product safety.

Personnel are protected from the dangers of hot detergent solutions being pumped around the plant by such devices as proximity switches, which ensure that vessel lids are properly closed and removable pipe sections are in place. The control system is programmed so that any fault on an operating CIP circuit will switch off the distribution pump and close the route in a safe manner.

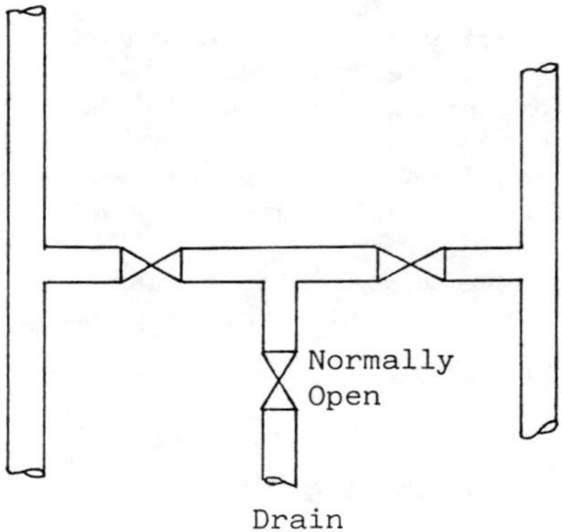

(a) Three valve arrangement

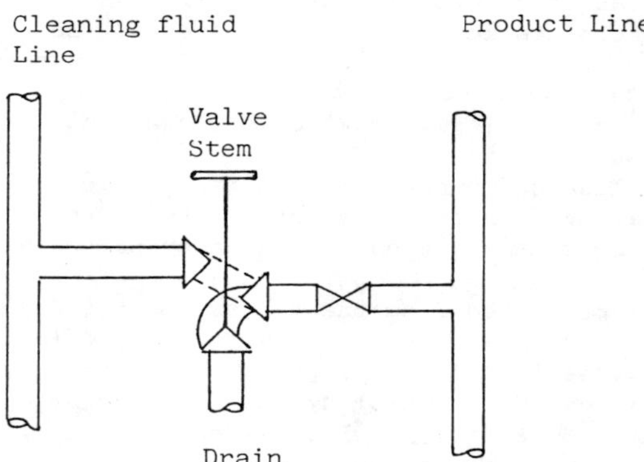

(b) 3-port valve arrangement

Figure 3. "Double block and bleed" arrangements to protect product from cleaning fluids

As it is necessary to clean parts of the plant, such as the medium make up tank, whilst product is elsewhere in the plant, we have engineered 'double block and bleed' protection of all interfaces between cleaning fluid and product. Thus if one valve seal should fail and pass cleaning fluid it will fall to drain rather than pressing up against a valve seat with product on the other side. Proprietary three way valves, with one port piped to drain, provide a neat method of achieving this.(See Figure 3).

DESIGN CHANGES

The new 1000L system incorporated a number of physical changes to enhance product protection. As the inoculum fermenter could be growing a different cell line to that in the main fermenter, the inoculum line contains a physically removable section to prevent cross contamination. The control system then checks that the piece is in the correct position for sterilizations, inoculation and cleans, and raises an alarm if an illegal connection is made.

Pharmaceutical process vessels require a double skinned jacket to prevent a pinhole leak causing contamination of the contents by service fluids. The disadvantage of this construction is poor heat transfer. To get around this we use a single skinned jacket with an attemperation loop filled with deionized water. A separate plate heat exchanger is used to heat or cool the water in the attemperation loop using steam or process water. This gives efficient heat transfer and an effective double barrier between products and services.

The prototype system still relied on a number of connections to sterile equipment being made by spearing through a septum to connect autoclaved laboratory glassware. It has been possible to remove the need for the majority of these operations by installing steam sterilizable stainless steel pipework with appropriate sterile barriers. These have the advantages of being easier to operate, introduce lower risks of contamination and are safer for the operator.

CONCLUSION

We now have nearly two years operating experience of fermenter systems with the design changes described above. The changes have enabled us to reduce the manpower allocation from 2 people/1000L system to 1.25 people / 1000L system with no operation requiring more than one person on the plant at a time.

Routine out-of hours working has now been eliminated along with many of the physically demanding or mundane activities. The turn-round time on the fermenter has been reduced from 4 days to 1.5 days, a saving in time which represents upto five additional production runs per system each year.

These improvements have been achieved by ensuring that the process engineers responsible for designing and specifying plant have personal experience of operating the plant, thereby understanding the problems encountered by the operators and seeing scope for improvements. By applying sound engineering principles to the problems, cost effective solutions were devised. Finally, a willingness by all parties, including operators, engineers and management to implement those changes ensures that all share the benefits.

REFERENCES

1. Birch, J R., Thompson, P W., Lambert, K and Boraston, R., The large scale cultivation of hybridoma cells producing monoclonal antibodies. Presented at the American Chemical Society, Philadelphia, 1984.

2. Thompson, P W., Wood, L A., Birch, J R., Lambert, K and Boraston, R., Antibody Production in Airlift Fermenters. Presented at the American Institute of Chemical Engineers Annual Meeting, Chicago, 1985.

INTERNATIONAL CONFERENCE ON BIOREACTORS AND BIOTRANSFORMATIONS
GLENEAGLES, SCOTLAND, UK: 9-12 NOVEMBER 1987

Sales talk

Paper C3

TWO CERAMIC MATRICES FOR THE LONG-TERM GROWTH OF ADHERENT OR SUSPENSION CELLS: SCALABILITY

Gordon G. Pugh, Guy J. Berg and Christopher H.J. Sear
Charles River Biotechnical Services, Inc.,
Wilmington, Massachusetts, USA and
Charles River Biotechnical Services Limited,
Margate, Kent, UK

ABSTRACT

Bioreactors have been developed for mammalian cell cultures using either a smooth-surfaced ceramic for the growth of adherent cells or a porous ceramic for the immobilization of suspension cells. Available surface areas range from 4,250cm^2 to 120,000cm^2 and up to 720,000cm^2 when run in multiples. The ceramics are used in the Opticell^{TM} culture system which is a closed-loop circumfusion system where the environmental parameters are computer controlled. For adherent cell growth, the scalability is nearly linear and dependent on the surface area provided. For immobilized hybridoma cultures, the productivity is scalable and when compared to cores of the same material, related to the volume of the porous ceramic matrix utilized.

INTRODUCTION

The use of large scale animal cell cultures in the production of viral vaccines is an established manufacturing procedure [1]. Additionally, the production of proteins for human diagnostic and therapeutic use has also begun to favor mammalian cells as a production tool [2,3,6]. The necessity for huge numbers of viable and productive cells has stimulated the development of many different approaches to the design of useful bioreactors [4,5]. While many have effectively addressed the need for a large surface area to medium volume ratio on a small scale, the problem of scaling up these bioreactors for fragile cells has proven to be more difficult than simply increasing the reactor volume or surface area available [9]. Problems such as cell distribution, oxygen transfer, longevity, nutrient supply, plugging, channeling, gradient formation, and shear have limited the efficiency of scale-up in systems such as stirred tanks, microcarriers, fluidized beds and hollow fibers.

A proven alternative to scalable mass cell culture is the use of bioreactors containing ceramic matrices. Immobilization of both adherent and suspension cells to these surfaces has resulted in the favorable production of both cells and cell derived proteins [7,8,9]. Scaling of these reactors for use in manufacturing settings has been achieved.

MATERIALS AND METHODS

In all of these studies, Opticell™ bioreactor systems were used with the ceramic Opticore™ growth chambers. The benchtop Opticell™ Model 5200R was used for small scale testing, the Model 5300 for intermediate scale work and the Model 5400 for the study on a $72m^2$ multireactor system. For adherent cells, the AD-51 Opticore™ was used to provide $4,250cm^2$, the AD-451 supplied $42,500cm^2$ and the AD1251 provided $120,000cm^2$ of usable surface. Two size chambers were used for suspension cell testing; the S-51 which provided $165cm^3$ of matrix and the S-451 which gave $1940cm^3$ of identical porous ceramic.

Two formulations of ceramic are utilized in the Opticores™. One has a smooth surface with little porosity and is appropriate for attachment dependent cells; the other has as much as as 50% porosity which is used to aid immobilization of suspension cells. Both ceramics resemble extruded cylinders and have hundreds to thousands of $1\text{-}2mm^2$ square channels running through their lengths. In this way, the available surface area is greatly increased in a given volume. The surface area of a single Opticore™ which is 30cm in length and 8cm in diameter is approximately $42,500cm^2$. The ceramic matrices are enclosed in a plastic cartridge with tubulated endcaps. A circular flow diverter is placed in each end to ensure a uniform distribution of flow across the entire core.

Cells are seeded into the ceramic chambers by recirculating a cell slurry several times through the core until a homogenous suspension has been reached. The flow is then stopped and with the reactor in the horizontal position, the cells are allowed to settle in each of the hundreds of channels onto one of the 4 flat sides. This procedure is repeated for each of the remaining 3 sides. The even distribution of cells both radially and longitudinally has been previously documented [7].

The Opticores™ are attached to the Opticells™ in a recirculating perfusion loop which contains electrodes for the measurement of dissolved oxygen, (DO_2), pH and temperature (See Figure 1). The microprocessor controller maintains the culture environment at desired setpoints as well as initiating medium flow rate changes or feed and harvest pumps. The oxygen consumption rate, (OCR), is used as the primary on-line monitoring device and is made possible with the use of DO_2 electrodes positioned both before and after the Opticore™. As the oxygenated medium flows through the electrode chambers, the computer calculates the DO_2 uptake of the culture by using the difference generated in the sensor readings. While not a direct measurement of cell number or absolute productivity, changes in this rate have been successfully used to monitor culture growth and productivity trends [8,9].

During the production of monoclonal antibodies from hybridoma cultures, the Opticell™ uses a procedure called the continuous feed and harvest system. This approach is similar to an automated chemostat where spent medium is removed and fresh medium added at the same rate. The overall addition and removal rates are based on the changes in the OCR and on glucose utilization.

Figure 1

Opticell Bioreactor Schematic

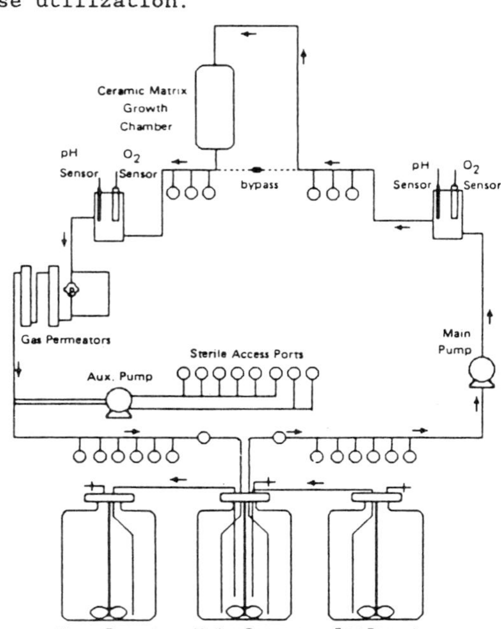

RESULTS

Scalability of Adherent Cell Cultures

The attachment dependent MDCK cells, (ATCC), were grown on Opticore™ matrices containing surface areas of 4,250cm^2, 42,500cm^2, and 120,000cm^2. The cell seeds were prepared by routine passaging on flasks and roller bottles and planted at similar densities on the ceramics (See Table 1). High glucose DMEM growth medium containing 8-10% Fetal Bovine Serum (FBS) was used in all cores at proportionally equivalent volumes. Fresh medium was added on day 3 and the cells were grown for a total of 5 days. The cultures were seeded at densitites of between 2.1-2.5 x 10^4 cells/cm^2. Table 1 summarizes the specifics of these tests and provides a comparison of yields. Figure 2 shows the relationship between the increase in cell number for these differently sized growth chambers and Figure 3 presents the individual OCR's. While the OCR curves appear proportional, the growth curves representing scalability of yield are strikingly similar. These parallel curves provide evidence of linear scale up of cell yields with surface area.

TABLE 1

SCALABILITY OF ADHERENT CELLS MDCK AND BHK-21

USING OPTICORE™ CERAMIC MATRICES

Cell Type	Number of Growth Chambers	Surface Area of Reactor (cm^2)	Starting Cell Number/cm^2	Total Media Used (L)	Days in Culture	Total Cell Yield	Ending Cell Number/cm^2	Index of Scala-bility*
MDCK	1	4,250	2.1x10^4	2.5L	5	10.4x10^8	2.4x10^5	100
MDCK	1	42,500	2.3x10^4	25L	5	10.9x10^9	2.5x10^5	104
MDCK	1	120,000	2.5x10^4	70L	5	38.3x10^9	3.2x10^5	142
BHK-21	1	4,250	2.8x10^4	3.45L	7	2.95x10^9	6.9x10^5	100
BHK-21	1	42,500	2.35x10^4	20L	7	30.6x10^9	7.0x10^5	101
BHK-21	1	120,000	2.5x10^4	60L	7	105x10^9	8.75x10^5	127
BHK-21	6	720,000	2.35x10^4	360L	7	52.5x10^{10}	7.3x10^5	106

* The index of scalability is calculated by dividing the cell yield of any size growth chamber by the cell yield of the control core which we selected as being 100 and equivalent to the yield from the smallest 4,250cm^2 cartridges. This index is intended to be an aid in comparing the relative efficiency of scalability.

Figure 2 MDCK Cell growth (seeding vs. final yield) on Opticores™ of increasing surface area

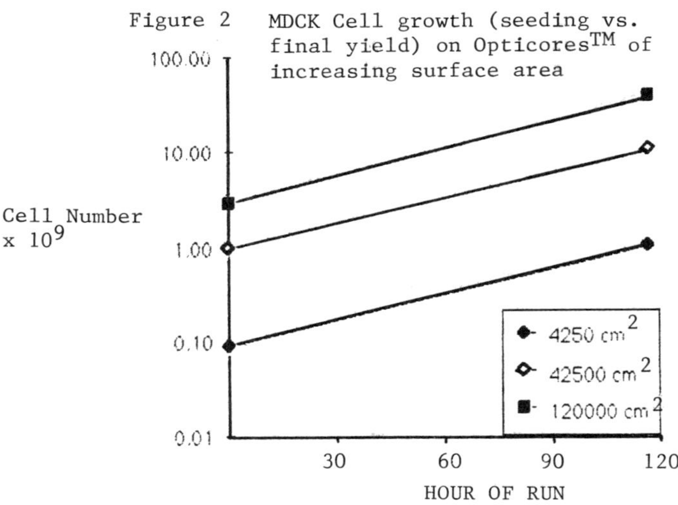

Figure 3 MDCK OCR's for Opticore AD-51, 451, 1251 Chambers

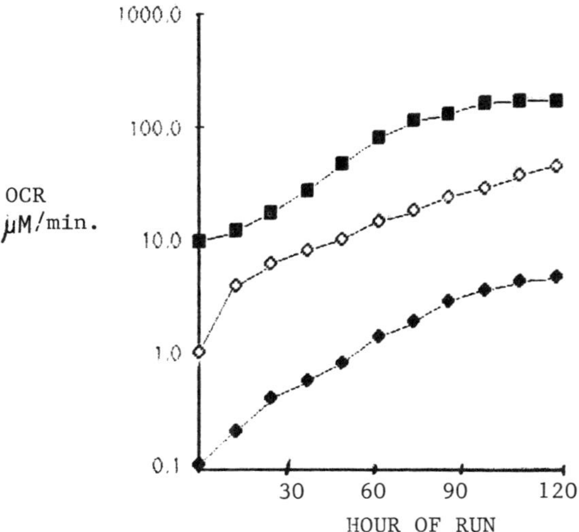

Another study on scalability of cell yields was carried out with the fibroblast-like BHK-21 (C-13) cell, (ATCC). The cells were again grown on bioreactors of increasing surface area, and additionally, in multiple cores connected in parallel. Total increase in scale from the smallest single reactor to the largest multiple chambered reactor was greater than 150x. As with the last tests, the ceramics were all planted at between 2 - 2.5 x 10^4 cells/cm^2, medium volumes and refeeding schedules were kept proportional and Opticell™ culture systems were used to develop OCR curves and regulate the culture environments. These cultures were all grown for 7 days and then harvested with 0.25% trypsin containing 0.01-0.02% EDTA.

Duplicate AD-51 cores having a surface area of 4,250cm^2 were seeded with an average of 1.2 x 10^8 BHK-21 cells. The average cell yield was 2.05 x 10^9. Similarly, AD451 chambers containing 42,500cm^2 were seeded with 1 x 10^9 cells, used 20 liters of DMEM with 8% FBS and yielded 3.0 x 10^{10} cells in 7 days. The Opticore™ 1251 containing 120,000cm^2 was planted with 3 x 10^9 cells, used 60 liters of media and yielded 10.5 x 10^{10} cells within 7 days. These results are summarized in Table 1. Additionally, six 1251 Opticores™, each containing 120,000cm^2 were connected for a total of 720,000cm^2 and run on an Opticell™ 5400 system. The chambers were attached in a parallel format and equal medium flows through each chamber were established prior to these studies. 17.6 x 10^9 cells were inoculated into this bioreactor which used 360 liters of DMEM medium containing 8% FBS over a period of 7 days. A total of 52.5 x 10^{10} cells were harvested from this multireactor system.

In summation, BHK-21 cell cultures were initiated at from 2.3 to 2.8 x 10^4 cells/cm^2 and yielded densities ranging from 6.9 to 8.75 x 10^5 cells/cm^2 in a seven day period. This suggests an average yield of 7.48 x 10^5 cells/cm^2 $\pm$ 15% regardless of the size or number of chambers used. Figures 4 and 5 show the growth curves and oxygen consumption rates of these BHK-21 cultures. As with the MDCK cells, these parallel lines provide additional proof that adherent cell yields are directly scalable in bioreactors utilizing ceramic growth supports.

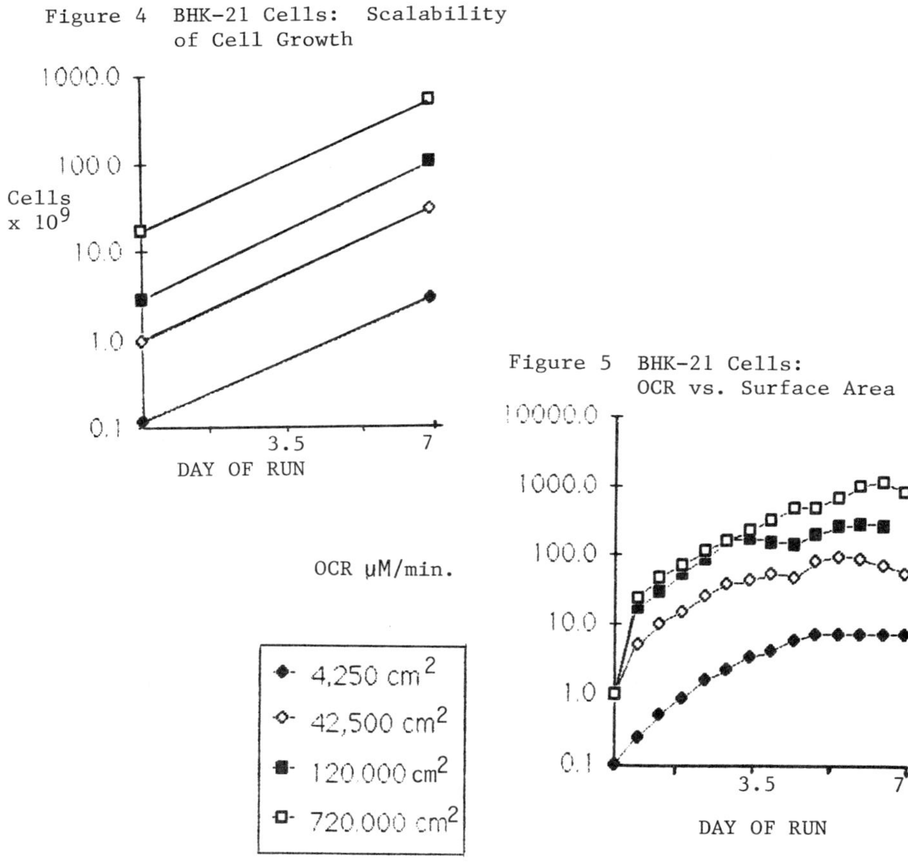

Scalable Productivity from Suspension Cells

As a general rule, hybridoma cells do not possess adherent characteristics and prefer to grow in the suspended state. The porous ceramics utilized for immobilization of these cells is scalable by the amount or volume of matrix presented for entrapment. Because the cells are physically entrapped in these irregularly shaped pores, it is difficult to remove cells after the growth phase to determine cell number. An indirect approach is to compare the amount of product secreted by the same cell lines, under similar conditions in the same type of matrix but of different size. For these tests, two sizes of porous ceramic were used. As the ceramics are extruded in cylinders with longitudinal channels of idential dimensions, one means of determining a ratio of expected scale is by comparing their volumes. The volume of the ceramic cylinder contained in the S-451 core is $1940cm^3$ and that of the S-51 core is $165cm^3$. Although only a fraction of these volumes actually contain matrix walls, they are directly proportional to matrix volume, which in the S-451 core is 12 times greater than the S-51. Two murine hybridomas 20-8-4S (ATCC HB 11) and L243 (ATCC HB 55) both secreting IgG_{2a} were used to examine scalability. In each case, Opticell™ systems were run in the continuous feeding and harvest mode throughout the production phase to maintain high productivity (8,9).

As an initial test, the production of IgG from 20-8-4S hybridoma cells was studied in the S-51 ceramic to be compared with previous work with the S-451 reactor system. 1 x 10^8 20-8-4S cells were immobilized in the S-51 in an Opticell™ 5200 containing DMEM supplemented with 10% FBS. After 6 days of growth in DMEM medium containing 10% FBS, the system was changed to the production phase employing the continuous feed and harvest with DMEM containing 1% FBS. While in production, the OCR was maintained at a constant state and the glucose level was controlled at between 500 and 1500mg/L. This system was allowed to produce antibody for 7 days in which time 8.1 liters of supernatant was collected. A total of 903mg of IgG was produced at a rate of 128mg per day.

Opticell™ 5300 systems using S-451 ceramics have been run with the 20-8-4S cell line in both serum-containing and serum-free production conditions (9). Systems using DMEM supplemented with 1% FBS yielded 50 grams of IgG in 30 days for an average of 1.7 grams per day. This bioreactor fitted with the $1940cm^2$ core, produced on a daily basis 13.3 times as much product as the S-51 containing system which is calculated to be 12 times smaller. As a feasibility test, these results are promising and support a scale index of 111 when going from an S-51 containing bioreactor to an S-451 system. An additional hybridoma was secured for testing to provide a further set of comparative data.

The IgG_{2a} secreting hybridoma L243 was grown in both the Opticell™ 5200R containing an S-51 core and in the Opticell™ 5300 utilizing an S-451 core. In each case, the cultures were allowed to grow for 7 days in DMEM containing 10% FBS. On day seven, a saturation density had been reached as indicated by the leveling OCR, and the production phase was begun. Having recirculating medium volumes of 3L and 30L respectively, the feed and harvest schedule was initiated at 30% per day or 1L/day for the S-51 and 10L/day for the S-451. DMEM supplemented with Biotain MPS was used as the production medium so that harvests would be serum-free. Glucose levels were monitored daily and adjustments made to the feeding rate to keep the concentration between 750 and 1250 mg/L. The oxygen consumption rate was kept high and steady and the pH, DO_2, and recirculating flow rates were kept proportional. Sampling for product was begun on day 10 and tested for IgG by ELISA techniques.

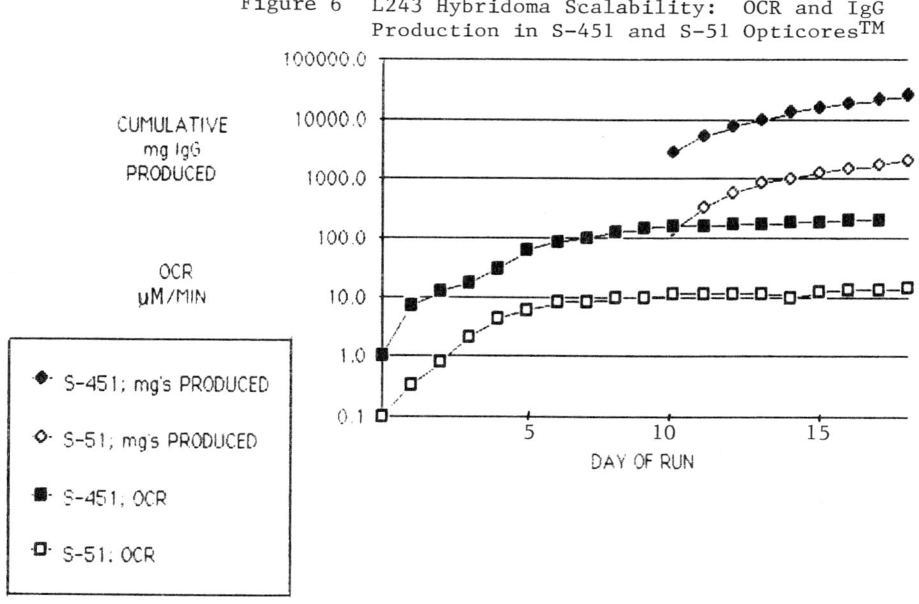

Figure 6 L243 Hybridoma Scalability: OCR and IgG Production in S-451 and S-51 Opticores™

It can been seen from Figure 6 that the respiration rates or OCR's of the two systems follow very similar trends and are consistent. During the production stage between days 10 through 16, the S-451 reactor maintained an OCR of 180um, while the S-51 reactor leveled at approximately 12um. This suggests that the 1940cm^3 ceramic is supporting a culture that is respiring at 15 times the 165cm^3 ceramic which we have estimated to be 1/12 the former's size. While this is very close to linear scalability, the OCR is only a monitoring tool, and actual productivity is of greater significance. Between days 10 and 18, the S-51 system produced an average of 233mg IgG per day for a total of 2.1 grams in 9 days. The S-451 system produced an average of 2850mg IgG per day or 25.7 grams in the same 9 day period. In this case, the S-451 ceramic, which is 12 times the size of the S-51 ceramic, produced 12.2 times the product in the given period. This result indeed indicates linear scalability.

CONCLUSIONS AND DISCUSSIONS

For attachment dependent mammalian cells, the use of Opticore^{TM} ceramic based growth chambers run in Opticell^{TM} systems has been shown to provide linearly scalable culture growth. These bioreactors were scalable both when increasing the chamber sizes and when using multiples. The production of proteins from cells entrapped on porous ceramics is scalable and directly related to the volume of ceramic used. The use of the oxygen consumption rate to predict scalability is useful if all other conditions can be kept proportionally consistent.

Future demand for cell derived biomolecules will eventually determine the necessary scale for animal cell bioreactors. In the studies discussed, there was no apparent loss of efficiency with scale-up using this technology. It has been reported (private communications) that cells in the Opticell^{TM} 5300 bioreactors have produced protein at levels equivalent to larger systems of 10 times greater capacitiy by volume and media used. Hence, it is difficult to compare these bioreactors to multi-thousand liter systems if their efficiencies are very different. Future work with immobilized hybridomas is planned on larger matrices and in multiple chambered bioreactors.

The authors would like to acknowledge the contribution of data by Mr. Ernest A. Bognar, Jr. Additionally, we would thank Ms. Kyle Wallace for her assay work and Ms. Joanne Lamond for her time in preparation of this manuscript.

REFERENCES

[1] Spier, R.E. 1980. Recent Developments in the Large Scale Cultivation of Animal Cells in Monolayers. Advances in Biochemical Engineering, 14: 119-162.

[2] Birch, J., Thompson, P., Lambert, K., Braston, R. In: Large-Scale Mammalian Cell Culture. Feder, J., Tolbert, W., (eds.) pp. 1-18, Orlando: Academic Press, Inc. 1985.

[3] Karkare, S., Phillips, P., Burke, D., Dean, R., In: Large-Scale Mammalian Cell Culture. Feder, J., Tolbert, W., (eds.) pp. 127-149, Orlando: Academic Press, Inc. 1985.

[4] Glacken, M.W., Fleishacker, R.J., Sinskey, A.J. (1983). Mammalian cell culture: engineering principles and scale-up. Trends in Biotechnology 1: 102-107.

[5] Margaritis, A., Wallace, J. (1987) Novel Bioreactor Systems and Their Applications. Bio/Technology 2: 447-450.

[6] Feder, J., Tolbert, W. (1983). The Large-Scale Cultivation of Mammalian Cells. Scientific American. 248: 36-43.

[7] Lydersen, B.K., Pugh, G.G., Paris, M.S., Sharma, B.P., and Noll, L.A. (1985). Ceramic matrix for large scale animal cell culture. Bio/Technology 3: 63-67.

[8] Lydersen, B.K., Putnam, J., Bognar E., Patterson, M. Pugh, G.G., Noll, L.A. In: Large-Scale Mammalian Cell Culture. Feder, J., Tolbert W. (eds.) pp. 39-58. Orlando: Academic Press, Inc. 1985.

[9] Putnam, J.E. In: Commercial Production of Monoclonal Antibodies. Seaver, S. (ed.) In press.

INTERNATIONAL CONFERENCE ON BIOREACTORS AND BIOTRANSFORMATIONS
GLENEAGLES, SCOTLAND, UK: 9-12 NOVEMBER 1987

Paper C4

PRODUCTION OF BIOMOLECULES BY CELLS CULTURED IN TRI-DIMENSIONAL COLLAGEN MICROSPHERES

Edward G. Hayman, Nitya G. Ray, Peter W. Runstadler, Jr.
Verax Corporation
HC 61, Box 6, Etna Road
Lebanon, NH 03766

ABSTRACT

Verax Corporation has developed and is manufacturing bioreactor systems intended for large scale culturing of eukaryotic cells. The key element of these systems is spherical three dimensional collagen lattices (microspheres) 500 microns in diameter. Collagen is a major constituent of extracellular matrix *in vivo* and has been used as a substratum for cultured cells for many years. Most cells are capable of attaching to collagen. Adhesion to a substratum is often a requirement for growth and survival *in vitro*. Twenty micron sections cut through the populated microspheres reveal tissue like morphology and densities within the sphere for hybridoma and fibroblastic cells. The microenvironment of the sphere has been useful in promoting the survival, growth, genetic stability and productivity of cells in serum free medium. This fluidized-bed bioreactor and its inherent scalability is described.

INTRODUCTION

The demand for the products of Biotechnology (monoclonal antibodies, tissue plasminogen activator, hepatitis B vaccine, erythropoietin, factor VIII and interleukin I - IV) is steadily growing [1]. Because of the requirement for post-translational modification of these products most must be made in eukaryotic cells [2]. To meet the demand for these products a new generation of bioreactors for mass cell culture is being developed [3].

The first wave of bioproducts, the monoclonal antibodies, could be produced in free suspension culture. At the research level, spinner culture flasks were suitable. In scale up and commercialization these were replaced by large stirred tanks or airlift bioreactors similar to what has been used for years to grow microorganisms. These can be run either in batch or continuous mode. The major draw back of these systems is the low concentration of antibodies compared to *in vivo* titers. To overcome the drawback, hybridoma cells were either encased in a semipermeable polymer membrane [4] or cultured on the shell side of hollow fiber membranes [5]. Practically these systems have not proved to be scalable.

The second wave of biotechnology products comes from attachment preferred mammalian cells. In order for these cells to grow and produce product *in vitro* these cells must be anchored to a surface. The problem was how to immobilize the cells in a scalable process. In 1967 van Wezel [6] had introduced the use of a microcarrier as a means to grow mass culture of anchorage dependent cells. While this is a scalable approach not all cells have proven adaptable to microcarrier cultures [7]. Verax has developed a process that is scalable, adaptable to both attachment dependent and independent cells and is continuous. The process uses a fluidized bed bioreactor. The material that is fluidized is a spherical collagen lattice.

MATERIALS AND METHODS

Cells and culture conditions: A mouse/mouse hybridoma and a transformed rat kidney (TRK) cell line were used to test microsphere loading. The hybridoma cells were cultured in a serum free medium developed by Verax. The rat kidney cell line was cultured in this medium supplemented with 1% fetal bovine serum. Cell counting was done using a Coulter Counter. Viability was established by dye exclusion and hemocytometer counts. Disruption of the collagen matrix was accomplished by treatment with 1 mg/ml bacterial type I collagenase.

Scanning electron micrographs: Samples of microspheres were fixed in 2% glutaraldehyde and viewed and photographed at the Dartmouth College Electron Microscope Facility, Hanover, NH.

Light micrographs of microsphere sections: Populated microspheres were fixed in 10% buffered formalin and embedded in 10% high gel strength agarose. Twenty micron sections were cut on a Vibratome 1000 sectioning system. The sections were stained with Masson Trichome stain.

RESULTS

The mass culture of mammalian cells presents some unique problems. In comparison to bacteria or yeast, mammalian cells are fragile, slow growing, have rigorous nutrient requirements and in many cases are anchorage dependent. In order to make the Verax system versatile enough to grow both attachment independent as well as attachment preferred (while these cells are capable of being adapted to suspension culture they grow and produce better when attached) cells, an immobilization substratum was required. Collagen is a major constituent of the extracellular structural matrix in higher animals. It is the most common protein in the animal world [8, 9]. It is immunologically benign, resistant to proteolysis and is a natural substrate for cell adhesion either directly or indirectly in conjunction with adhesion promoting molecules such as fibronectin, vitronectin and laminin [10, 11]. Extensive literature describes the effects of collagen substrates on cell growth, migration differentiation as well as adhesion. Collagen in the form of a three dimensional lattice appears to be most active in promoting these effects [12, 13].

The basic elements of the Verax system is a spherical three dimensional lattice (microspheres) composed primarily of native type I collagen. The microspheres are manufactured from bovine hide collagen by a proprietary process (U.S. patents applied). The finsihed microspheres are highly crosslinked for structural robustness and attrition resistance and are sterilized. By precisely controlling the various steps in this process Verax can produce microspheres of various degrees of openness. Two examples of microspheres are shown in figure 1.

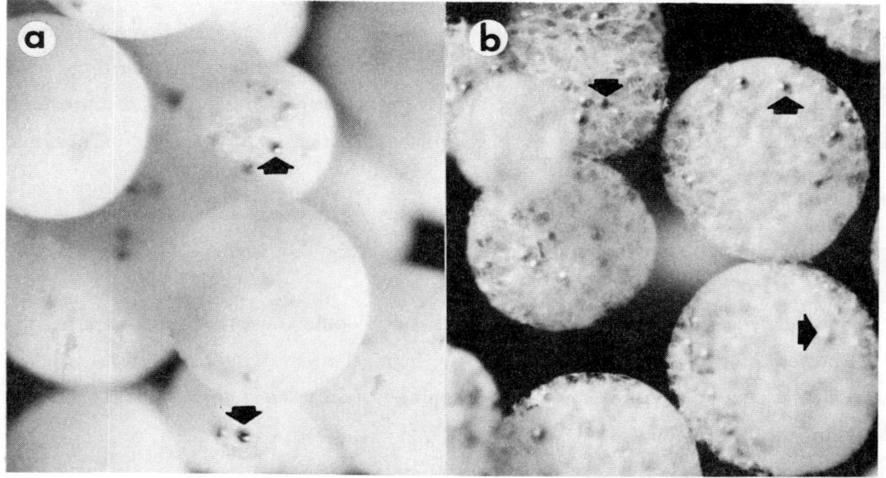

Fig 1 Collagen microsphere a) Example of tight lattice structure. b) Example of open lattice structure with 20 - 40 micron pores/channels. Arrows point to weighting material that is used to vary specific gravity of microsphere.

The microspheres in figure 1a have a very tight lattice structure. Cells grow mainly on the surface of such microspheres. The microspheres in figure 1b are examples of the lattice structure that has given the best results to date. These microspheres are 500 - 600 microns in diameter and have interconnecting pores and channels in the order of 20 -40 microns. These photographs also reveal another unique feature of the Verax microspheres. The small spheres scattered about the microsphere are weighting elements. These are added to regulate the specific gravity of the microspheres. Presently we can vary the specific gravity from 1.05 - 3.0. The importance of this feature is discussed below.

Figure 2 is a scanning electron micrograph of the microsphere type shown in fig 1b. The sheet like nature of the collagen is seen, surfaces on which cells can easily attach, yet the structure is open for good mass transfer.

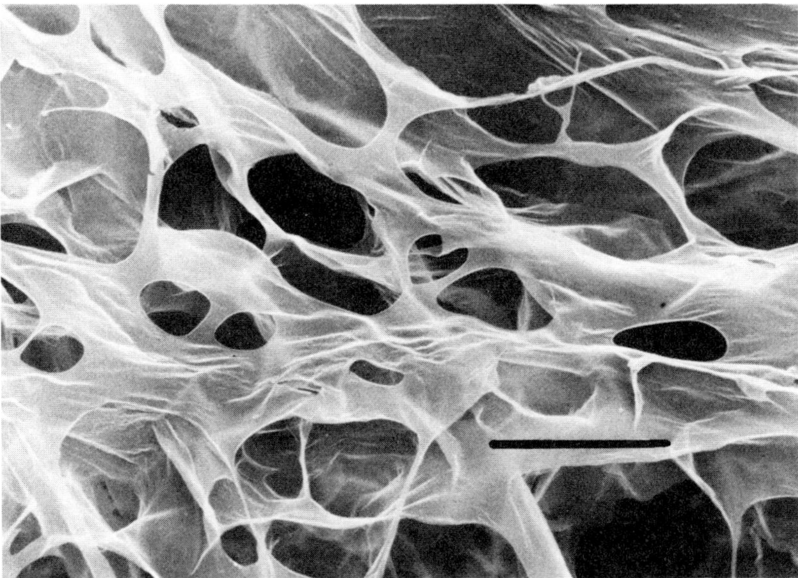

Fig 2 Scanning electron micrograph of microsphere similar to figure 1b showing the sheet like structure of the collagen. Bar = 10 microns.

Microsphere loading tests were done to determine which matrix configuration was most suitable for cell growth and protein production. The microspheres were inoculated by placing 10 ml of microspheres in a flask and adding a suspension of single cells of either an attachment independent cell, a murine hybridoma, or an attachment preferred cell, a transformed rat kidney cell. After 24 hrs the free cells were removed and the number of loaded cells determined. This was done by removing a sample of the microspheres, and then mixing the sample with collagenase and counting the liberated cells. A composite graph of this data is shown in figure 3.

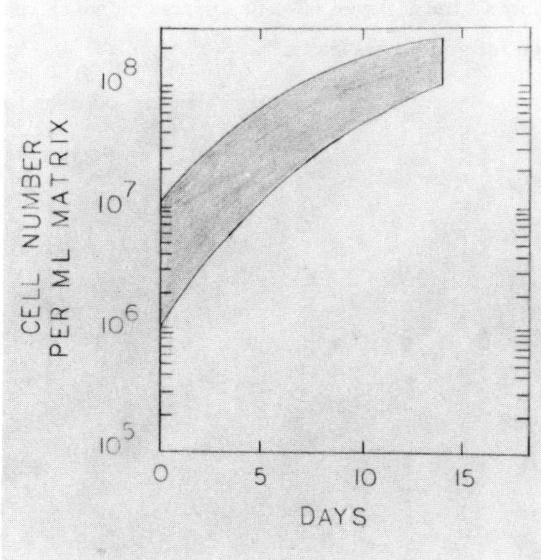

Fig 3 Population of microspheres. Number of viable cells, either hybridoma or rat kidney cells, was determined after disruption of the microspheres with collagenase. Graph is composite of 30 samples run in duplicate.

On day zero, 1 ml of microsphere contained 10^6 - 10^7 cells per ml of matrix. Within two weeks 1 ml of matrix contained between 1 to 2.5 x 10^8 viable cells. Medium consumption was 100 - 125 ml per day.

Samples of populated microspheres were taken, fixed in formalin, embedded in agarose, sectioned and stained. An example of such a section populated with TRK cells is shown in Figure 4a.

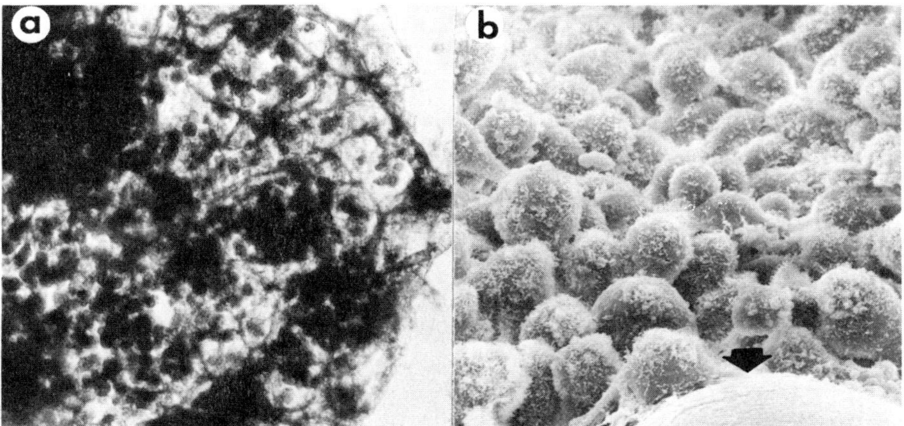

Fig 4 a) Stained 20 micron sections of populated microspheres. Cell can be seen throughout the matrix. b) Scanning electron photo micrograph of cells on surface of populated matrix. Arrow points to weighting element.

When examined microscopically such sections reveal that the cells have populated the interior as well as the surface of the microspheres. The cells appear to be firmly attached to the matrix. Hybridoma cells also appear to be able to attach to the matrix but not as firmly as fibroblastic cells. Scanning electron micrographs of a populated microsphere such as seen in figure 4b reveal that the surface of the microsphere is populated by a dense layer of cells.

Matrix Application

A wide variety of cell types have been successfully cultured in microspheres. Success is monitored not only by high cell number but by productivity measurements. Among the cells that have been cultured in microspheres are chinese hamster ovary cells making 5 different genetically engineered products, genetically engineered african green monkey kidney cells, human embryonic kidney cells and mouse mammary tumor cells. Non engineered normal rat kidney cells, transformed rat kidney cells and human hepatoma cells have also been successfully cultured. More than 30 different hybridomas encompassing mouse/mouse, mouse/rat, mouse/human and human/human have also been cultured in the microsphere to high cell densities.

The Bioreactor

The central feature of the Verax bioreactor system is its fluidized bed of microspheres. A fluidized bed occurs when particles of a given size are put in a vertical vessel and a fluid of lower density is caused to flow upward through them with sufficiently high velocity that the particles no longer rest on one another but are suspended in the fluid and are free to move about. Some of the advantages of a fluidized bed reactor are: scalability, superior mass transfer capabilities, good mixing and low shear.

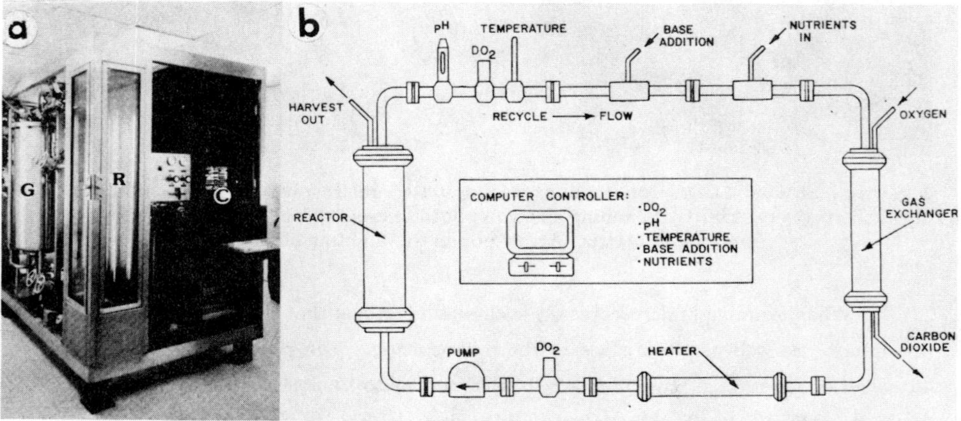

Fig 5 a) Photograph of Verax System 2000, a 24 liter bioreactor capable of maintaining 10^{12} viable cells. C, computer control, R, reactor, G, gas exchanger. b) Schematic of bioreactor elements.

Figure 5a is a photograph of the Verax production scale bioreactor. The schematic figure 5b shows that the system consists of two independent liquid flow paths that operate at two different flow rates. One flow path is used to supply fresh medium and to remove harvest and waste. For the system shown this will be in the range of 150 - 900 liters per day. This rate will support between 1 - 4 x 10^{12} cells. The second flow path is used to recycle medium through a tube and shell gas exchanger and a heater then back through the bottom of the reactor so that the weighted matrix is fluidized into a homogeneous bed. The flow rate in the recycle loop is in the range of 4 - 30 liters per minute. The bioreactor is computer controlled. Dissolved oxygen, pH and temperature are monitored and controlled at the desired levels by the computer. The bioreactor is designed to operate in the continuous mode which allows sophisticated adjustment of the steady state environment in order to maximize productivity.

Bioreactor Performance

Table 1 displays productivity data using the collagen lattice, fluidized bed bioreactor system. The cell specific productivity and yields are equal to or better than other process systems for which equivalent data are available. The hybridoma data were obtained using serum free media; the lymphokine and interferon data were with 1% fetal bovine serum. Because of the nature of the fluidized bed/microsphere system the process can be scaled from bench top to production using the same basic process. Maximum cell density in the fluidized bed is typically achieved within 10 to 15 days starting from an inoculum of 2.5 x 10^5 cells per milliliter of reactor volume.

TABLE 1

Bioproduct	Cell type	Concentration in harvest	Units/liter reactor/day	Units/10^6 viable cells/day
IgG	Murine hybridoma	40 - 70 μg/ml	300 - 500 mg	24 μg
Lymphokine	CHO	2700 - 3400 units/ml	17 - 21 x 10^6 units	5000 units
Mouse interferon	CHO	10,000 - 13,000 units/ml	9.5 - 12.0 x 10^7 units	4400 units

DISCUSSION

Verax has incorporated the unique cell culture features of a three dimensional collagen matrix in a fluidized bed with the steady state environment obtainable in a continuous culture process. The result is a bioreactor system for mammalian cells that not only is versatile, in that it can handle both attachment independent cells and attachment dependent cells, but is also scalable. We have found that a collagen sphere 500 microns in diameter, with an open structure so that cells can populate the interior as well as the surface is most suitable for a variety of cell types. The use of a tridimensional collagen lattice in a fluidized bed has many advantages: post inoculation of cultures; protective environment for fragile cells; direct sampling of matrix associated and free cells; ease of separation of biomass from product and waste; and scalability. Furthermore the open structure of the matrix allows for good mass transfer. Cells *in vivo* live a very protected life. They live in an environment in which pH, temperature, oxygen tension etc. are maintained within well defined set points. Fresh nutrients are constantly being supplied to them and wastes removed. The Verax bioreactor process tries to duplicate this environment.

The authors wish to thank the following for technical assistance: Sandy Warner, Cheryl Grabe, Ben Turner, Steve Parker, Kathy Maruk and Dave Dugdale. The authors thank Baar Graf for preparing the manuscript.

REFERENCES

1. McCormick, D., Pharmaceutical Markets for the 1990's. Bio/Technology, 1987, 5, 27.

2. Van Brunt, J., Immobilized Mammalian Cells: The Gentle Way to Productivity. Bio/Technology, 1985, 4, 505-510

3. Hu, W.S. and Dodge, T., Cultivation of Mammalian Cells in Bioreactors. Biotechnology Progress, 1985, 1, 209-215.

4. Nilsson, K., Scheirer, W., Merten, O., Ostberg, L., Liehl, E., Kallinger, H. and Mosbach, K., Entrapment of Animal Cells for Production of Monoclonal Antibodies and Other Biomolecules. Nature, 1983, 302, 629-630.

5. Hopkinson, J., Hollow Fibre Cell Culture: Applications in Industry. In Immobilized Cells and Organelles, ed. B. Mattiason, CRC Press, Boca Raton, 1983, pp. 88-89.

6. Van Wezel, A.L., Growth of Cell-Strain and Primary Cells on Microcarriers in Homogenous Culture. Nature, 1967, 216, 64.

7. Margaritis, A. and Wallace, J.B., Novel Bioreactor Systems and Their Applications. Bio/Technology, 1984, May, 447-453.

8. Yang, J. and Nandi, S., Growth of Cultured Cells Using Collagen as Substrate. International Review of Cytology, 1983, 1, 249-286.

9. Sabelman, E., Biology, Biotechnology and Biocompatability of Collagen. In Biocompatability of Tissue Analogs, CRC Press, Boca Raton, 1985, pp. 27-66.

10. Kleinman, H., Klebe, R.J. and Martin, G., Role of Collagen Matrices in the Adhesion and Growth of Cells. J. Cell Biol., 1981, 88, 473-485.

11. Hayman, E.G., Pierschbacher, M.D., Suzuki, S. and Ruoslahti, E., Vitronectin - A Major Cell Attachment - Promoting Protein in Fetal Bovine Serum. Exp. Cell Res., 1985, 160, 245-258.

12. Schor, S.L., Schor, A.M., Winn, B. and Rushton, G., The Use of Three-Dimensional Collagen Gels for the Study of Tumor Cell Invasion In Vitro: Experimental Parameters Influencing Cell Migration Into the Gel Matrix. Int. J. Cancer, 1982, 29, 57-62.

13. Hall, H.G., Farson, D. and Bissell, M., Lumen Formation By Epithelial Cell Lines in Response to Collagen Overlay: A Morphogenetic Model in Culture. Proc. Natl' Acad Sci USA, 1986, 19, 4672-4676.

INTERNATIONAL CONFERENCE ON BIOREACTORS AND BIOTRANSFORMATIONS
GLENEAGLES, SCOTLAND, UK: 9-12 NOVEMBER 1987

Paper D1

A NOVEL METHOD FOR THE PRODUCTION OF DEXTRAN AND FRUCTOSE

PE Barker & I Zafar
Department of Chemical Engineering and Applied Chemistry,
Aston University
Aston Triangle
Birmingham, B4 7ET

and

RM Alsop
Fisons Pharmaceuticals plc
London Road
Holmes Chapel
Crewe, Cheshire, CW4 8BG

ABSTRACT

A batch chromatographic reactor-separator was used to synthesise dextran and fructose from sucrose. The stationary phase consisted of an ion exchange resin in the calcium form and the eluent was a dilute solution of dextransucrase. The simultaneous separation of products not only resulted in pure dextran but also in high purity fructose. Moreover this separation of fructose from the dextran produced dextran of a different molecular weight distribution (MWD) than that obtained from a conventional batch reactor.

To obtain high conversions and separation, the chromatographic reactor generally used more enzyme than the conventional batch fermentation reactor, although there are possibilities of attaining parity or even smaller consumptions of enzyme.

INTRODUCTION

Ever since the patents taken out by Dinwiddie (1), Magee (2) and Gaziev (3), the chromatographic reactor has been the subject of much research for its analytical as well as its chemical engineering applications. The industrial reactions studied have included the dehydrogenation of butanol (4) and butenes (5,6), the Haber ammonia synthesis reaction (7) and the hydrolysis of methyl formate (8). These reactions were all reversible and it was conclusively shown that conversions greater than those predicted by thermodynamics can be obtained. However, despite these favourable results, the chromatographic reactor has not found industrial applications like the other conventional reactors such as plug flow, well mixed and packed beds. The reason for this is mainly due to the extent of conversion, being used as a performance index (9,10). No emphasis was placed on the reactors ability to separate the products and hence justify its use on economic grounds.

In this paper we experimentally demonstrate that the separation of products is also an important criteria for assessing a chromatographic reactor. The reaction studied was the enzymatic conversion of sucrose to fructose and dextran by dextransucrase.

Principles of Operation

The principle of operation of the chromatographic reactor involves the injection of a reactant or reaction mixture into the reactor column as a pulse. The pulse is then swept down the column by the mobile phase. On passing through the column, the reactant is converted to products which have a different affinity for the stationary phase and they travel through the column at different velocities so that separation tends to occur.

There are two possible ways of operating a chromatographic reactor (Figure 1). The first method involves the reaction taking place in the stationary phase which may be liquid supported or solid catalyst. The stationary phase acts as a catalyst as well as an adsorber. Where this is not possible eg. with an immobilised enzyme and adsorbent a mixed bed can be used. The mobile phase can be a reactant or it can be an inert phase.

The second method involves the reaction taking place in the mobile phase (gas or liquid). The mobile phase includes the reactant and the catalyst whilst the stationary phase acts only as an adsorbent.

These types of reactors offer three major advantages as compared to other conventional reactors:

(1) Inclusion of two unit operations, reaction and separation into one thus helping to reduce operating and capital costs.

(2) The instantaneous separation of products and reactants allows greater conversions to be attained with reversible reactions.

(3) The interference of inhibitors or unwanted by products in the reaction process can be lessened by the separation process.

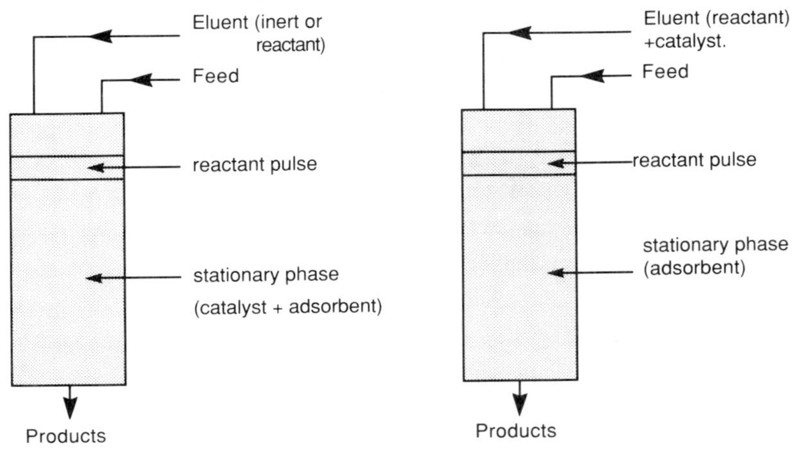

(a) Stationary phase acting as catalyst and adsorber.

(b) Stationary phase acting as adsorbent only

Figure 1- Operation of a chromatographic reactor

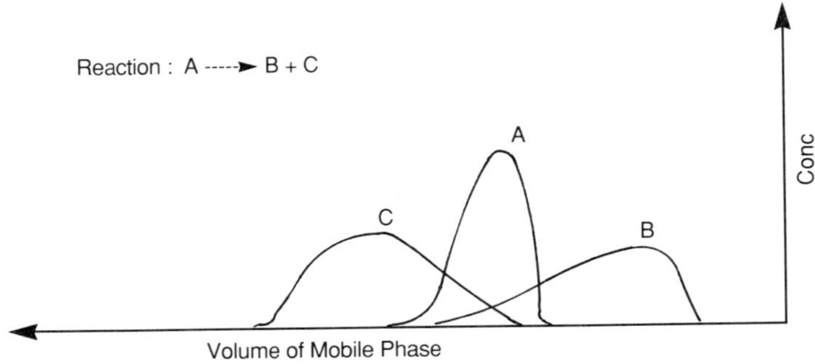

Fig 2 Profile of products and reactant $K_{dB} < K_{dA} < K_{dC}$

To attain the above advantages it is essential for the values of the distribution coefficient (Kd) of the reactant to lie between those of the two products (see Figure 2). If however the Kd value of the reactant is less or greater than the products, longer reactor lengths will be needed to attain the required conversion and purity.

Reactor Characterisation

There were two different reactor sizes used in this work. Both were packed with Lewatit TSW 40 which was supplied in the calcium form by Bayer. The criterion used for comparing their performance was the height equivalent to a theoretical plate (HETP). This was calculated by using the Gluekauf equation (11).

$$\text{HETP} = \frac{L}{8\left(\frac{t_r^{\,i}}{W_{h/e}}\right)^2} \qquad \ldots\ldots\ 1$$

The distribution coefficient (Kd) at infinite dilution was determined by the following equation:

$$\text{Kdi} = \frac{V_i - V_o}{V_T - V_o} \qquad \ldots\ldots\ 2$$

From the distribution coefficients the columns separation potential from the reactant sucrose in terms of the separation factor ($\alpha_s i$) was obtained:

$$\alpha_s{}^i = \frac{K_{di}}{K_{ds}} \qquad \ldots\ldots\ 3$$

The results obtained for each reactor column are presented in Table 1. The larger diameter reactor was found to have a higher voidage even though the same packing technique was used. This discrepancy is expected, because it is known that an increase in diameter can lead to packing inefficiencies. These inefficiencies were shown up in the HETP values for glucose, sucrose and fructose. Also the separation factors indicate that the larger diameter column gives better separation between sucrose and fructose, although the band spreading is greater.

TABLE 1
Column characterisation for the two reactors

		Column 1	Column 2
Diameter (cm)		9.7	19.4
Packing height (m)		1.5	1.75
Elution volumes (cm^3)	V_d	37.28	254.04
	V_s	53.25	299.28
	V_g	61.43	327.12
	V_f	76.7	428.04
Distribution coefficients	K_{ds}	0.192	0.16
	K_{dg}	0.29	0.26
	K_{df}	0.474	0.62
Number of theoretical plates	WRT sucrose	349.2	80.2
	WRT glucose	422.5	81.2
	WRT fructose	494.5	104.2
HETP (cm)	WRT sucrose	0.43	2.18
	WRT glucose	0.36	2.16
	WRT fructose	0.3	1.68
Separation factor	$\alpha^d{}_s$	-	-
	$\alpha^g{}_s$	1.51	1.63
	$\alpha^f{}_s$	2.47	3.88
Voidage		0.31	0.475

d = dextran g = glucose f = fructose s = sucrose

Equipment Description

The equipment (see Figure 3) consisted of a jacketed reactor, which was maintained at 25°C during experimentation and at 4°C, when not in use. This lowering of temperature reduced the loss of activity of enzyme.

The eluent consisted of diluted dextransucrase at pH 5.2, which was pumped by a peristaltic pump. The reactor output was allowed to flow into a reservoir maintained at 70°C. A high temperature was maintained to stop the reaction proceeding further and hence enabling analysis of products to be performed. The products were tracked by a differential refractometer and then analysed using conventional HPLC and gel permeation chromatography techniques. The dextran due to its larger molecular weight distribution (MWD) could not be analysed by HPLC, but was successfully characterised using gel permeation chromatography.

The first product to elute from the reactor was dextran and then sucrose. The fructose eluted last due to the formation of a loose complex with the calcium ions on the resin.

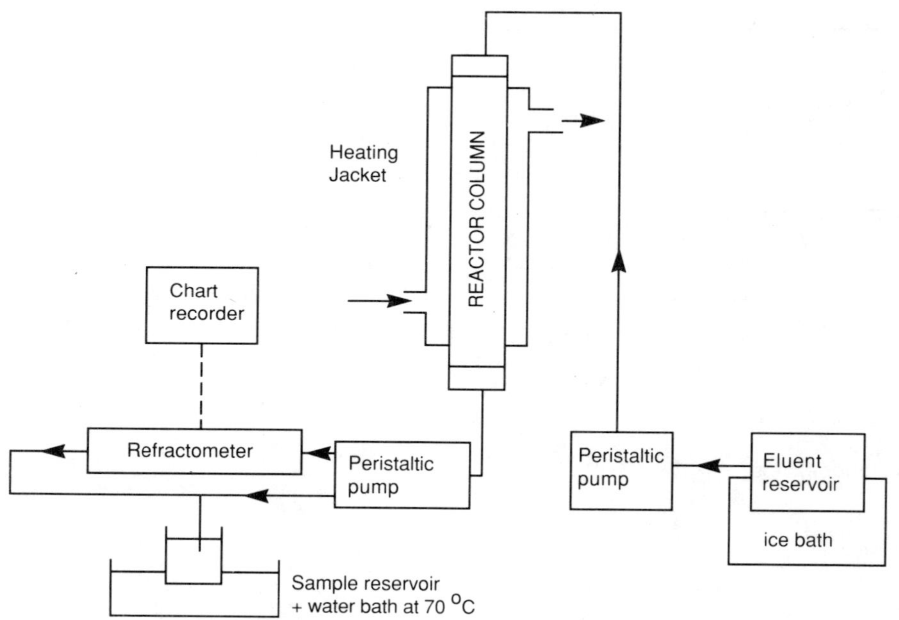

Fig 3 Schematic Diagram for Equipment Layout

RESULTS

Effects of Enzyme Eluent Concentration

The results for the effect of enzyme eluent concentration for the 1.94cm diameter column on % conversion are presented in Figure 4. It was noted that the reactor initially does not produce very high conversions at economic enzyme concentrations. Only after some usage was consistency and higher conversions attained. This behaviour was attributed to gradual adsorption of enzyme on to the resin and therefore enabling greater contact with the sucrose substrate. The adsorption was a very loose one and the enzyme was washed off when the eluent was switched to deionised water. Secondly, there could be gradual displacement of water from the void volumes and resin pores.

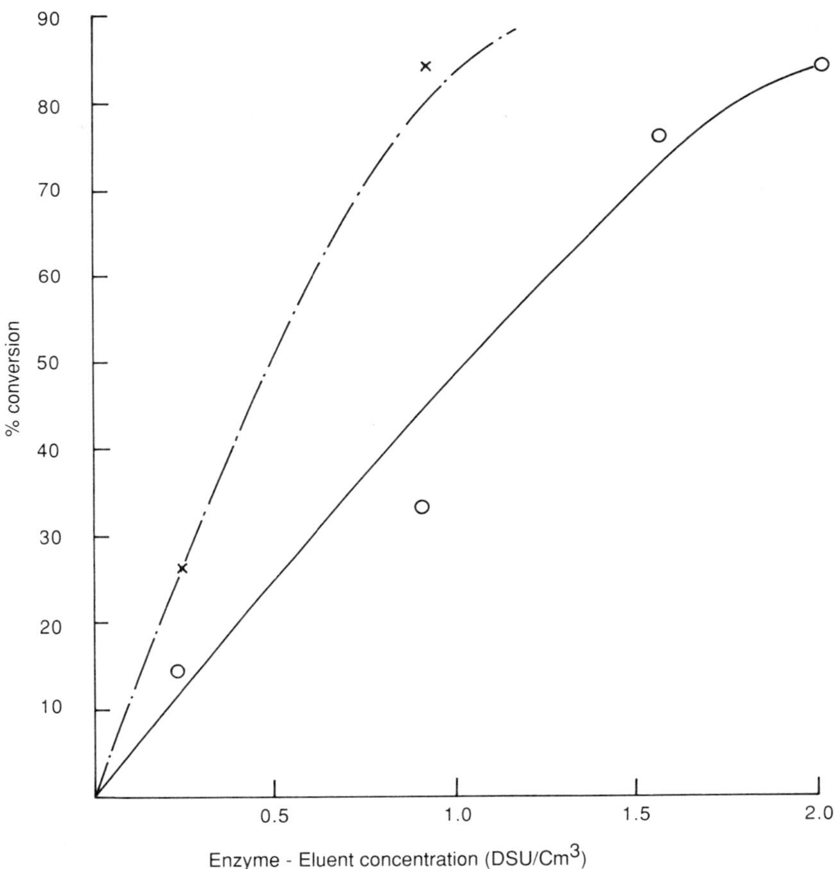

Fig 4 Effect of Enzyme - Eluent concentration on % conversion

Effect of Temperature

The effect of temperature on the reactor performance was investigated on the 0.97cm diameter reactor by varying the temperature between 19°C and 26°C (see Figure 5). The results indicated that for every 1°C drop in temperature there is a reduction of 2.14 in percentage conversion. This behaviour was similar to that obtained for a conventional batch fermentation reactor. At temperatures greater than 26°C a reduction in conversion is expected. Moreover at 30°C the enzyme is at its most active state, but this is offset by it being stable for only a short period of time.

Another aspect of temperature change is the ability of the resin to separate the reactant from the products. It is known that a change in temperature affects the Kd values, but this effect is expected to be very minor over the temperature change under investigtion.

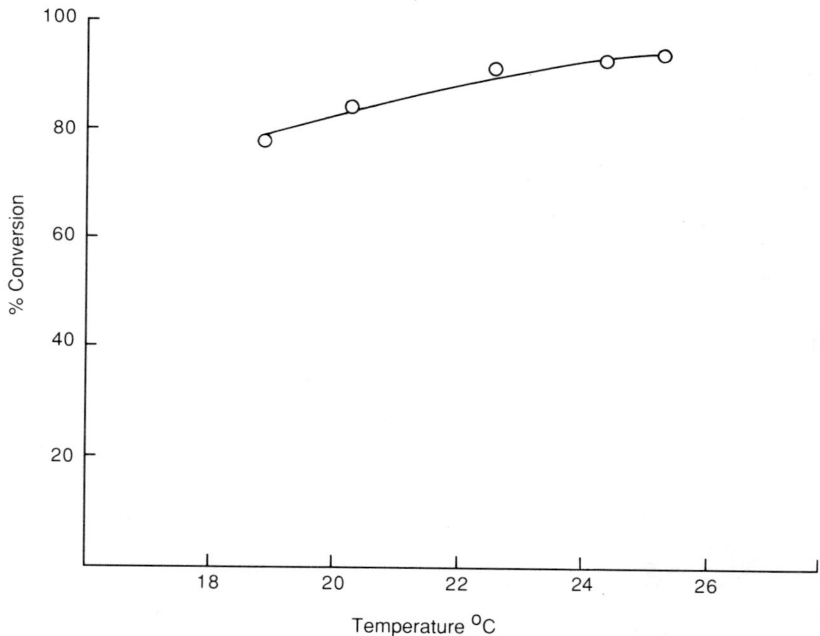

Fig 5 Effect of temperature on %conversion

Effect of Pulse Size at a Concentration of 2% w/v

The pulse size of sucrose in both reactors was varied between 3%-13% of empty column volume. In both reactors similar trends were observed, and the results for the larger diameter reactor are presented in Table 2. As expected an increase in pulse size led to a decrease in percentage conversion, but not in absolute yield of products. The fructose produced reached a plateau value at a sucrose pulse size greater than 9.3% of empty column volume. Further increase in pulse size does reduce productivity and moreover creates purity problems. This is due to the dilution effect of the enzyme free pulse. However this can be compensated for by including the enzyme in the pulse at the same concentration or at a higher concentration than in the reactor.

TABLE 2
Effect of pulse size on percentage conversion and purity

Run No	Charge size cm^3	% of total column volume	Mass of fructose formed (x 10^{-3}g)	% con-version	Excess enzyme usage Actual (DSU)/ Theoretical (DSU)[a]	Res-olution	% recovery of fructose recovered	Amount of pure fructose ($x10^{-3}$g)
20-2-1.38-0.8-25	20	3.7	166.75	79.25	2.34	0.54	67.4	112.07
30-2-1.38-0.8-25	30	5.6	245.65	77.8	1.59	0.46	49.3	121.11
50-2-1.38-0.8-25	50	9.3	360.5	68.5	1.08	0.36	45.99	165.78
70-2-1.38-0.8-25	70	13.1	340.01	46.2	1.15	0.12	32.55	110.68

(a) The enzyme consumption in a conventional reactor based on the residence time in a chromatographic reactor

Also presented in Table 2 are the enzyme usage figures. It can be seen that the chromatographic reactor uses slightly higher amounts of enzyme than a conventional reactor. This slight disadvantage can be overcome by selecting a stationary phase which will adsorb greater amounts of enzyme hence enabling greater contact with the sucrose.

A comparison of the amount of enzyme used in the two reactors is given in Figure 6. Clearly the larger diameter reactor seems to be consuming less enzyme as compared to the smaller one. This is due to the larger diameter reactor having a higher voidage. Since the reaction takes place mainly in the void volumes an increase in voidage can be advantageous.

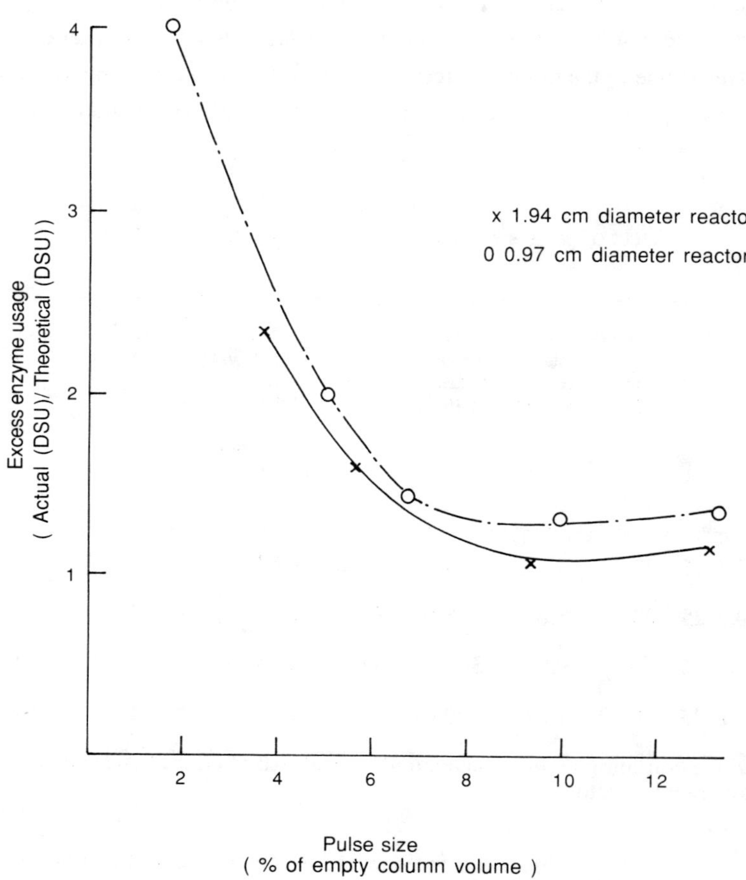

Fig 6 Comparison of enzyme consumption between the two reactors

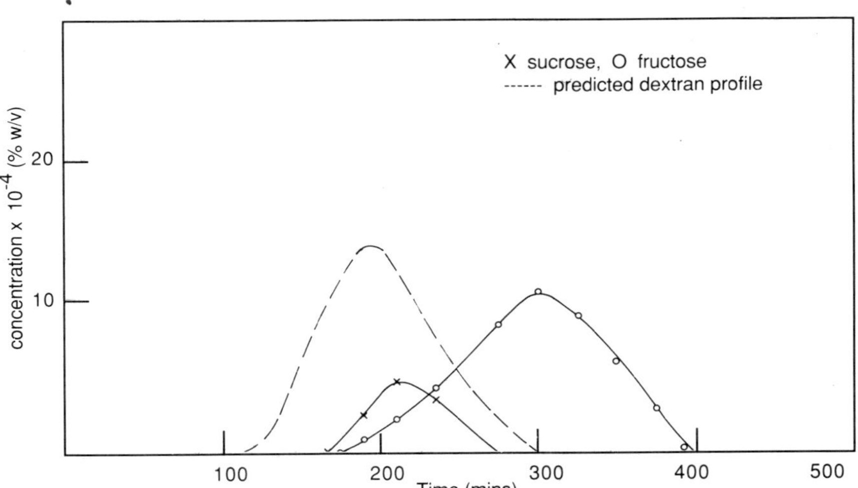

Run number was defined in the following manner 20-2-1. 38-0. 8-25 where 20 =pulse size (Cm^3), 2=pulse concentration % w/v , 1.38 = eluent flowrate ($Cm^3 min^{-1}$),0.8=enzyme eluent concentration (DSU/cm^3) 25 =temperature (^oC).

Fig 7 Experimental concentration profile for run 20-2-1. 38-0. 8-25

However this has to be balanced against the loss in separation efficiency (increase in zone spreading) due to the increase in the void volume of the reactor. The chromatograms for some of the runs carried out at different pulse sizes are given in Figures 7-10. There are three major observations that can be made from them. Firstly, as the pulse size increases the fructose peak time decreases from 300 minutes to 215 minutes. This in turn resulted in a drop in resolution between sucrose and fructose. A similar trend is also expected for dextran, although the degree of resolution between dextran and sucrose will be worse. Secondly, although the degree of resolution worsens with pulse size, the maximum recovery of pure fructose occurs at 9.3% of empty column volume. Any further increase in pulse may yield the same quantity of fructose, but a greater overlap between peaks leads to a drop in yield of pure fructose. Thirdly in an industrial environment, the unreacted sucrose may have to be recycled or allowed to reach 100% completion in a separate vessel and then added back to the dextran produced from the chromatographic reactor.

Molecular Weight Distribution (MWD) of Dextran

The MWD of dextran produced by dextransucrase at different sucrose concentrations for a conventional batch reactor were obtained. The MWD was obtained from the Lichrosphere DIOL columns (E Merck) and the results obtained are presented in Table 3. From the results it can be noted that as the initial concentration of sucrose is increased, the amount of low molecular weight dextran increases and that its concentration becomes significant at sucrose levels greater than 10% w/v. This low molecular weight dextran is unusable in the manufacture of clinical dextran and therefore is rendered as a loss in yield. The production of this low molecular weight dextran is due to an increase in concentration of fructose acceptor molecules (12,13) at high sucrose concentrations. These acceptor molecules cause early termination of the polymer chain. One method of overcoming this problem of fructose interfering with the polymerisation is to carry out the reaction in a chromatographic reactor. Here the fructose produced in the reaction is partially removed from the reaction zone thus enabling greater production of high molecular weight material. The results obtained are presented in Table 4. Clearly the partial separation of fructose does have an effect on the MWD of dextran produced from the chromatographic reactor. At high sucrose concentrations a greater proportion of high and intermediate molecular weight dextran is produced. However at a sucrose concentration of 2% w/v the conventional batch reactor produces greater amounts of high molecular weights. This could be explained in terms of band broadening effects. Due to the large quantity of the sucrose pulse injected, it is possible that the fructose as it was formed interfered with the rear end of the dextran peak. However at a sucrose concentration of 2% w/v the conventional batch reactor produces greater amounts of high molecular weight dextran.

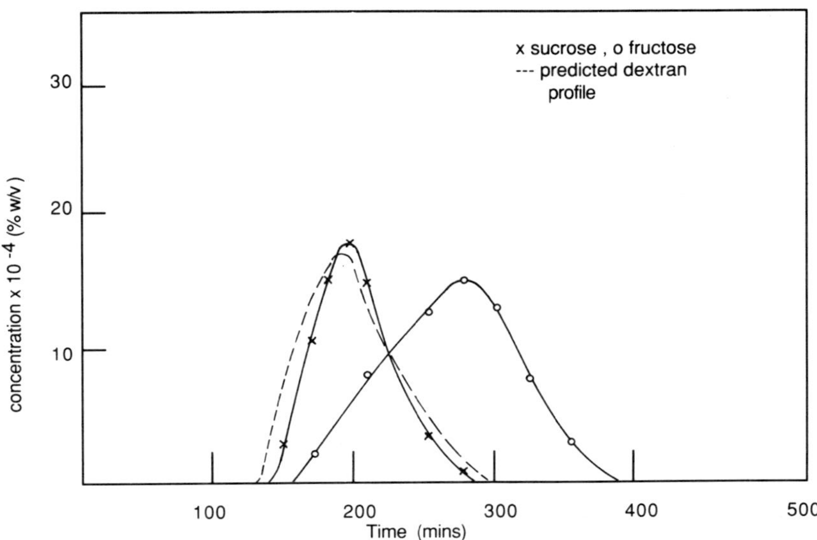

Fig 8 Experimental concentration profile for run 30-2-1. 38-0. 8-25

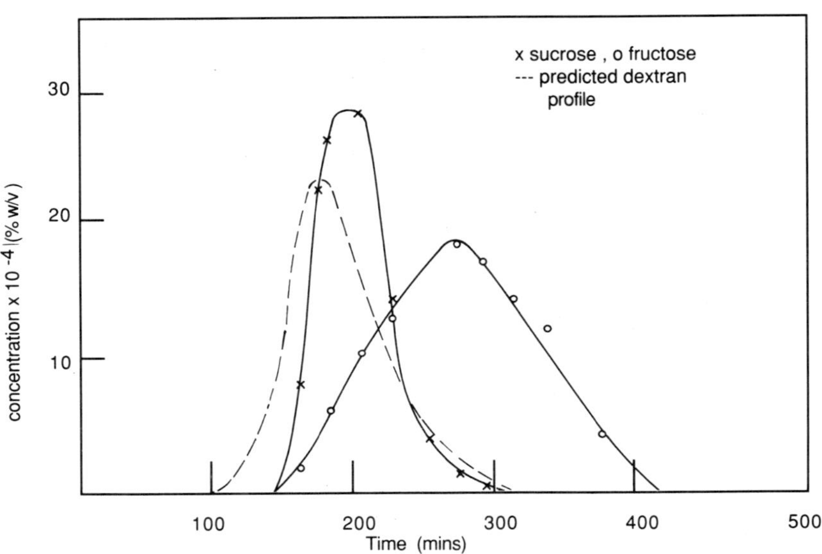

Fig 9 Experimental concentration profile for run 50-2-1. 38-0. 8-25

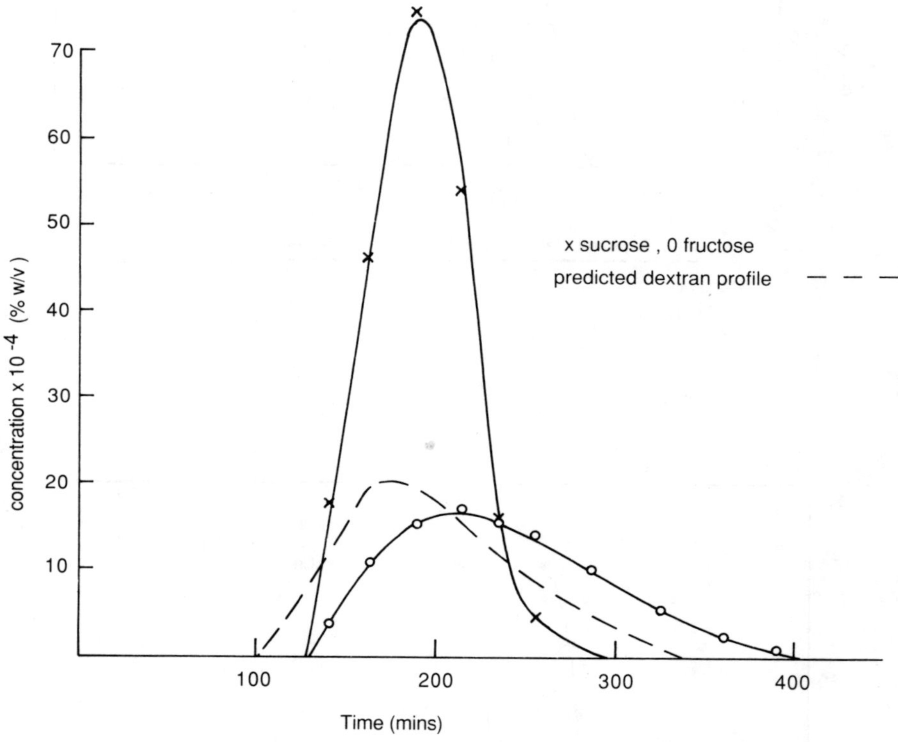

Fig 10 Experimental concentration profiles for run 70-2-1.38-0.8-25

TABLE 3
Molecular weight distribution of dextran at different sucrose concentrations for a conventional batch reactor

	% Dextran in each molecular weight range							
Sucrose conc %w/v	> 160000	88000 to 160000	59000 to 88000	34000 to 59000	20000 to 34000	9000 to 20000	4500 to 9000	1000 to 4500
2	97.62	1.06	0.6	0.48	0.16	0.08	-	-
5	94.66	1.21	0.86	0.88	0.42	0.52	0.32	1.13
10	88.85	2.0	1.4	1.55	0.77	1.12	2.05	2.26
15	68.98	2.49	1.77	2.55	2.12	5.56	9.13	7.4
20	43.65	2.46	2.34	4.48	5.4	15.28	17.74	8.29

TABLE 4
Molecular weight distribution of dextran at different sucrose concentrations for a chromatographic reactor separator

	% Dextran in each molecular weight range							
Sucrose conc %w/v	> 160000	88000 to 160000	59000 to 88000	34000 to 59000	20000 to 34000	9000 to 20000	4500 to 9000	1000 to 4500
2	89.04	3.32	2.78	2.06	0.84	1.17	0.79	-
15	80.7	2.4	3.27	8.06	5.04	0.44	0.07	-
20	77.35	3.65	4.24	8.97	2.79	1.83	1.17	-
20	74.52	3.08	4.29	10.37	5.85	1.6	0.25	-
20	64.34	4.02	4.78	10.52	9.24	5.87	0.92	-

CONCLUSION

Dextran and fructose have been produced for the first time in a chromatographic reactor (14). The parameters for the production have been identified as the enzyme-eluent concentration, eluent flowrate, voidage, pulse size and concentration. The reactor needs some time to reach steady state, whereby a certain amount of adsorbtion of enzyme on the resin surface occurs. The dextran produced from the chromatographic reactor gave a different MWD of dextran than that for a conventational batch reactor at high sucrose concentrations. This difference was attributed to partial removal of the fructose acceptors from the reaction zone. Moreover by including other acceptor molecules in the eluent a certain amount of tailoring of dextran (MWD) is envisaged. Apart from dextran, the process gives a useful source of high purity fructose which is currently not produced in the UK.

NOMENCLATURE

DSU	Dextran Sucrase Unit. One DSU is the amount of enzyme to convert 1 milligram of sucrose to dextran at 25 °C and a pH of 5.2.
e	Logarithmic base
h	Chromatogram peak height
K_{di}	Distribution coefficient of component i
L	Bed height
MWD	Molecular Weight Distribution
S	Sucrose
t_r^i	Retention time of component i
V_i	Elution volume of component i
V_o	Column void volume
V_T	Total column volume
$W_{h/e}$	Peak width at height h/e
$\alpha_s i$	Separation factor for component i with respect to sucrose

ACKNOWLEDGEMENTS

The authors would like to thank the directors of Fisons Pharmaceuticals plc and the SERC for financial support to enable this project to be undertaken.

REFERENCES

1 Dinwiddie J and Morgan W, 1961, US Patent 2, 976, 132

2 Magee E, 1961, Canadian Patent, 631, 882

3 Gaziev G, Roginskii S and Yanovskii MI, 1963, USSR Patent, 149, 398

4 Roginskii SZ, Zimin RA and Yanovskii MI, 1965, Dok Akad Nauk, SSSR, 164(1), 144-146

5 Roginskii SZ, Semenenko EI and Yanovskii MI, 1963, Dok Akad Nauk, SSSR, 153(2), 383-385

6 Semenko EI, Roginskii SZ and Yanovskii MI, 1964, Kinet Katal, 5(3), 490-495

7 Unger B and Rinker R, 1976, Ind Eng Chem Fund, 15(3), 225-227

8 Wetherold RG, Wissler E and Bischoff K, 1969, Ind Chem Eng, 61(4), 11-21

9 Schweich D and Villermaux J, 1978, Ind Eng Chem Fund 17(1), 1-7

10 Schweich D and Villermaux J, 1982, Chem Eng J, 24, 99-109

11 Gluekauf E, 1955, Trans Farad Soc, 51, 34

12 Alsop RM, 1983, Industrial production of dextrans, Progress in industrial microbiology, Vol 18, Elsevier, 1-44

13 Robyt JF, 1980, Mechanisms of saccharide polymerisation and depolymerisation, Proc Symp, Miami Beach, Academic Press, 43-54

14 Alsop RM, Barker PE, Zafar I 'Process'. UK patent application No 86/05975 (1986).

INTERNATIONAL CONFERENCE ON BIOREACTORS AND BIOTRANSFORMATIONS
GLENEAGLES, SCOTLAND, UK: 9-12 NOVEMBER 1987

Paper D2

BIOREACTORS FOR ANAEROBIC BACTERIA AND GENE-ENGINEERED BACTERIA WITH CROSS-FLOW FILTRATION

Takeshi Kobayashi

Department of Chemical Engineering
Nagoya University
Chikusa-ku, Nagoya 464,
Japan

ABSTRACT

A new fermentation system with continuous separation of inhibitory metabolites by cross-flow filtration was developed. With this fermentation system, high concentration cultivations of anaerobic bacteria were carried out for improving productivity of cell-mass itself and vitamin B_{12}. In the case of lactic acid bacteria, this system made it possible to remove continuously lactate while retaining cells completely in the fermenter. *Streptococcus cremoris*, *Lactobacillus casei* and *Bifidobacterium longum* were cultivated up to high concentrations of 81.5, 49.0 and 54.4 g-dry cell weight per liter, respectively. The cell productivities obtained were 19, 9 and 7-times respectively, as high as that of each corresponding conventional batch cultivation. This system was also effective for cultivations of *Propionibacterium shermanii* and *Butyribacterium methylotrophicum*. Vitamin B_{12} accumulated to 52 and 93 mg per liter, respectively. This system was also applied for on-off regulation of gene expression from tryptophan promoter. Recombinant plasmid containing β-galactosidase gene fused to *trp* promoter was introduced in *Escherichia coli* C600. By controlling tryptophan concentration, very high biomass was achieved and the amount of produced β-galactosidase was about 10% of total cellular proteins.

INTRODUCTION

In fed-batch culture, a high concentration of biomass was obtained [1-3] when nutritional conditions in the broth were kept at optimal values. At a high biomass concentration, however, accumulation of metabolites caused sharp growth retardation [4]. To achieve high concentrations of microorganisms which excrete metabolites inhibitory to growth, several fermentation processes with continuous removal of inhibitory metabolites have been studied, including dialysis culture, extractive fermentation and flash fermentation. Although productivities of cell-mass and metabolic products were improved by these fermentation processes, each has disadvantages associated with the separation of inhibitory products. Filtration is one of the most promising methods of removing inhibitory metabolites. For separating inhibitory metabolites from culture broth, ceramic filter was used for cultivations of lactic acid bacteria and vitamin B_{12} producers.

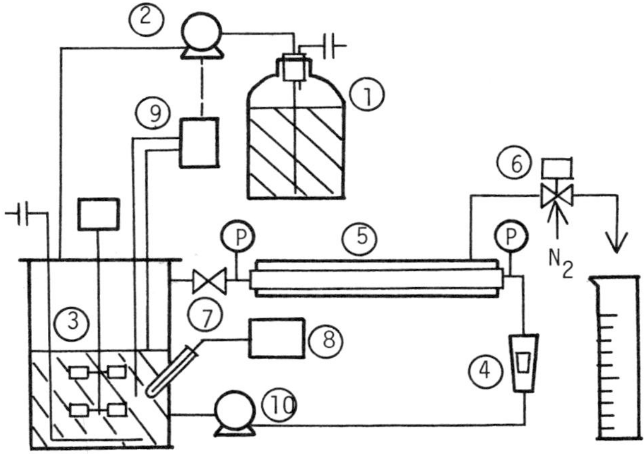

Figure 1. Schematic diagram of a fermentation system with cross-flow filtration. 1, reservoir for fresh medium 2, pump for feeding fresh medium 3, fermenter 4, flow meter 5, filtration unit 6, solenoid valve 7, ball valve 8, pH controller 9, level controller 10, pump for recycling culture broth.

Overproduction of useful proteins for medical and industrial uses is one of the most important targets in genetic engineering. Various techniques have been developed for overproduction of cloned gene products. We have already reported on the usefulness of the two-stage cultivation [5], which can be divided into the cell growth phase and the recombinant protein-production phase. In the case of expression vectors using *trp* promoter, decreasing tryptophan concentration in a medium becomes important. The filtration system was applied for removal of tryptophan in the medium with full induction of a cloned gene.

MATERIALS AND METHODS

A schematic diagram of the fermentation system with cross-flow filtration is shown in Fig. 1. A multihole type (17 holes, 4 x 750 mm) of Membralox, available from Ceravail Co., was used throughout this study. The ceramic filter has an average pore size of 0.2 μm and a filtration area of 0.18 m^2. The filtration area was enlarged by increasing the number of the ceramic filter when necessary. The circulation rate of the culture broth was 10 l/min, which corresponds to a superficial liquid velocity of 0.7 m/s. Fresh medium of the same volume as the filtrate was supplied via the feeding pump, which was coupled to a level controller to maintain a constant working volume in the fermenter. The time and interval for backflush was regulated with a timer connected with a solenoid valve. The backflush was usually performed for 0.1 s at an interval of 5 min with nitrogen gas at a pressure of 5 atm.

For the growth of *Streptococcus cremoris*, *Lactobacillus casei* and *Bifidobacterium longum*, MRS medium [6,7] was used. The cross-flow filtration was started at the late-logarithmic phase. *Propionibacterium shermanii* and *Butyribacterium methylotrophicum* were cultivated in PZ medium [8]. Vitamin B_{12} content was measured by the bioassay method using *Escherichia coli* 215. *E. coli* C600 was used as a host strain. The hybrid plasmid pMCT98 in which *lacZ* gene was fused to *trp* promoter in frame was constructed [9] and was introduced into the host strain. FB medium [10] supplemented with 0.5 g/l of leucine, 3 g/l of threonine, 5 mg/l of thiamine, 100 mg/l of ampicillin and 0.25 % of casamino acids was used, and tryptophan was added to the medium for the preculture and the fed-batch culture. Fresh medium which was used for replenishment due to cross-flow filtration did not contain tryptophan. Growth of the organism was monitored by measuring the optical density at 570 nm (OD_{570}).

RESULTS AND DISCUSSION

In a batch cultivation of *S. cremoris* with pH control at 6.5, final cell concentration was 2.8 g-dry cell/l. Lactate concentrations of more than 22 g/l resulted in rapid decrease of the growth rate and no growth was observed in the medium containing 35 g/l of lactate. By separating lactate, it was possible to prolong the logarithmic growth period and to

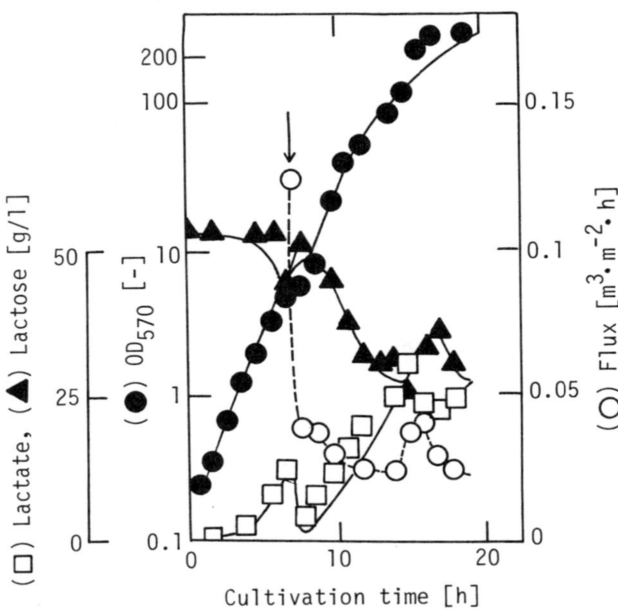

Figure 2. Experimental results of cultivation of *S. cremoris* in the fermenter with cross-flow filtration. The arrow indicates the start of filtration.

achieve high concentration of biomass as shown in Fig. 2. Cross-flow filtration was started at 6 h of cultivation, when the growth rate began to decrease with accumulation of lactate. The final cell concentration was 81.5 g-dry cell/l, 29-fold that without cross-flow filtration. The cell productivity was 19-fold that of conventional batch cultivation.

In the case of *L. casei*, the final cell concentration was 49 g-dry cell/l in the fermentation with the filtration, 22-fold that of conventional batch cultivation.

Main metabolic products are acetate and lactate in the culture of *B. longum*. Removal of acetate and lactate could be expected to prolong logarithmic growth and to improve productivity of cell mass. Figure 3 shows high concentration cultivation of *B. longum* in the fermenter with cross-flow filtration using three ceramic filters connected in series. Cross-flow filtration was started at a cultivation time of 7 h. The final cell concentration reached 54.4 g-dry cell/l, seven times that

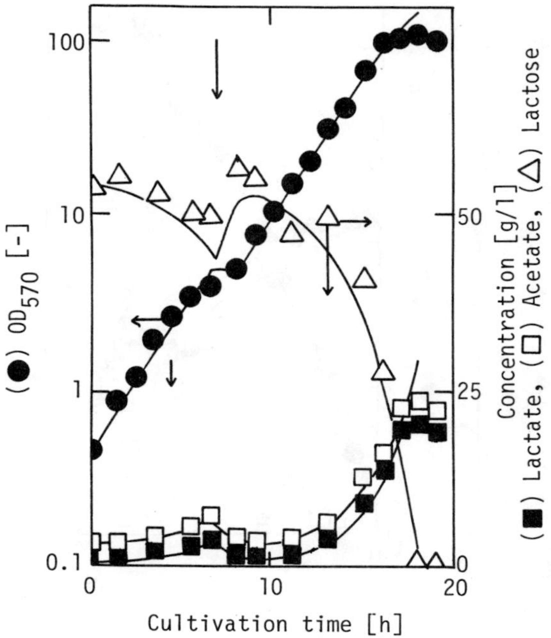

Figure 3. Experimental results for the cultivation of *B. longum* in the fermenter with cross-flow filtration.

achieved without the filtration. Cell productivity per unit cultivation time in the fermenter with the filtration was 7-fold that of the batch culture.

In the cases of *P. shermanii* and *B. methylotrophicum*, inhibitory metabolites are propionate and butyrate, respectively. By the removal of the metabolite in the cultivation of *P. shermanii* with the filtration, the cell concentration of 227 g-dry cell/l and 52 mg-vitamin B_{12}/l were obtained. The cell and vitamin concentrations were 35-fold and 24-fold those in the batch cultivation. Figure 4 shows cultivation result of *B. methylotrophicum* in the filtration system. The final cell and vitamin concentrations were 33.4 g-dry cell/l and 93 mg/l, respectively.

The filtration system was also applied for the two-stage cultivation of gene-engineered microorganism as shown in Fig. 5. During the initial stage of the cultivation, tryptophan concentration was kept 170 mg/l. The filtration was started when OD_{570} became 10. After starting the filtration, tryptophan concentration dropped rapidly and biosynthesis of β-galactosidase was induced. At the end, the specific activity of β-galactosidase was 30 U/mg protein, which was estimated to correspond to about 10% of total cellular proteins. Plasmid was maintained stably and the final cell concentration was 66 g-dry cell/l.

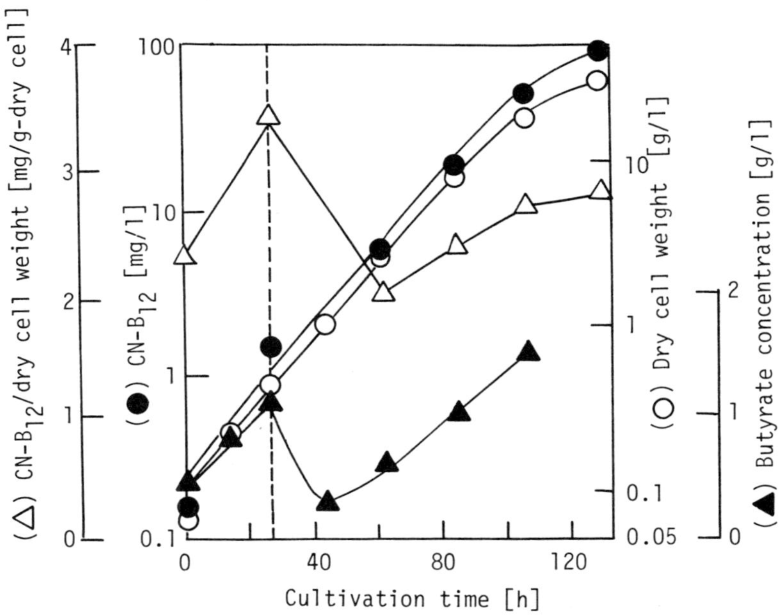

Figure 4. Cultivation of *B. methylotrophicum* with filtration

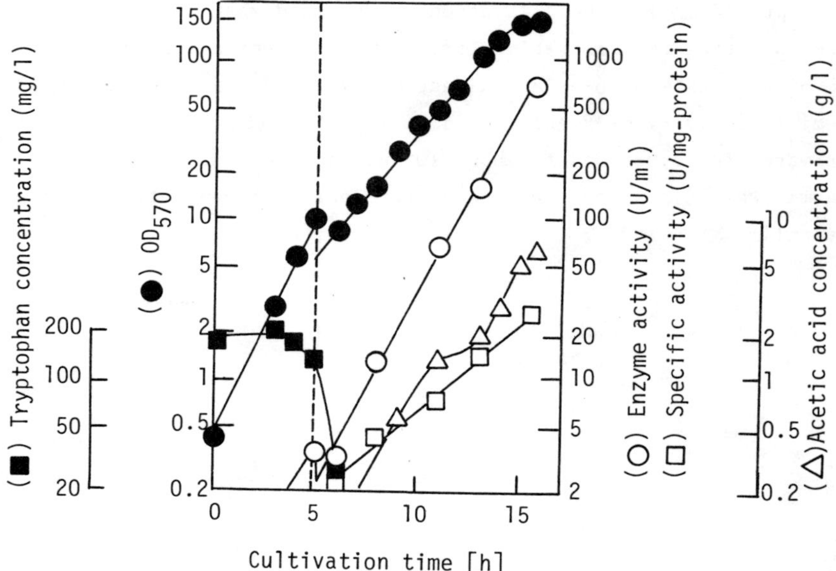

Figure 5. On-off regulation of *trp* promoter by the cross-flow filtration. Tryptophan concentration was kept at 170 mg/l for the first 5 h, then feeding of the amino acid was stopped and filtration was started. Just after starting, the cells were diluted 1.75-fold since 1.5 l of the fresh medium was added to fill the dead space of the filtration apparatus (pump etc.). The broken line indicates start of the filtration.

REFERENCES

1. Yano, T., Kobayashi, T. and Shimizu, S., Silicone tubing sensor for detection of methanol. *J. Ferment. Technol.*, 1978, **56**, 421-427

2. Mori, H., Yano, T., Kobayashi, T. and Shimizu, S., High density cultivation of biomass in fed-batch system with DO-stat. *J. Ferment. Technol.*, 1979, **12**, 313-319

3. Yano, T., Kobayashi, T. and Shimizu, S., High concentration cultivation of *Candida brassicae* in a fed-batch system. *J. Ferment. Technol.*, 1985, **63**, 415-418

4. Yano, T., Mori, H., Kobayashi, T. and Shimizu, S., Reusability of broth supernatant as medium. *J. Ferment. Technol.*, 1980, **58**, 259-266

5. Mizutani, S., Iijima, S. and Kobayashi, T., Fed-batch culture of *Escherichia coli* harboring a runaway-replication plasmid. *J. Chem. Eng. Japan*, 1986, **19**, 111-116

6. Taniguchi, M., Kotani, N. and Kobayashi, T., High concentration

cultivation of lactic acid bacteria in fermenter with cross-flow filtration. *J. Ferment. Technol.*, 1987, **65**, 179-184

7. Taniguchi, M., Kotani, N. and Kobayashi, T., High concentration cultivation of *Bifidobacterium longum* in fermentation with cross-flow filtration. *Appl. Microbiol. Biotechnol.*, 1987, **25**, 438-441

8. Nanba, A., Nukada, R. and Nagai, S., Inhibition by acetic and propyonic acids of the growth of *Propionibacterium shermanii*. *J. Ferment. Technol.*, 1983, **61**, 551-556

9. Kawai, S., Mizutani, S., Iijima, S. and Kobayashi, T., On-off regulation of tryptophan promoter in fed-batch culture. *J. Ferment. Technol.*, 1986, **64**, 503-510

10. Mizutani, S., Iijima, S. and Kobayashi, T., Effect of amino acid supplement on cell yield and gene product in *Escherichia coli* harboring plasmid. *Biotechnol. Bioeng.* 1986, **28**, 204-209

INTERNATIONAL CONFERENCE ON BIOREACTORS AND BIOTRANSFORMATIONS
GLENEAGLES, SCOTLAND, UK: 9-12 NOVEMBER 1987

Paper D3

A NOVEL MEMBRANE MODULE FOR USE IN BIOTECHNOLOGY THAT HAS HIGH TRANSMEMBRANE FLUX RATES AND LOW FOULING

J.M. Wyatt[1,2], C.J. Knowles[2] and B.J. Bellhouse[3]

(1) BMP Ltd. Abingdon, Oxfordshire, (2) Biological Laboratory, University of Kent, Canterbury, (3) Medical Engineering Unit, Oxford University, Oxford.

ABSTRACT

A novel cross-flow membrane separator developed for blood treatment has been assessed for its biotechnological potential. Applications of this system are described particularly with regard to harvesting microorganisms. A five fold greater transmembrane flux has been achieved compared to conventional cross-flow filtration systems.

INTRODUCTION

Membranes and membrane separation technology is being widely utilized today in biotechnology. Two modes of filtration are used, dead-ended and tangential or cross-flow filtration. In dead-ended mode the fluid is passed perpendicular to the plane of the filter whereas in cross-flow the fluid is passed across the filter surface. The filtration of cells by dead-ended filtration is limiting with material rapidly accumulating on the membrane surface. The build up of a layer on the membrane is termed concentration polarization. In cross-flow filtration the effect of concentration polarization is greatly reduced as the fluid moves parallel to the membrane surface sweeping the membrane relatively clear of retained material.

In this paper we describe a novel membrane device utilizing the principle of cross-flow filtration.

DESCRIPTION

Membrane devices have been developed for blood treatment. These consist of membrane lungs for oxygen and carbon dioxide transfer [1] and blood-plasma separators for harvesting human plasma through a microporous membrane [2]. The present paper reports the further development of these membrane modules for use in biotechnology.

The membrane separator consists of a sandwich of two membranes between which the feed (for example microbial cells) flows and two outer compartments, one on either side of the feed channel. Substances that are able to permeate the membranes (for example growth medium) are collected in the outer two compartments when a hydrophilic microporous membrane is used. The devices differ from conventional hollow fibre or simple plate and frame devices in that the membranes consist of a large number of small, part-spherical dimples, concave to the fluid channel. Alternatively the membranes may be furrowed, like corrugated cardboard. By the use of a suitable simple pumping arrangement, oscillatory (pulsatile) flow is provided, together with a superimposed overall flow through the system (Figure 1).

At each forward phase of flow, vortices form in all the concave dimples, are affected when flow reverses, and are then replaced immediately by a set of vortices rotating in the opposite direction. This vortex mixing provides mixing of high efficiency in mass transfer, with low shear which causes negligible damage to delicate cells such as blood cells. The high rates of trans-membrane fluxes, up to 5-fold greater than for conventional separation systems mean that lower levels of recycling are required for cell harvesting or other concentrative processes, thereby enhancing volumetric efficiency and reducing overall shear effects. The vortex mixing reduces concentration polarization effects thereby enabling more stable flux rates to be attained. These vortex mixing membrane devices have already proven their value for biomedical applications but until now little attention has been paid to their biotechnological potential.

FIGURE 1

Diagrammatic representation of the pulsatile cross-flow filtration device

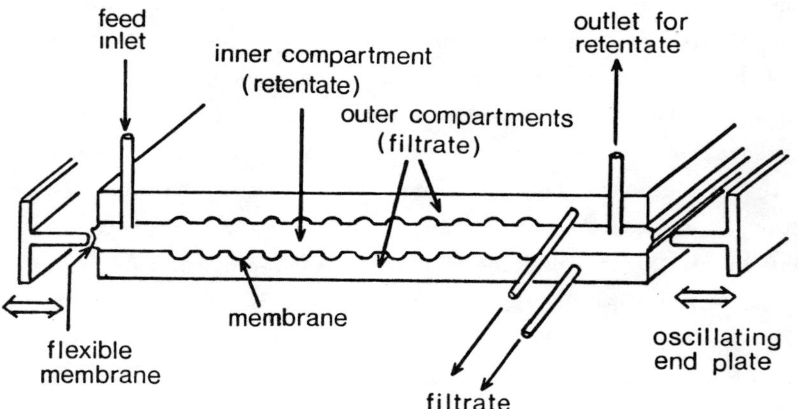

RESULTS

The pulsatile cross-flow filtration system has been intensively studied to assess its potential for the filtration of microorganisms. The system used was as described previously by Bellhouse et al. [2] for use in plasmaphoresis.

The test microorganism used in the majority of our studies was *E. coli* as the Gram negative microbe is renowned for its ability to rapidly blind conventional filtration equipment. The microbial culture is separated into two streams by the filtration system; the retentate, the effluent from the centre of the membranes in which all the biomass is retained; and the permeate, the cell free effluent that has passed through the membranes.

Permeate Flux: The rate of permeate flux has been assessed under a variety of conditions using commercially available 0.2μ pore size polysulphone membranes formed into a dimpled shape using a technique developed at the Medical Engineering Unit, University of Oxford. The rate of permeate flux with this device (membrane area ≈200cm^2) has been compared to a number of commercially available filtration systems such as other cross-flow equipment, pleated cartridges and hollow fibre systems. In each case the volumetric efficiency (permeate flux (lhr^{-1} m^{-2}) of the pusatile cross-flow filtration was greater than the commercial equipment as determined by experimentation and analysis of published data [3]. The volumetric efficiency of the system under the conditions used and the membrane utilised being of the order of 300-400 l hr^{-1} m^{-2} with no pressure applied to the system.

Optimal permeate flux was shown to be achieved using a dimpled membrane with pusatile flow, which gave a volumetric efficiency considerably greater than with flat membranes both with and without pulsatile flow or a dimpled membrane without pulsing (Table 1).

TABLE 1

Membrane	Pulsing 5 Hertz	Volumetric Efficiency 1 hr^{-1} m^{-2}	Clean water Flux % of initial rate
Polysulphone-Flat	-	144	25
Polysulphone-Flat	+	195	51
Polysulphone-Dimpled	-	140	25
Polysulphone-Dimpled	+	345	75

The volumetric efficiency and fall in clean water flux after filtration of an E. coli suspension (30 minutes continuous operation) using flat and dimpled membranes with and without pulsing.

The pulsatile action with a dimpled membrane structure results in a reduction of concentration polarization at the membrane surface. Table 1 demonstrates this showing the volumetric efficiency of the system after 30 minutes continuous operation concentrating a suspension of E. coli. The degree of concentration polarization differed as shown by the reduction in clean water flux across the membranes from before use and after the experiment.

The oscillation speed used to pulse the system was found to affect the efficiency of filtration. The rate of permeate flux increased with an increase in the oscillation speed between 2 and 5 Hertz; an oscillation speed of 5 Hertz was used routinely in this study.

Mode of Operation: The filtration system because of its greater volumetric efficiency compared to other systems can be used efficiently in at least two modes.

(i) Single pass filtration. The feedstock, whether it be a microbial culture, animal cell culture or whatever can be considerably concentrated using a single pass through the filtration system. Only a small peristaltic pump is required with flow rates generally being less than 100ml min^{-1}. The advantage of this mode of operation is the very low shear stress imposed on the cells, a factor which can be important with regard to use of plant or animal cells. A ten-fold concentration can be achieved, 80-90% of the fluid being removed as the permeate in a single pass.

(ii) Recycle filtration. In this mode the retentate is recycled to the feedstock continuously and pumped back through the filtration system. Higher flow rates can be used than in the single-pass mode thereby increasing the possible permeate flux. The pump speeds required to obtain comparative volumetric efficiencies in recycle mode using the vortex mixing system are much less than those used in commercial systems, and therefore pumps of lower shear can be used for this mode of operation.

Pressure: The volumetric efficiency of the filtration system can be improved by applying low pressures to the retentate line. Increasing the pressure correspondingly increases the efficiency in both single-pass and recycle modes of operation. Table 2 gives an example of the typical affect of pressure on the percentage of an *E. coli* suspension obtained as the permeate in a single pass experiment. Greater pressures can be exerted when the system is used in recycle mode due to the higher flow rates that can be used.

TABLE 2

Pressure (Hg mm)	Permeate (%)
0	49
8	66
16	81
40	88
56	94

The affect of pressure on the percentage of permeate obtained from an *E. coli* suspension using repeated single pass filtration.

SUMMARY

The pulsatile cross-flow filtration device has been shown to have a higher volumetric efficiency than other cross-flow devices and to be applicable to a wide range of microorganisms as well as plant and animal cells, achieving rapid filtration under low shear conditions. The system can therefore be used, for example, to reduce filtration time in downstream processing or comparable rates of permeate flux can be obtained using a smaller area of membrane.

Any membrane that can be formed into a dimpled or corrugated shape could be used in the filtration system to give a higher flux rate compared to the same membrane used as a flat sheet and/or without pulsatile flow.

APPLICATIONS IN BIOTECHNOLOGY

The pulsatile cross-flow filtration device has a number of potential applications in biotechnology. These include:

(a) Rapid microfiltration for the harvesting of microbial cells or cell recycle

(b) Removal of cell debris from cell-free extracts after breakage of the cells

(c) Harvesting shear sensitive cells, in particular animal or plant cells. The system could be used for harvesting or for cell recycle to a bioreactor.

(d) Prefiltration of water and media

(e) Ultrafiltration e.g. to separate enzymes

(f) Clarification of solvents.

REFERENCES

1. Dorrington, K.L.D., Ralph, M.E., Bellhouse, B.J., Gardaz, J-P. and Sykes, M.K. J. Biomed. Eng. 7 (1985) 89-99.

2. Stairmand, J.W., Bellhouse, B.J., Jamal, Z., Lewis, R.W.H., Urban J.P. and Entwistle, C.C. Life Support Systems 4 (1986) 193-204.

3. Kroner, K.H., Schutte, H., Hustedt, H. and Kula, M.R. Cross-flow filtration in the downstream processing of enzymes. Process Biochemistry, April (1984) 67-74.

INTERNATIONAL CONFERENCE ON BIOREACTORS AND BIOTRANSFORMATIONS
GLENEAGLES, SCOTLAND, UK: 9-12 NOVEMBER 1987

Paper D4

EFFECTS OF BIOMASS RECYCLE ON THE PERFORMANCE OF BIOTRANSFORMATION FERMENTATIONS USING MUTANT ORGANISMS

well presented

J. Hubble and S. Joyce
School of Chemical Engineering,
University of Bath,
Claverton Down, BATH BA2 7AY
UK

ABSTRACT

While mutant organisms are commonly used to promote biotransformations otherwise impossible with wild type strains, severe problems can result from the limited stability of the mutant. In many cases there is a tendency for reversion, to the faster growing wild type which can subsequently take over the fermentation. In practice, strain instability often rules out any consideration of continuous operation as a means of production. While batch wise fermentation may be possible, the potential productivity advantages of continuous operation have been widely documented and are clearly attractive. This is particularly true where the substrate and product of the conversion are both inhibitory and unable to sustain microbial growth. In these cases biomass must be produced at the expense of a second substrate and subsequently used as little more than a simple catalyst. During the catalytic phase addition of the inhibitory raw material must be closely matched to the rate of conversion which will decline as the inhibitory product accumulates.

The complexity of a batch process operating within the constraints outlined above obviously presents a number of problems. However, many of these may be overcome if continuous operation is possible. The frequency of strain reversion during a fermentation will be a function of the number of cell divisions occurring. In the fermenter this will will be a function of biomass concentration and specific growth rate. The growth rate is in turn a function of substrate and inhibitor concentrations. Therefore it is clear that the stability of the fermentation with respect to mutant revision is a function of the operating parameters in addition to the biological stability of the mutant.

Recent developments in membrane technology now offer the potential for a self-contained module for cell recovery and recycle during continuous fermenter operation. Biomass recycle would allow cell densities in the fermenter to be maintained at lower specific growth rates than is possible in a simple chemostat. By adjusting the ratio of feed to recycle streams and the degree of concentration achieved in the separator, it is possible to control the net biomass formation rate, and hence the sensitivity of the fermentation to microbial reversion.

This paper reports the results obtained from dynamic modelling studies carried out for a recycle fermenter where an unstable mutant is used to convert an inhibitory substrate to an inhibitory product. Particular attention is paid to the effects of operating conditions and effective run durations.

Power utilisation horrendous – Beechams

INTRODUCTION

Mutation has been used to improve the yield of a number of fermentation based processes. One of the most impressive examples is that of penicillin production where the use of mutation and screening techniques led to highly productive strains giving yields orders of magnitude higher than those obtainable with wild type organisms.[1] However, a major problem associated with the use of mutant organisms arises from spontaneous mutations causing reversion to the wild type. While spontaneous mutations will occur in all fermentations the consequence of mutation is usually to give a slower growing organism. In fermentations with mutant organisms the growth rate may be expected to be impaired by a reduced efficiency imposed by the desired product formation. When this is the case, reversion to the wild type might be expected to give a faster growing organism, incapable of producing the desired product but which would tend to competitively take over a continuous fermentation by virtue of its higher growth rate.[2]

Bioconversions based on the metabolism of xenobiotic compounds are becoming more common and are often seen as alternatives to traditional organic synthesis. Many of these reactions are oxidase mediated and have a requirement for the reduced form of nicotinamide adenine dinucleotide (NAD(P)H).[3] For the conversion to proceed at an acceptable rate the NAD(P) produced by the primary reaction must be regenerated at the expense of the cell's metabolism. Hence the bioconversion is imposing a strain on the cell such that mutants showing an increased conversion rate will be at a growth disadvantage to the wild type organism. This situation is typified by the work reported by Dalton[4] on the conversion of benzene to benzene–cis–glycol, and, although yet to utilise mutant organisms, work undertaken at Bath on the N–Demethylation of pharmocological intermediates is also expected to show similar constraints.[5] The instability of fermentations showing reversions has led to a general avoidance of continuous fermentation in favour of fed batch processes. However, as observed by Dalton, such fermentations are limited by the accumulation of toxic products unless specific product recovery methodologies can be made available.

A similar constraint is imposed on fermentations with unstable recombinant cultures where loss of plasmid can result in a lowering of productivity and a takeover of the fermentation by the wild type organism. The theoretical implications of plasmid loss is considered by Ollis[6] for both batch and continuous fermentations.

The object of this report is to consider the advantages of semi–continuous fermentations as an alternative to fed batch for fermentations employing mutant organisms. During continuous operation product removal is automatically accounted for, and although the fermentation duration may be limited by mutant reversion the productivity may be enhanced by a reduction of product inhibition. More significantly the availability of microfiltration membranes and their use in recycle fermentations suggests that cell recycle is now a viable option for this type of fermentation.[7,8] The use of biomass recycle reduces the cell formation rate required to maintain a given productivity and hence reduces the excess biomass production. As the frequency of mutation will be a function of the number of cell division, biomass recycle might also be expected to increase the stability of the fermentation.

THEORY

In many biotransformations the biomass is required solely as a catalyst to promote a specific conversion. In some cases this conversion would represent the first stage in the complete catabolism of the compound by the organism. Hence the effective utilisation of the biotransformations requires the use of mutant organisms, unable to degrade further the desired product.

In a continuous fermentation based on a conversion of this type we must expect to provide a second carbon source for growth and maintenance energy. In the event that the mutant organism is unstable and able to revert to a fully functional 'wild type' a continuous fermentation will be inherently unstable. If the wild type organism grows faster than the mutant or is able to utilise the reactant or product of the desired conversion as a growth substrate then the duration of the fermentation effectively would be limited.

For the initial modelling study a situation was chosen where the wild type organism grows faster than the mutant but is unable to utilise the additional carbon sources present in the fermentation medium. The fermentation is modelled assuming that there are no wild type organisms present at time zero but that spontaneous mutations can occur. The frequency for spontaneous biological mutations is commonly reported as 1 in 10^6 cell divisions. Taking an average figure of 10^{12} microbial cells per gram this gives a formation rate of wild type organisms of 10^{-6} grams per gram of biomass produced.(9)

Simple Chemostat

In a chemostat, steady state is maintained by matching the growth rate of the microorganism to the rate at which they are lost from the fermenter. For this case biomass production is linked to both dilution rate and substrate concentration and so there is little scope for improving the stability of the fermentation. However, as mutant biomass is being used as a catalyst for the desired conversion, the rate of production is also dependent on the operating conditions of the fermenter. This suggests that, within the constraints of the limited stability productivity can still be optimised.

At steady state in the chemostat there is a balance between the specific growth rate of the organism, the dilution rate and the endogenous respiration rate

$$\mu = D + K_e \qquad (1)$$

Assuming Monod kinetics for μ we can solve this expression to give the steady state substrate concentration(10)

$$S_1 = \frac{K_s(D+K_e)}{\mu_m - (D+K_e)} \qquad (2)$$

The relationship between biomass concentration and substrate used is given by astoichiometric yield coefficient

$$X = Y(S_1^0 - S_1) \qquad (3)$$

Substituting for S in equation (3) allows the steady state biomass concentration to be calculated:

$$X = Y\left[S_o - \frac{K_s(D+K_e)}{\mu_m-(D+K_e)}\right] \qquad (4)$$

The rate of conversion of reactant to product can most simply be described in terms of a first order rate constant which is a function of the biomass concentration. Solving the balance for the reactant gives

$$S_2 = \frac{D\,S_2^0}{D + KX} \qquad (5)$$

Similarly solving the mass balance for product (assuming a reaction stoichiometry of 1:1) gives

$$S_3 = \frac{K\ S_3 X}{D} \qquad (6)$$

So the productivity of the fermentation at steady state can be calculated from the mass balances of each component.

The stability of the fermentations can be assessed by numerically integrating the differential equations describing the fermentations. This obviously requires that an additional equation be introduced to describe the wild type biomass. The two resulting differential equations can be written

$$\text{Mutant} \qquad \frac{dX_1}{dt} = X_1\left[\mu_1 - (K_e + D + F\mu_1)\right] \qquad (7)$$

$$\text{Wild type} \qquad \frac{dX_2}{dt} = \mu_2 X_2 + F\mu_1 X_1 - X_2(D+K_e) \qquad (8)$$

Where F is the fraction of biomass showing reversion.
These equations together with those describing substrate reactant and product can be numerically integrated from a given set of initial conditions to show the stability of the fermentation.

Fermenters with biomass recycle

As the stability of the fermentations considered here depends on the rate of biomass production, it might be expected that the stability would be improved by reducing net biomass production. For this to be achieved, while maintaining both the desired biomass concentration and the specified dilution rate biomass must be recycled. In addition to improving the stability of the fermentation biomass recycle offers the advantage of allowing higher dilution rates and also leads to a reduction in the demand for the growth substrate.

When considering biomass recycle it is necessary to distinguish between types of separator and their performance. Generally two modes of operation are possible. Firstly, operation with a fixed ratio of output to input solids concentration (settler); secondly, operation with a constant output solids concentration (centrifuge).[11] More recently there has been interest in the use of membranes for biomass recycle and these can also be used with a constant output solids concentration. In the case of processes requiring sterile operations membranes offer a potentially simpler solution to the problem of biomass retention and so the system to be simulated here is based on this approach.

The reactor layout assumed is described in figure 1. A constraint on this system is that it is not possible to recycle more biomass than is entering the membrane separator. If we consider a balance across the separator we can show that this constraint is satisfied when

$$R \leqslant \frac{1}{\beta - 1}$$

where R = the recycle ratio, and $\beta = Xr/X$.

As the membranes will only be selective for the microorganisms in the system the mass balances for substrate reactant and product will be the same as for the simple chemostat. However the biomass balances for the recycle fermenter must be amended to account for the recycle stream.

$$\text{Mutant} \qquad \frac{dX_1}{dt} = D(X_1^r - (1+R)X_1) + X_1(\mu_1 - K_e - F\mu_1) \qquad (9)$$

$$\text{Wild type} \qquad \frac{dX_2}{dt} = D(X_2^r - (1+R)X_2) + \mu_2 X_2 - K_e X_2 + F\mu_1 X_1 \qquad (10)$$

Fig.1. Proposed reactor layout

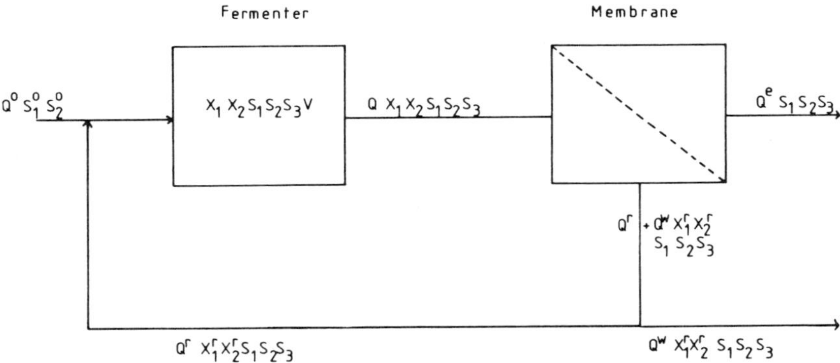

The steady–state mass balance for the mutant can be combined with the mass balance for substrate and solved to give the steady state substrate concentration and hence the steady state biomass concentration. As with the chemostat model the steady state solution can be used to investigate the effect of changes in the operating conditions on productivity while the dynamic simulation can be used to ascertain the stability of the fermentations.

Calculations

Steady state solutions for fermenter performances were calculated using a commercial spreadsheet program (VP–planner, Paperback Software). Dynamic simulations were carried out using the interactive simulation language ISIM (Simulation Sciences). All packages were run on an IBM compatible personal computer.

Initial conditions for dynamic simulations were taken from the spreadsheet steady state solutions. Typical fermentation profiles are shown in figure 2. In addition to the effects of operating conditions the time taken for the reverted organism to dominate the fermentation is a function of both the spontaneous mutation rate and the ratio of the organism's specific growth rates. In this work these values were unchanged throughout as they represent biological rather than physical properties.

RESULTS

Simulations were carried out using data obtained in batch fermentations using *Pseudomonas putida* cultured on glucose. The kinetic constants used were $\mu max = 0.77h^{-1}$ and $Ks = 0.5g\ell^{-1}$. The observed yield of biomass for substrate consumed was 0.15.(12) An endogenous respiration rate of $0.01h^{-1}$ was also assumed.

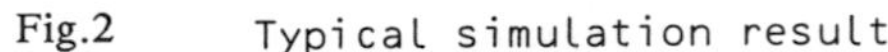

Fig.2 Typical simulation result

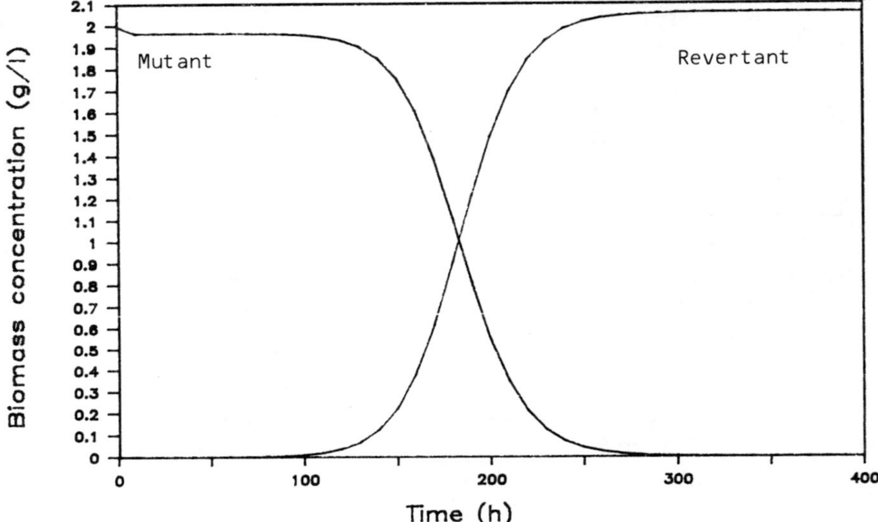

Chemostat

Using the mass balances described in the theory section for continous fermentation and assuming a specific reactant to product conversion rate of $1.5g\ell^{-1}\ h^{-1}$ the effect of dilution rate on reactor productivity was studied (fig.3). In this simulation biomass was kept constant by varying the substrate concentration. The results show that the maximum theoretical productivity approaches $(\mu m - K_e)S_3$ however this value could not be approached in practice as it would require unrealistic substrate concentration. An additional constraint in many cases is the effect of catabolite repression on the microbial conversion of xenobiotic compounds which limits the dilution rate which can be used. The situation represented in figure 3 simply shows the consequences of changes in the steady state growing conditions for the fermentation. In the case of processes using an unstable mutant the length of fermentation will be related to the rate of biomass formation and hence the dilution rate. This situation is shown in figure 4 where the mutant is assumed to spontaneously revert to the wild type which has a μmax value 10% higher. The stable length of the fermentation is taken to be the time elapsed before the product concentration falls to 75% of the steady state value. Taking the results presented in figures 3 and 4 and considering the achieved productivity over the stable fermentation time, allows the effect of dilution rate on total product fermentation to be assessed (figure 5). The results obtained suggest that a maximum product fermentation is asymptotically approached as dilution rate is increased. However in practice the effects of catabolite repression might be expected to lead to a discrete optimal dilution rate.

Fig.3 Effect of dilution rate on fermenter productivity (X = 2g/ℓ)

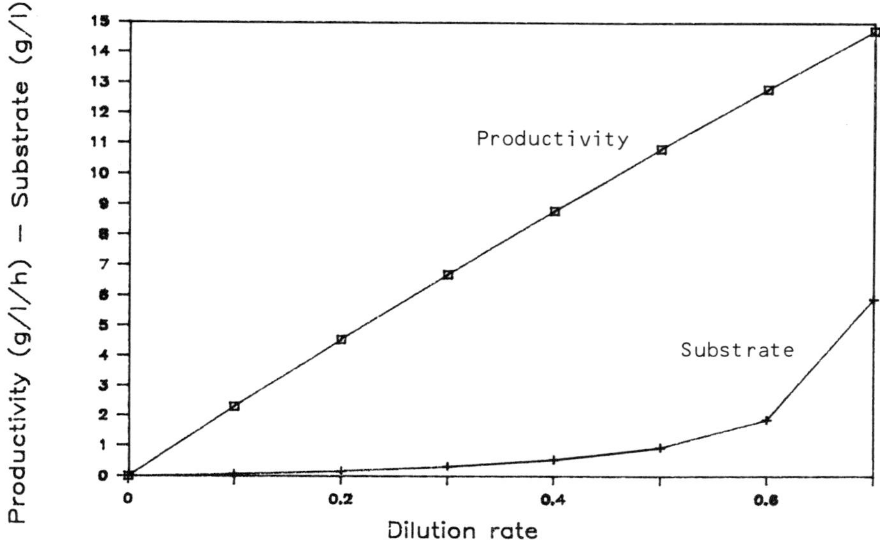

Fig.4 Effect of dilution rate on stable fermentation time

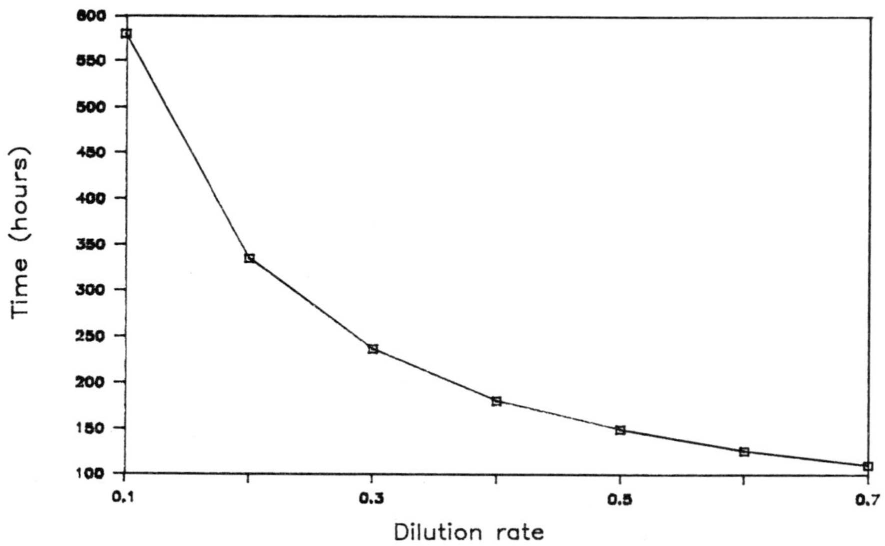

Chemostat with Recycle

Although the results obtained with simulations of a simple chemostat suggest there is some scope for optimisation. the major constraint is the relationship between dilution rate and net biomass fermentation.

The problem of maintaining fermenter biomass concentrations at high dilution rates can be minimised by employing cell recycle such that the requirement for net biomass formation is reduced. As described in the theory section biomass recycle is constrained by biomass availability but within this limit the biomass formation rate can be varied independently of dilution rate by changing the recycle rate and the concentration of biomass in the recycle stream. For a recycle fermenter biomass production is given by $(\mu - Ke)X$ rather than DX. Figure 6 shows the effect of dilution rate on productivity and substrate concentration. As with the example for a simple CSTR the biomass concentration is maintained at a constant $2g\ell^{-1}$ by varying S^0. The concentration of recycled biomass was set at $10g\ell^{-1}$ with a recycle ratio of 0.2.

Fig.5 Effect of dilution rate on total reactor productivity in a semi-continuous fermentation

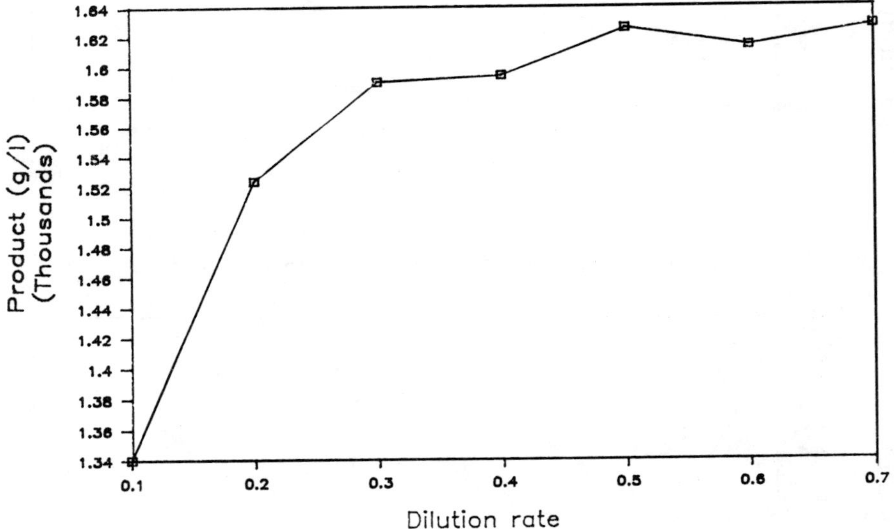

In the recycle fermenter the increased residence time of cells in the reactor leads to a significant loss of biomass from endogenous respiration at low dilution rates. This can be observed as a higher substrate feed requirement at low dilution rates. A comparison of figure 6 with figure 1 clearly shows the increased productivity for a given biomass production rate which can be obtained with cell recycle. The effect of dilution rate on the stability of a recycle fermenter is shown in figure 7. As with the simple CSTR the stable fermentation time decreases with increasing dilution rate, however, cell recycle increases the stable fermentation time considerably. At a dilution rate of $0.1h^{-1}$ the recycle fermenter gives a fermentation time of nearly 1600 hours compared with a CSTR time of less than 600 hours. Similar increases in total product are observed as a result of increases in both productivity and stable fermentation time. (Figure 8)

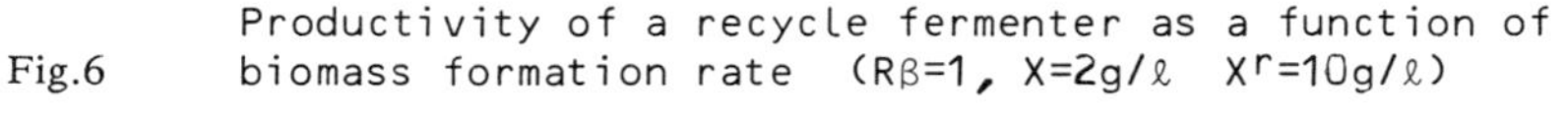

Fig.6 Productivity of a recycle fermenter as a function of biomass formation rate (Rβ=1, X=2g/ℓ X^r=10g/ℓ)

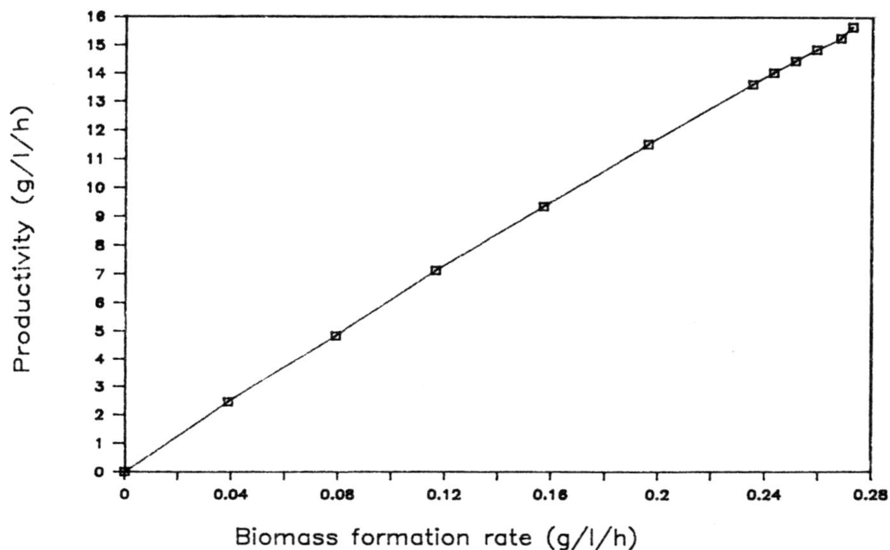

Fig.7 Effect of dilution rate on stable fermentation time (with biomass recycle)

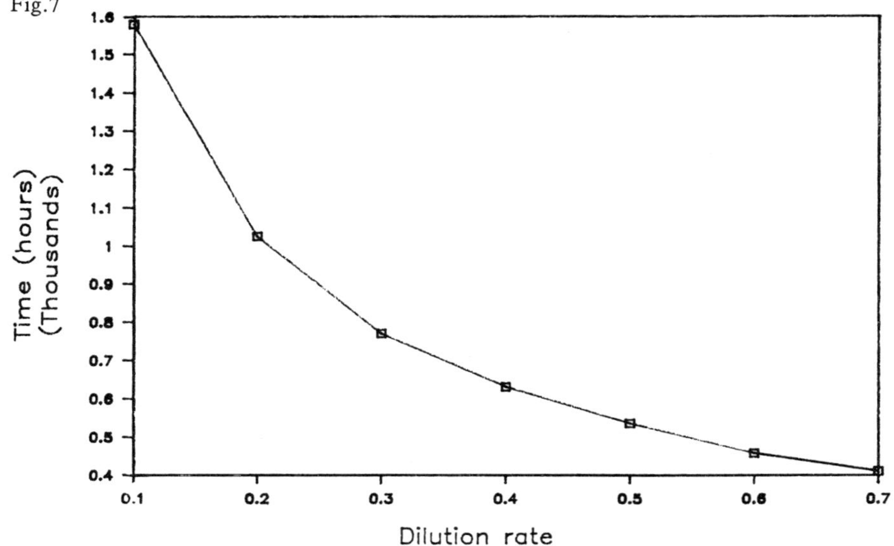

Fig.8 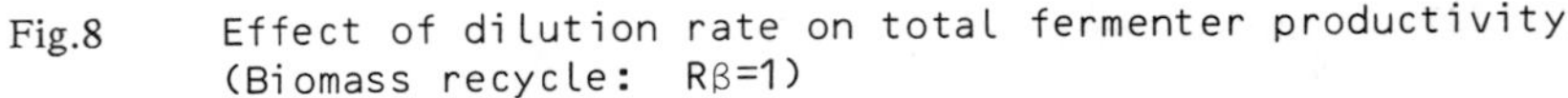

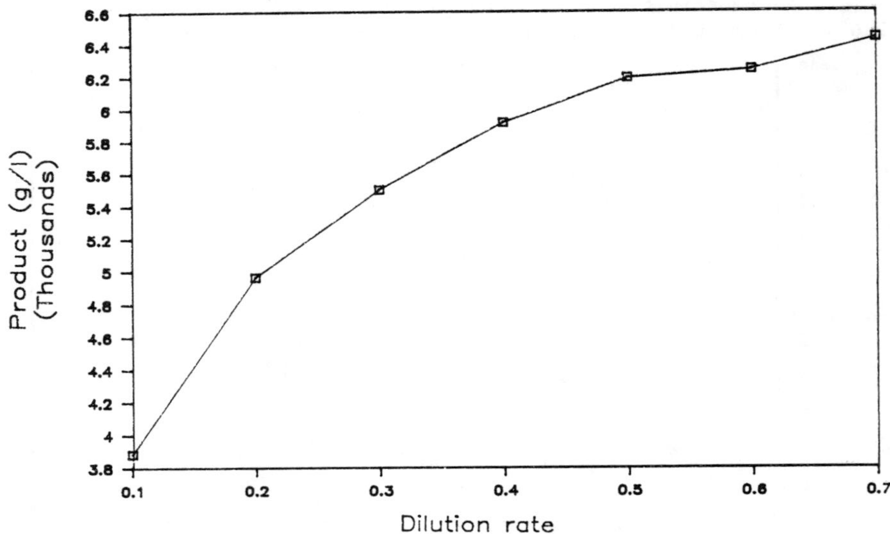

The biomass recycle has a clear effect on the the reactor performance suggesting that the combination of separator performance and recycle ratio will be critical to reactor performance. Figure 9 shows the effect of various operating protocols on the relationship between biomass fermentation rate and productivity for a recycle fermenter. As the biomass recycle is reduced by lowering the recycle ratio net biomass formation increases for any given fermenter productivity. Similarly the stable fermentation time decreases towards the value obtained with a simple CSTR.

Effects of Inhibition

In the earlier simulations the conversion was assumed to be described by a first order rate constant to show general trends, however in a more rigorous examination the conversion must be considered in terms of saturation kinetics taking account of inhibition effects.

Fermentations of the type considered here will be significantly affected by inhibition. This falls into three distinct types, firstly cell growth will be modified by the presence of both the reactant and product of the bioconversion. Secondly the enzyme system responsible for the conversion might be expected to show catabolite repression in the presence of the growth substrate. Finally, the enzyme system promoting the conversion may be expected to show reactant and product inhibition. In fed-batch fermentations these phenomena can be isolated by allowing growth to occur prior to the addition of reactant. In continuous fermentation the situation is more complex and inhibition effects must be considered collectively. While each individual fermentation will be different we can make some simplifying assumptions to show the general consequences of inhibition.

Taking a continuous fermentation where the initial steady state conditions are set to allow optimal conversion, the inhibition effects on growth will be most apparent when the fermentation becomes unstable. In the context of the conversion being simulated these effects can be ignored for initial assessments. The major effects of interest concern catabolite repression and enzyme system inhibition. Although these are the consequence of fundamentally different mechanisms they can most conveniently be modelled in terms of classical enzyme kinetics.[13]

Fig.9 Relationship between productivity and biomass formation rate

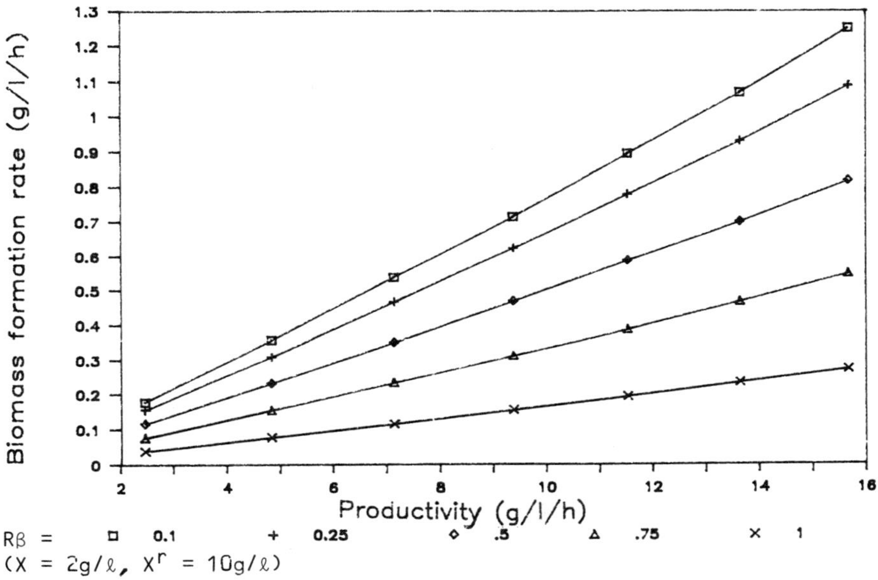

(X = 2g/ℓ, X^r = 10g/ℓ)

In the case of catabolite repression synthesis of the required enzyme system is inhibited at the operon level. This suggests that enzyme concentration will be a function of inhibitor concentration but that the kinetic properties of the enzyme will remain unchanged. This situation is anologous to non competitive inhibition where:

$$Vmax' = \frac{Vmax}{\left[1 + {}^{I}/_{Ki}\right]} \qquad \text{Km unchanged}$$

For the general case of substrate and product inhibition the mechanism of inhibition are chosen as uncompetitive and competitive respectively, on the basis that substrate cannot compete for a single enzyme site, but that binding of two substrate molecules could block the active site. In the case of product inhibition structural similarities between substrate and product could be envisaged as leading to competition for the active site.

Substrate inhibition (uncompetitive)

$$\text{Vmax}' = \frac{\text{Vm}}{\left[1 + {}^{I}/_{Ki}\right]} \qquad \text{Km}' = \frac{\text{Km}}{\left[1 + {}^{I}/_{Ki}\right]}$$

Product inhibition (competitive)

Vmax unchanged $\qquad \text{Km}' = \text{Km}\left[1 + {}^{I}/_{Ki}\right]$

Modification of the kinetic expressions by these factors allows the effects of inhibition on steady state reactor performance to be investigated.

Catabolite Repression

As can be seen from the results presented in figure 3 biomass fermentation rate is a function of substrate concentration in a CSTR. The desirability of a high dilution rate in terms of productivity has been shown for a non catabolite repressed system (fig 4). However, as stated previously, repression of this type may be expected to lead to a discrete productivity optimum. Figure 10 shows that as the substrate concentration in the fermenter increases the rate of conversion and hence the productivity falls. In practice this could be additionally compounded by the effect of reactant inhibition and the inhibitory effect of concentrated reactant on microbial growth. While catabolite repression would occur in a fermenter with recycle the substrate concentrations required to maintain cell mass production are considerably lower. For the results presented in figure 6 the substrate concentration required to maintain the highest dilution rate was $0.12\text{g}\ell^{-1}$ as opposed to $6\text{g}\ell^{-1}$ in the simple CSTR. Clearly these are significant advantages to cell recycle in terms of minimising catabolite repression.

Fig.10 Effect of catabolite repression on the relationship between productivity and dilution rate

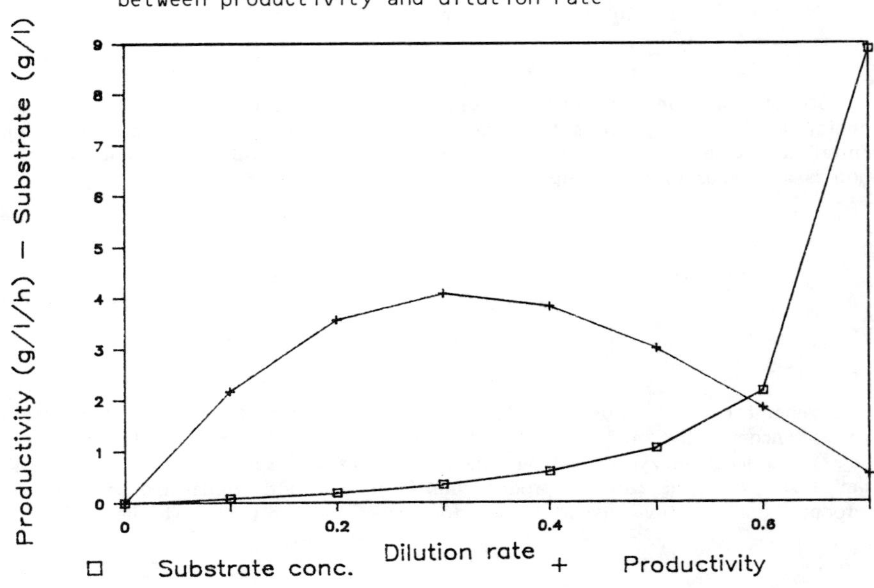

Reactant and Product Inhibition

Simulations of both CSTRs and CSTRs with recycle show increased productivity with increasing dilution rate. However, while productivity increases, the fraction of reactant converted to product goes down, altering the balance between product and substrate inhibition. In the simple CSTR this effect is likely to be swamped by the consequences of catabolite repression, and so this type of inhibition might be expected to be of greater significance in the case of recycle fermenters. The steady state effects of this are predicted in figure 11 using the following values for the enzyme conversion.

Km = 0.5 $g\ell^{-1}$ Vm = 4 $g\ell^{-1}$ h^{-1} g^{-1} (biomass)
Ki (reactant) = 2 $g\ell^{-1}$ and Ki (product) = 15 $g\ell^{-1}$

Fig.11 Effect of dilution rate on the conversion reaction

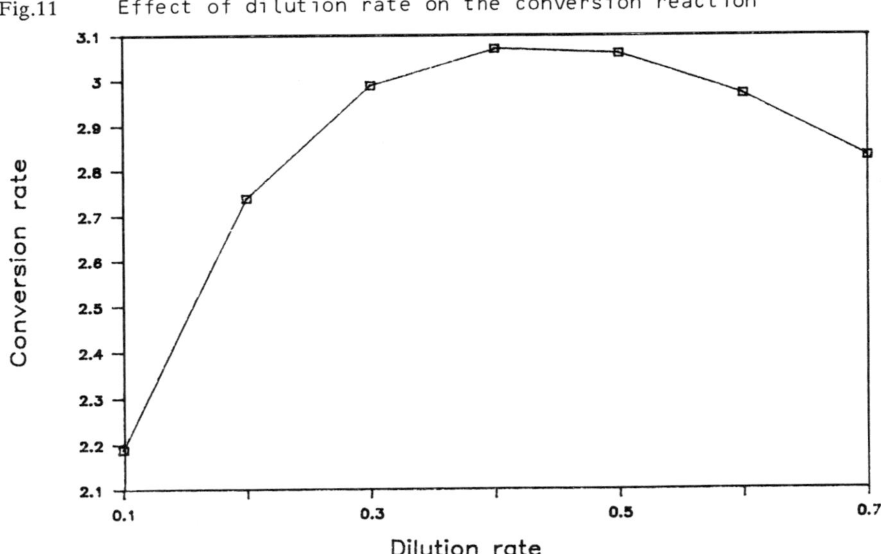

The results show that as the dilution rate is increased the concentration of unconverted reactant increased. This leads to an elevation of the rate due to the nature of the dependence of V on S. However, as the reactant concentration increases inhibition effects lead to a decrease in the rate. As different patterns of reactant and product inhibition could be seen with different organisms, optimisation of this aspect of the fermentation is critical to the performance of recycle fermenters.

CONCLUSIONS

The aim of this report was to consider the effects of fermenter operating parameters on the performance of fermentations using mutant organisms. While every individual fermentation may be expected to be unique it is apparent that there are common underlying constraints which may be expected to affect performances in similar ways.

In the simulation results presented kinetic constants were taken from work conducted with *Pseudomonas putida*, as an example of an organism which has been used for conversion of this type. The results obtained show the limited scope for optimising a simple CSTR fermenter in terms of productivity and stable fermentation time. The dependence of biomass fermentation rate on dilution rate means that the stability of the fermentation is totally dependent on the biological properties of the cells. In practice catabolite repression might be expected to impose severe limitations on realistic dilution rates which can be utilised. Taken in conjunction these factors suggest that there will be an optimum dilution rate for a given fermentation but that the productivity and stability may be too low to allow profitable operation.

In the case of cell recycle biomass formation can be uncoupled from dilution rate. As shown in figure 9 for any given dilution rate biomass formation is a function of the amount of biomass recycled. This suggests that fermentation stability will also be a function of biomass recycle, allowing much greater control over the fermentation performance. As the rate of biomass formation is reduced the substrate concentration required is also reduced minimising the effects of catabolite repression. Reduction of biomass formation has the additional attraction of reducing effluent problems and minimising substrate costs compared with simple chemostat operations.

The consequences of reactant and product inhibition on growth are complex and are not considered in detail here as they will vary considerably from organism to organism. In the case of conversion rate, reactant and product inhibition may be expected to influence the choice of dilution rate chosen. (Figure 11). However, the optimal dilution rate must also take account of desired productivity as well as fractional conversion.

To summarise, the results presented here suggest that semi–continuous fermentation could be considered as an alternative to some fed batch fermentations based on mutant organisms. When used in a cell recycle mode this approach offers the opportunity to increase fermentation duration by reducing product accumulation in the reactor.

The complexity of any given fermentation requires that each case be considered on its individual merits taking into account additional factors not considered here (start up/time to reach steady state and stability of the cells as catalysts). While little discussion has been aimed at the separation unit required for cell recycle, the potential of results presented here (based on a reactor biomass concentration of $2g\ell^{-1}$ and a recycle concentration of $10g\ell^{-1}$) suggests that current membranes capable of filtering cell suspensions of 20 wt%[14] would not represent a limiting factor in the adoption of this approach.

In the assessment of the cell recovery stage no consideration has been given to the possibility of selectively recovering the desired organism. Given the likely physical similarity of mutants of the same strain the feasibility of a totally selective separation is questionable, however, Ollis[6] points out that the enrichment required to stabilise a fermentation is not great. Clearly if selective recycling is possible the fermentation can be run for extended periods without the problem of revertant takeover.

NOMENCLATURE

D	Dilution rate	reciprocal time
K_e	Endogenous respiration rate	reciprocal time
K_s	Saturation constant monod equation	concentration
Km	Michaelis constant	concentration
F	Fraction of biomass reverted	–
Ki	Inhibitor constant (reactant or product)	concentration
X_1	Mutant biomass	concentration
X_2	Wild type biomass	"
S_1	Growth substrate	"
S_2	Reactant	"
S_3	Product	"
R	Recycle ratio	–
Vmax	Maximum rate of enzyme catalysed reaction	concentration/time
μ	Specific growth rate	concentration/time/ biomass concentration
μmax	Maximum specific growth rate	"
β	Ratio of recycle to reactor biomass concentration (X^r/X)	–

REFERENCES

1. Perlman, D. (1979) 'Microbial Production of Antibiotics' in Microbial Technology vol.1, p241 (Peppler & Perlman eds) Academic Press, New York.

2. Slater, J H (1985) 'Microbial Growth Dynamics' in Comprehensive Biotechnology vol.1, p189 (Moo-Young ed) Pergamon Press, New York.

3. Smith, R V, Davis, P J (1980) 'Induction of Xenobiotic Monooxygenases' Advances in Biochem. Eng. 14, 61, Springer-Verlag, Berlin.

4. Dalton, H (1986) 'Biotransformations for Industry' SERC Bulletin 3, 8.

5. Sewell, G (1982)
Ph.D thesis, University of Bath.

6. Ollis, D F (1982) 'Industrial Fermentations with (unstable) Recombinant Cultures', Phil. Trans. R.Soc. Lond. B,297, 617–629.

7. Khorakiwala, K H, Cheryan, M, Mehaia, M A (1985) 'Ethanol Production in a membrane recycle bioreactor' Biotech. Bioeng. Symp.15, 249–262.

8. Cheryan, M, Mehaia, M A (1984) 'Ethanol Production of glucose using *Saccharomyces Cerevisiae*' Process Biochem. 19, 204–208.

9. Atkinson, B, Mavituna, F (1983) Biochemical Engineering and Biotechnology Handbook, Macmillan, Byfleet.

10. Bailey, J E, Ollis, F (1986) Biochemical Engineering Fundamentals 2nd Ed., McGraw Hill, New York.

11. Sundstrom, D W, Klei, H E (1979) Wastewater Treatment, Prentice Hall, N.J.

12. Crossley, S (1985) 'Influence of Culture History on the growth of *Psuedomonas putida* on glucose in batch culture' MSc Thesis U.C. Swansea.

13. Palmer, T (1981) Understanding Enzymes, Ellis Horwood, Chichester.

14. Tutunjau, R S (1985) 'Cell Separation with Hollow Fibre Membranes' in Comprehensive Biotechnology vol.2, p367, (Moo-Young ed) Pergamon Press, Oxford.

INTERNATIONAL CONFERENCE ON BIOREACTORS AND BIOTRANSFORMATIONS
GLENEAGLES, SCOTLAND, UK: 9-12 NOVEMBER 1987

Paper E1

BIOCATALYSIS IN LOW WATER, PREDOMINANTLY ORGANIC REACTION MIXTURES

Peter J Halling
Department of Bioscience and Biotechnology,
University of Strathclyde, Glasgow G1 1XW,
UK.

ABSTRACT

A variety of enzyme and whole cells are catalytically active in systems containing high proportions of water-immiscible organic liquids; such reaction mixtures offer several advantages for the industrial application of biocatalysis. The system splits into two or more phases, but in some cases the amount of water is so low that even the polar phase around the biocatalyst is far from a dilute aqueous solution. Such systems, characterised by a thermodynamic water activity significantly less than 1, have been formed from dried whole cells, dried enzyme powders, or enzymes dried onto support particles. Many lipases, some proteases and a few other enzymes have been used. With lipases, reaction rates can be comparable to those in higher water systems, though may be lower, and depend on the precise water content and hydration history. Hydrolytic equilibria are shifted further towards synthetic products, and attractive packed bed reactors may be operated with water adsorbed in the catalyst bed.

BIOCATALYSIS IN ORGANIC MEDIA

A wide variety of enzyme and whole cell catalysts have now been shown to remain active in predominantly organic reaction mixtures. Such reaction systems are of considerable industrial interest, as they overcome some of the disadvantages cited against biological catalysis, which result from the use of dilute aqueous solutions.

Firstly, the increased solubility of non-polar species allows higher concentrations of many potential reactants and products to be used in the process, with more efficient use of the catalyst and less expense in solvent removal. Secondly, cheap and readily available hydrolytic enzymes may be used as efficient catalysts for the reverse, synthetic reactions; though the equilibria catalysed by these enzymes favour hydrolysis in dilute aqueous solution, in organic media they are shifted to give useful yields of synthetic products (over 99% in some cases). Thirdly, the biocatalyst may be protected from non-polar starting materials or products that can inactivate it, since these mainly partition into the organic phase. Similarly side reactions (chemical or biocatalysed) that occur in the aqueous phase may be reduced by extraction of reactants into the organic phase. Finally, since a biocatalyst may often be used for just one step in a synthetic sequence, organic media will usually aid integration with chemical steps, for which organic solvents are the norm.

Some biocatalysts can be used in mixtures of water with polar organic solvents such as methanol or dimethylsulphoxide. However, most of them are inactivated when the solvent concentration exceeds around 30% [1], which limits the application of these systems. There are, nevertheless, some attractive processes, such as in peptide synthesis [2].

There is more general potential in the use of reaction systems containing water-immiscible organic liquids (which may be solvents or the reactants themselves). In this case, the system splits into two or more phases, and the biocatalyst remains in a relatively water rich phase that includes only low concentrations of the non-polar organic species. Hence, inactivation of the catalyst is often less of a problem than with water-miscible solvents. In these "biphasic" or "two-phase" systems, many biocatalysts can remain catalytically active with total water contents of 1% or less. I have recently reviewed the whole area of biocatalysis in two-phase systems [3].

Enzymes have also been shown to act in predominantly organic media where the aqueous phase is only present at the molecular level, with the catalyst apparently dissolved. These systems, such as microemulsions or reversed micelles, will not be discussed in this paper.

VERY LOW WATER BIPHASIC REACTION MIXTURES

We would expect the behaviour of a biocatalyst in a two-phase system to depend on the microenvironment in which it is present, that is the nature of the aqueous or polar phase. In many cases this remains a dilute aqueous solution, even when the total water content is very small; hence it is not surprising that the properties of the biocatalyst are not greatly affected. However, it is possible to use reaction mixtures in which the polar phase becomes very concentrated, or even consists of no more than adsorbed water on catalyst particles. This paper will be concerned just with these very low water systems, in which significant changes in biocatalyst properties might be expected.

It is important to be able to distinguish those reaction mixtures that fall into the "very low water" category. Unfortunately, this is not as easy as might be hoped. The water content of the whole system is a poor guide: with a non-polar organic phase having little capacity to dissolve water, less than 1 % of water can be sufficient to separate as a dilute aqueous solution around the biocatalyst. An ideal approach would be to measure directly the composition of the polar phase, but sampling this is usually impossible as it is restricted to the pores of support particles. I would suggest that measurement of the thermodynamic water activity (a_w) is a useful method for the characterisation of these systems. Though it does not tell us the precise nature of the polar phase, the a_w will be significantly less than 1 if and only if there is no dilute aqueous solution phase present.

A little theory is helpful to understand the use of the water activity [4]. It is usual to define it as a relative activity, with pure water having a value of 1. As an activity it has advantages over concentration or content, because by definition: i) It is equal in all phases at equilibrium. One useful consequence of this is that a_w may in principle be measured straightforwardly via the relative humidity of an equilibrated gas phase. ii) It determines the mass action effect of water on any equilibria in which water is a reactant.

In a biphasic system, a low total water content is not a sufficient condition to obtain an a_w value significantly less than 1. To achieve a reduction in a_w, the organic phase must be less than saturated with dissolved water; hence if water concentrations are to be measured, that in solution in the organic phase is the most informative value. As noted, reduced a_w also requires an aqueous or polar phase that is very concentrated, or just consists of water adsorbed to the biocatalyst. Considering the likely nature of the polar phase is usually the best means of estimating likely a_w values when measurements are not available. Thus systems where the total water content is no more than the weight of the biocatalyst are likely to have a low a_w, while systems made from a water-saturated organic phase and a catalyst containing a dilute aqueous solution will be of a_w close to 1, however low the proportion of catalyst in the system, the solubility of water in the organic phase, or the total water content.

So far, three approaches have been used to produce low water two-phase reaction mixtures. The simplest employs nothing more than dried enzyme particles suspended in an organic phase. Alternatively the enzyme may first be dispersed over the surface of solid support particles, usually inorganic materials such as celite or silica. Finally, the catalyst may consist of whole microbial cells containing an enzyme of interest. In each case a small quantity of water may be added back before use. Fig. 1 shows the sorts of structure that can be produced.

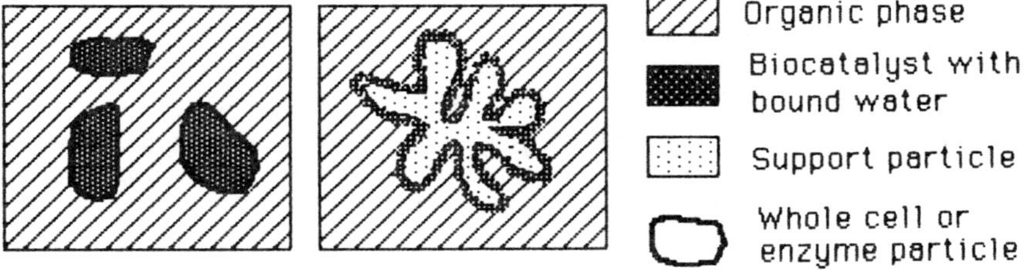

Fig. 1. Structures of very low water biphasic reaction mixtures.

BIOTRANSFORMATIONS DEMONSTRATED IN VERY LOW WATER BIPHASIC SYSTEMS

Lipases

Lipase activity has been clearly demonstrated in several systems of measured a_w significantly less than 1 [5-7], and in many more where low water activity may be confidently predicted from published compositions [3]. The catalyst can be the isolated enzyme (either simple particles or dispersed over a support surface), or whole microbial cells containing lipases. A wide variety of esterification, transesterification and ester hydrolysis reactions have been shown to be possible; as expected the equilibrium in these reaction mixtures favours synthesis compared with conventional emulsion systems. Particular interest has surrounded the 1,3-specific transesterification (or "interesterification") of triacylglycerols in oils and fats, where industrial application looks likely [6]. This reaction can be exploited to produce mixtures of triacylglycerols (triglycerides) similar to those present in expensive natural fats, from cheaper starting materials. The most common catalyst consists of lipase dispersed over the surface of celite particles, and is used to treat a solution of the fat in a non-polar solvent like hexane.

Proteases

Most predominantly organic reaction mixtures used with proteases prove on close examination to contain sufficient water to give a dilute aqueous solution phase. However, some probably or certainly fall into the very low water category [8-12] (see also Fig. 2 below), covering 7 enzymes in total, of which chymotrypsin and thermolysin have been used in 2 or more studies each. Hydrolysis, peptide synthesis and transesterification have all been demonstrated; dried enzyme powders and supported catalysts have been used.

Other biocatalysts

It is clear that phospholipases within fungal cells can be hydrolytically active in very low water systems [13], though only limited studies have been made. Similarly xanthine oxidase powder acting while suspended in dried solvents must have had a_w well below 1 [14]. A number of other enzymes have been used in systems of low water content, but this may still have been sufficient to form a dilute aqueous solution phase; in many cases it was shown to be essential to add water back to the dried catalyst. However, experiments with horseradish peroxidase in dioxane [15] almost certainly generated low a_w values (because of the solvent's capacity to dissolve water), and these may possibly have occurred in studies with polyphenol oxidase [16] and cholesterol oxidase [17,18] .

REACTION RATES IN VERY LOW WATER SYSTEMS

Clearly if these reaction mixtures are to be applied in industry, it is important that catalyst activities are comparable with those in higher water systems. Unfortunately, such a comparison is difficult to make. Besides the problems of unambiguously identifying very low water systems, essential rate data is missing in many publications. Furthermore the reaction(s) followed under low water conditions are often different (e.g. esterification instead of hydrolysis), and the catalysts have a wide range of purities and specific activities.

I believe, however, that a reasonably fair comparison may be made, at least for lipases [19]. It is possible to find rate data for synthetic reactions under higher water conditions, and hydrolytic activity of the catalysts may be used as a measure of the number of active lipase molecules present. Table 1 shows rate data for lipases from a range of sources, normalised in this way to give reaction rates per unit hydrolytic assay activity.

Despite the considerable uncertainties involved in making these comparisons, I believe the following conclusions are justified by the data collected in Table 1 (and some other data discussed in reference [19]).

1. Rates in low water systems can be at least as high as with the same amount of lipase catalysing the same reaction in a high water emulsion.
2. Supported (and possibly cell-bound) enzymes tend to give faster rates than simple powders.
3. Reaction rates per unit weight of low water catalyst are similar, even when the lipase activity varies widely. It may be that physical factors are limiting, such as the interfacial area.

SHIFT OF HYDROLYTIC EQUILIBRIA BY LOW WATER ACTIVITY

As noted, the equilibria catalysed by hydrolases may be shifted in organic media to give useful yields of synthetic products. One factor that contributes is the prefential solvation of the synthetic product in the organic phase, relative to the hydrolytic fragments. In very low water systems, it can be predicted that as a_w is reduced, mass action effects of water will contribute a further shift in equilibrium [4].

Shifts in equilibrium positions in low water systems as a function of water activity have been demonstrated experimentally. Table 2 shows some results for lipase-catalysed transesterification [5]. The equilibrium position of the hydrolytic side reaction that generates diacylglycerols is further shifted towards synthesis if a_w is reduced substantially by water removal from the headspace gases. We have also demonstrated such an effect during peptide synthesis catalysed by thermolysin dried down onto the surface of silica particles [20]. The catalyst has only low quantities of adsorbed water, and is suspended in an organic solution of the reactants. Table 3 shows the effect of a_w on the equilibrium position.

TABLE 1

Rates of lipase-catalysed esterification or transesterification on good substrates.
See reference [19] for full details of sources of data.

Enzyme source	Reaction rate per unit weight (given only for low water catalysts) 10^{-3} mole.s^{-1}.kg^{-1}	Reaction rate per unit activity 10^{-3} mole.s^{-1}.kat^{-1}
Dried enzyme powders		
Pancreas	0.68	8.3
Pancreas	1.4	1.5
Pancreas	1.4	6.7
M. meihei	2.5	*
M. meihei	0.094	0.28
R. japonicus	3.6	130
C. cylindraceae	1.8	0.18
Cell bound enzyme		
R. arrhizus	0.58	59
Enzyme on dispersant		
R. niveus (celite)	8.9	440
R. delemar (celite)	*	2.8
R. japonicus (perlite)	2.1	280
A. niger (celite)	1.1	87
M. meihei (celite)	2.5	*
M. meihei (duolite)	0.63	3.9
M. meihei (resin)	5.2	*
Emulsion systems with dissolved enzyme		
A. niger		45
Pancreas		5.2

1 Kat(al) = amount of catalytic activity which converts 1 mole in 1 sec.

TABLE 2

Effect of water removal during enzymic interesterification.

		Water removal	Control
Fat product composition (%)	Triacylglycerols	68.2	51.7
	Diacylglycerols	1.2	8.7
	Fatty acids	30.6	40.0
Final a_w		0.07	0.87

Reaction mixture : 228 ml petroleum ether, 120 g fat, 12.5 g celite-lipase, 1 g water.

TABLE 3

Position of peptide synthesis equilibrium with silica-thermolysin at different water levels

	Final a_w	$\frac{[\text{Z-Phe-Phe-OMe}]}{[\text{Z-Phe}][\text{Phe-OMe}]}$ mMolar^{-1}
Control	1.0	2.4
Control, then remove water and re-equilibrate	0.7	3.3
	0.3	7.1
Add less water to catalyst at start	0.7	2.7
	0.5	5.0

EFFECT OF WATER CONTENT ON REACTION RATE

Within the range giving very low water systems, the precise water content usually influences reaction rate. A variety of lipases show increasing activity as water is added, up to about 1 g per g dried catalyst (ie. enough to give a_w of about 0.9 if the solvent dissolves little water) [6,21,22]. Nevertheless, these catalysts show significant activity (5-50% of the maximum) in the initial "dry" state, which will of course contain some water molecules never removed. Fig 2 shows the effect of varying water activities on the reaction rate achieved with a silica-thermolysin catalyst [20]. In contrast, a number of other biocatalysts have been found to be completely inactive until water contents reach about 1 g per g or more, when a clear aqueous solution phase will usually begin to develop.

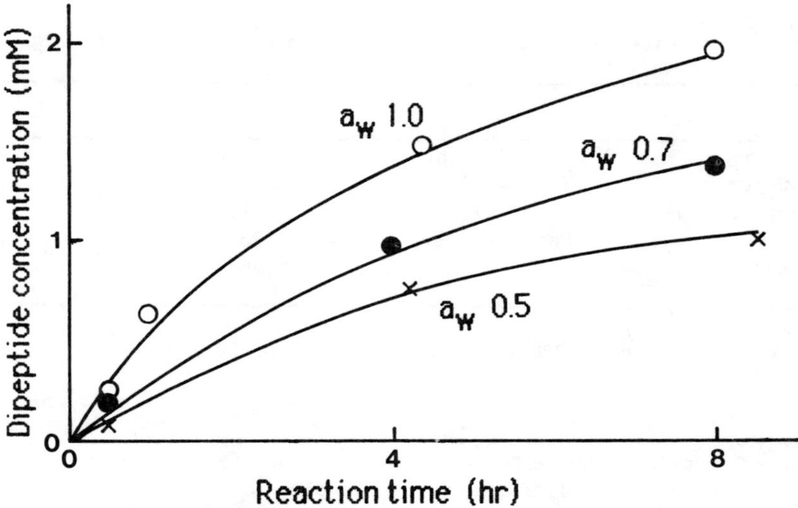

Fig. 2. Rates of peptide synthesis with silica-thermolysin catalysts of different water contents.

Catalysts were hydrated with different quantities of water before addition to an ethyl acetate solution of reactants.

Interestingly, it seems that the hydration history of the catalyst can affect its activity, as well as the final water content. Table 4 shows that celite-lipase catalysts of a given water activity are more active if they are initially hydrated to a higher level and then dried once in the reaction mixture [5]. We have recently observed similar effects with silica-thermolysin catalysts [20]. Similarly Table 5 shows that different drying regimes can markedly affect the activity of a celite-lipase catalyst that is not re-hydrated before use [23].

OTHER INFLUENCES ON BIOCATALYST ACTIVITY AND STABILITY

Nature of the solvent

The identity of the solvent or organic liquid(s) seems to influence activity and stability in the same general way as in higher water systems, with the most non-polar species being least damaging [21,22].

Temperature

Much attention was given to the demonstration by Zaks & Klibanov [24] of enhanced thermal stability of pancreatic lipase powders under very low water conditions in a predominantly organic system. A similar observation has now been reported by Ueda *et al.*[25] for whole cells of *Rhodococcus equi* containing lipase activity.

TABLE 4

Transesterification rates with celite-lipase catalysts hydrated by different methods

The first two reaction mixtures had identical compositions, but the water had been added initially to either the celite-lipase catalyst or to silica gel particles present in the reactor.

Reaction mixture	Transesterification rate (increase in stearate % in 500 min)	Final a_w
Wet catalyst + dry silica	12.9	0.28
Dry catalyst + wet silica	4.6	0.34
Wet catalyst control	14.5	0.87

TABLE 5

Transesterification rates with celite-lipase catalysts dried by different methods

The catalysts had the same enzyme and final water contents, and were used without adding back water.

Drying procedure	Rate (10^{-3} mole.s^{-1}.kg^{-1})
Freeze drying	0.84
Rapid vacuum drying	0.96
Slow initial rate of drying	8.9

PACKED BED REACTORS FOR VERY LOW WATER BIOCATALYSTS

A number of studies [6,7,26] have demonstrated attractive reactors for the application of very low water catalysts, with the solid particles (with adsorbed water) packed into a fixed bed through which an organic solution of the reactants is pumped, as illustrated in Fig. 3. Fig. 4 shows that such a packed bed reactor, for lipase-catalysed fat interesterification, may be operated continuously for many days [6].

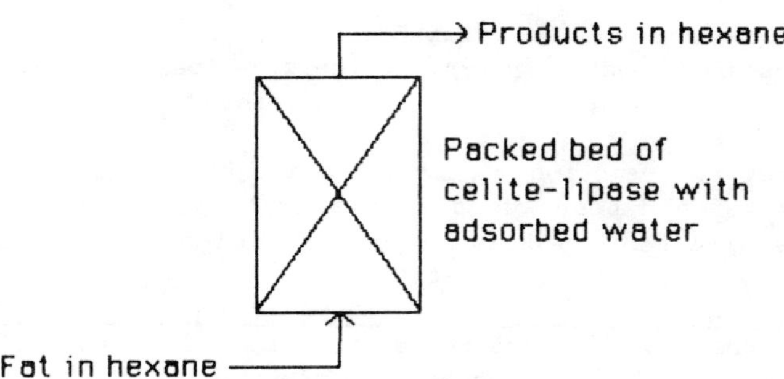

Fig. 3 Continuous interesterification reactor

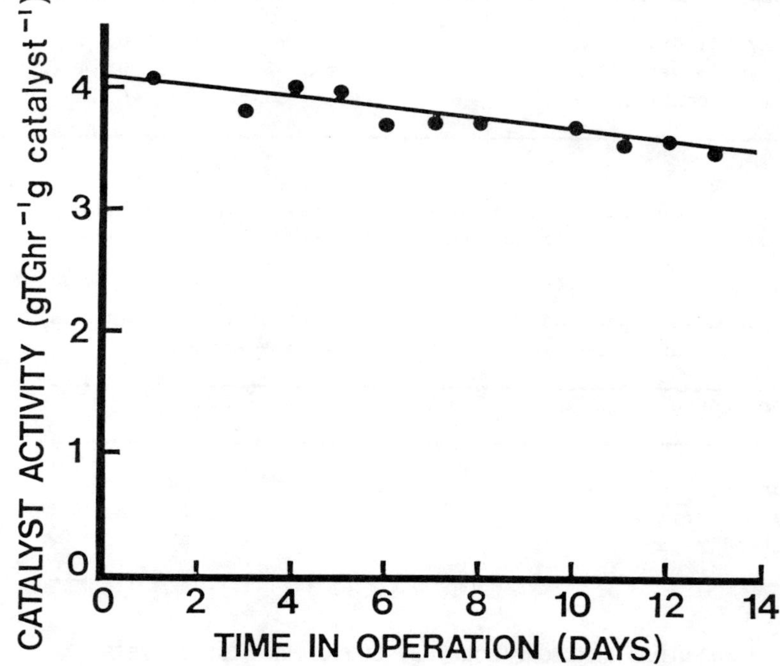

Fig. 4 Operational stability of celite-lipase catalyst in a packed bed reactor.

REFERENCES

1. Butler, L.G., Enzymes in non-aqueous solvents. Enzyme Microb. Technol., 1979, 1, 253-259.

2. Jakubke, H.D., Kuhl, P. and Konnecke, A., Basic principles of protease-catalysed peptide bond formation. Angew. Chem. Int. Ed., 1985, 24 85-93.

3. Halling, P.J., Biocatalysis in multi-phase reaction mixtures containing organic liquids. Biotechnol. Adv., 1987, in press.

4. Halling, P.J., Effects of water on equilibria catalysed by hydrolytic enzymes in biphasic reaction systems. Enzyme Microb. Technol., 1984, 6, 513-517.

5. Halling, P.J. and Macrae, A.R., Fat processing. Eur. Patent Appl., 1982, 64,855.

6. Macrae, A.R., Interesterification of fats and oils. In Biocatalysts in organic synthesis, ed. J. Tramper, Elsevier, Amsterdam, 1985, pp. 195-208.

7. Crook, S., Macrae, A.R. and Moore, H., Edible fats. Eur. Patent Appl., 1986, 170,431.

8. Dastoli, F.R., Musto, N.A. and Price, S., Reactivity of active sites of chymotrypsin suspended in an organic medium. Archiv. Biochem. Biophys., 1966, 115, 44-47.

9. Isowa, Y., Kakutani, M. and Yaguchi, M., The influence of water content in organic solvents on yields of peptide formation using proteases. In Peptide Chemistry 1981, ed. T. Shiori, Protein Research Foundation, Osaka, 1981, pp 25-30.

10. Kuhl, P., Walpurski, J. and Jakubke, H.-D., Studies on the influence of reaction conditions on the alpha-chymotrypsin-catalysed peptide synthesis in an aqueous-organic two-phase system. Pharmazie, 1982, 37, 766-768.

11. Ooshima, H., Mori, H. and Harano, Y., Synthesis of aspartame precursor by solid thermolysin in organic solvent. Biotechnol. Lett., 1985, 7, 789-792.

12. Zaks, A. and Klibanov, A.M., Substrate specificity of enzymes in organic solvents vs. water is reversed. J. Amer. Chem. Soc., 1986, 108, 2767-2768.

13. Blain, J.A., Patterson, J.D.E., Shaw, C.E. and Akhtar, M.W., Study of bound phospholipase activities of fungal mycelia using an organic solvent system. Lipids, 1976, 11, 553-560.

14. Dastoli, F.R. and Price, S., Catalysis by xanthine oxidase suspended in organic media. Archiv. Biochem. Biophys., 1967, 118, 163-165.

15. Dordick, J.S., Marletta, M.A. and Klibanov, A.M., Peroxidases depolymerise lignin in organic media but not in water. Proc. Nat. Acad. Sci. US, 1986, 83, 6255-6257.

16. Kazandjian, R.Z. and Klibanov, A.M., Regioselective oxidation of phenols catalysed by polyphenol oxidase in chloroform. J. Amer. Chem. Soc., 1985, 107, 5448-5450.

17. Kazandjian, R.Z., Dordick, J.S. and Klibanov, A.M., Enzymatic analyses in organic solvents. Biotechnol. Bioeng., 1986, 28, 417-421.

18. Saunders, R., Cheetham, P.S.J. and Hardman, R., Microbial transformation of crude fenugreek steroids. Enzyme Microbial Technol., 1986, 8, 549-555.

19. Halling, P.J., Rates of enzymic reactions in predominantly organic, low water systems. Biocatalysis, 1987, in press.

20. Cassells, J.M. and Halling, P.J., manuscript in preparation.

21. Blain, J.A., Patterson, J.D.E. and Shaw, C.E.L., The nature of mycelial lipolytic enzymes in filamentous fungi. FEMS Microbiol. Lett., 1978, 3, 85-87.

22. Zaks, A. and Klibanov, A.M., Enzyme catalysed processes in organic solvents. Proc. Nat. Acad. Sci. US, 1985, 82, 3192-3196.

23. Matsuo, T., Sawamura, N., Hashimoto, Y. and Hashida, W., Producing a cacao butter substitute by transesterification of fats and oils. GB Patent Appl., 1980, 2,035,359.

24. Zaks, A. and Klibanov, A.M., Enzymatic catalysis in organic media at 100 ^{o}C. Science, 1984, 224, 1249-1251.

25. Ueda, M., Mukataka, S., Sato, S. and Takahashi, J., Conditions for the microbial oxidation of various higher alcohols in isooctane. Agric. Biol. Chem., 1986, 50, 1533-1537.

26. Patterson, J.D.E., Blain, J.A., Shaw, C.E.L., Todd, R. and Bell, G., Synthesis of glycerides and esters by fungal cell-bound enzymes in continuous reactor systems. Biotechnol. Lett., 1979, 5, 211-216.

INTERNATIONAL CONFERENCE ON BIOREACTORS AND BIOTRANSFORMATIONS
GLENEAGLES, SCOTLAND, UK: 9-12 NOVEMBER 1987

Paper E2 Not presented

HALO-FERMENTATION, A NOVEL LOW WATER PROCESS FOR THE PRODUCTION OF ORGANIC CHEMICALS AND ENZYME PROTECTIVE AGENTS

E.A. Galinski
Institut für Mikrobiologie, Meckenheimer Allee 168, 5300 Bonn 1
Federal Republic of Germany

ABSTRACT

Salt lakes and solar evaporation ponds (salinas) are natural environments of low water activity (down to 0.75) and form ecological niches for halophilic microorganisms. To cope with the osmotic pressure of the surrounding medium the majority of halophilic/halotolerant eubacteria have evolved mechanisms which rely on the synthesis of organic osmotica. These "compatible solutes" are accumulated in the cytoplasm to molar concentrations and exhibit both osmoregulatory and enzyme protective functions. A screening project carried out in our laboratory has permitted the selection of a number strains which are potentially useful for the mass production of organic osmotica including the newly discovered novel amino acid, ectoine. The major problems associated with biological fermentation processes at high salinity are the inherent technical difficulties such as increased material corrosion. These technical obstacles may be solved either by using salt resistant fermenter materials (NaCl-tolerant Cr-Ni-Mo alloys) or by employing non-corrosive substitutes (e.g. polyethylene glycol) to maintain a low water activity.

INTRODUCTION

Concentrated salt solutions are naturally occurring extreme environments, being found as coastal lagoons, salt and soda lakes, and solar evaporation ponds. Strongly saline habitats are primarily a domain of microorganisms [1]. To cope with the physiological stress of low water activity (high osmolality) of the surrounding medium effective mechanisms of osmoadaptation have been evolved. With few exceptions the halophilic and halotolerant eubacteria so far examined produce high concentrations of organic osmotica (2-3 mol/kg of cytoplasmic water), which can be grouped into four typical classes of compounds: POLYOLS, SUGARS, AMINO ACIDS, and BETAINES [1]. These compounds have been named "compatible solutes" because they do not interfere with enzyme action [2].

TABLE 1
Spectrum of organic compounds ("compatible solutes") described so far in halophilic, halotolerant and osmophilic microorganisms

Polyols	Sugars	Amino Acids	Betaines
Glycerol	Glucose	Glycine	Glycine betaine
Erythritol	Fructose	β-Alanine	
Arabitol	Sucrose	γ-Aminobutyric acid	
Mannitol	Trehalose		
Sorbitol		Proline	Proline betaine
Altritol		Glutamate	Glutamate betaine
Cyclitol			
Glycosyl glycerol (gluco-, manno-, galactoside)			

In addition a novel cyclic amino acid was detected in and subsequently isolated from extremely halophilic species of the bacterial genus *Ectothiorhodospira* [3]. This new compound (Fig.1) has since then been found in a variety of different organisms and without doubt plays a major role in haloadaptation.

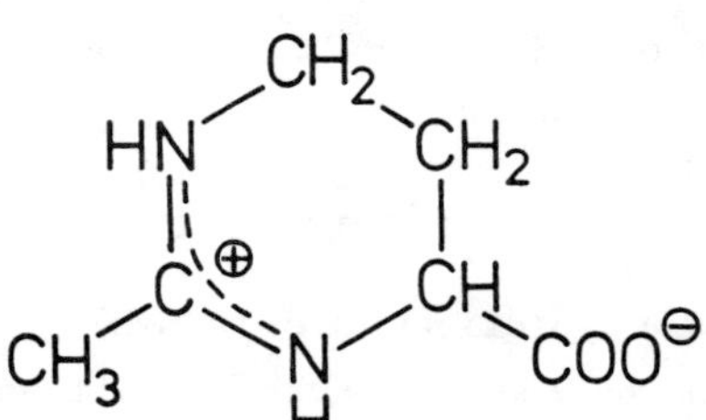

Figure 1: 1,4,5,6-Tetrahydro-2-methyl-4-pyrimidine carboxylic acid, a novel cyclic amino acid, which was given the trivial name "ectoine" due to its discovery in the bacterial genus *Ectothiorhodospira*.

We believe that the organisms' ability to produce and accumulate high concentrations of organic compounds (up to 20% of the cells' dry weight) makes them potentially useful for the biotechnological production of simple organic substances. In addition to their function as an osmoticum compatible solutes are required to maintain the enzymes' hydration shell in an environment of low water activity and to counterbalance the effect of elevated ionic strength within the cytoplasm. In view of this second physiological function compatible solutes may well find a technological application as stabilizers and protective agents in enzyme technology.

Large scale fermentation of halophilic microorganisms requires special materials. Although "normal" austenitic Cr-Ni-steel is resistant to neutral salt solutions (due to the protection by a chromium oxide passive layer) and corrosion is, therefore, negligible on plane surfaces, localised corrosion becomes a critical factor. The chloride anion effectively destroys the protecting passive layer locally and gives rise to pitting. This type of corrosion drastically increases above a critical temperature and, therefore, prevents the use of standard "stainless steel" for halo-fermentation (up to 50^oC). A suitable fermenter for aerobic fermentation of halophilic organisms must conform to the following requirements: High pitting potential and a minimium danger of intercrystalline corrosion (e.g. at welding zones), crevice corrosion (e.g. in anaerobic crevices under seals and encrustations), and stress corrosion cracking. Similarly deleterious effects on submersed measuring and regulating devices must be kept to a minimum. The use of non-corrosive substitutes (e.g. polyethylene glycol) could also provide a conceivable alternative to maintain a low water activity (at a minimum salt concentration) provided the organic media components like polyethylene glycol can be recovered after fermentation.

MATERIALS AND METHODS

Screening Method:

Samples from the salinas of Alicante (Spain), which were supplied by A. Wohlfarth (Dipl. biol.) of our laboratory, were subjected to a screening program for moderately halophilic and halotolerant aerobic bacteria. A typical enrichment medium was composed as follows: 1 l of demin. water, 1.0 g NH_4HCO_3, 1.0 g KNO_3, 0.5 g K_2HPO_4, 0.1 g $MgSO_4 \times 7H_2O$, 0.05 g $CaCl_2$

x $2H_2O$, 0.01 g $FeSO_4$ x $7H_2O$, 100 g NaCl, pH 7.5, 0.1 g yeast extract. The basal medium was then supplemented with 10 g/l of a variety of carbon sources including polymeric naturally occurring materials like starch, cellulose, and chitin. Pure cultures were then routinely grown in liquid culture on an incubator shaker and harvested in mid logarithmic phase to yield a minimum of 200 mg of cell material (wet weight) for subsequent HPLC analysis on a 200 x 8 x 4 Nucleosil-5NH_2 column (Macherey & Nagel, Düren) [4]. Solvent: 65% acetonitril/water (v/v); flow rate: 1 ml /min; detection: refractive index monitor (LDC/Milton Roy). Strains which proved to be potent producers of organic compounds were then fermented to yield large quantities of cell material for NMR analysis and further identification of yet unknown organic compounds.

Fermentation Vessels:

Fermentation on a small scale was carried out using Biostat M bench top fermenters (B.BRAUN, D-3508 Melsungen, FRG) with working volumes of 1.5 l. This type of all-glass fermenter offers the advantage of little contact between metal parts and corrosive culture medium, but is still subjected to corrosion on the top lid and immersed metal parts over a prolonged period of halo-fermentation. For laboratory scale fermentations up to 13 l working volume we have, therefore, employed a BIOENGINEERING fermenter (type L 1523) which was specially modified to meet our requirements (Bioengineering AG, CH-8636 Wald, Switzerland). The fermentation vessel comprises heated mantle piece, glass cylinder, and top plate. The metal parts of the system are made of austenitic Cr-Ni-Mo-Cu steel (type 1.4539) (see Table 2 in the RESULTS section). Other metal fermenter components like impeller, pH- and O_2-electrode housings, aeration tube, sparge ring etc. are protected by teflon or halar layers. To avoid undue wear of the bearing assembly through contact with crystalline salts the fermenter is equipped with overhead stirring. Despite the provision of in-situ sterilisation facilities, the fermentation medium is always autoclaved seperately in a glass container.

Culture conditions:

The fermentation medium for starch degrading actinomycetes was essentially the same as described for screening purposes except that KNO_3 and yeast extract were omitted and that NH_4HCO_3 was raised to 3 g/l. The temperature of fermentation did not exceed 40°C. Water activity, i.e. osmolality, was varied using different concentrations of NaCl, polyethylene glycol or combinations of both depending on the organisms salt requirement. Bacterial growth was monitored by colorimetric determination of starch degradation (quantitative iodide method) and gravimetric determination of bacterial cell mass. In order to reveal the whole spectrum of compounds cells were harvested in the exponential growth phase.

Identification of Compatible Solutes:

Using a modification of the method of Bligh & Dyer [5] a minimum of 10 g of cell material (wet weight) was extracted with 57 ml of a mixture of methanol/chloroform/water (10 : 5 : 2.2 v/v). Phase separation was achieved by adding 16.5 ml of chloroform and 16.5 ml of water. The aqueous top phase (40 ml of a methanol-water mixture) containing cytoplasmic components, soluble proteins, and extracellular salts was concentrated (to 1/4 of volume), deproteinized using perchloric acid, and desalted on an ion retardation column (BIORAD AG 11A8). The remaining non-ionic compounds (including betaines) were collected for NMR identification. 1.5 ml of an aqueous solution containing at least 10 mg of organic material was supplemented with 0.3 ml D_2O to provide an internal lock signal, and 10 μl of acetonitril were added as internal standard. Spectra were recorded in the pulsed Fourier transform mode with proton noise decoupling, using a BRUKER (model WH-90) spectrometer operating at 22.628 MHz and 90.02 MHz for the decoupling channel.

Replacement of NaCl in Fermentation Media:

Various types of polyethylene glycol (average molecular mass 200, 400, and 600) were examined for their potential use as osmotica. Of the physical parameters which have to be considered in choosing a substitute for saline fermentation media density, viscosity, and osmolality (water activity) are the most important. Viscosity was measured using a BROOKFIELD (model LVT) rotary viscosimeter which was equipped with an adapter for small sample volumes. Osmolality (Osm) was determined on a KNAUER vapour pressure osmometer at a temperature of 40°C. Osmolality (Osm), osmotic pressure (π), water potential (ψ), and water activity (a_w) are interconvertible according to the following equations: π (MPa) = 2.602 x Osm (at 40 C); ψ (MPa) = - π (MPa); ln a_w = - 18.05 x 10^{-3}x Osm.

RESULTS

Choice of fermenter material:

Table 2 compares the chemical composition of different types of steel considered as fermenter materials. Whereas AVESTA 17-12-2.5 closely resembles the V4A-fermenter material commonly supplied with high quality fermenters on the laboratory scale, the other types of steel represent materials which have been successfully used in halofermentation, or have been designed for salt water applications (254SMO). The increasing resistance (from top to bottom) can be explained by the elevated proportion of chromium, nickel, and molybdenum, which affects both the stability of the passive layer and dissolution tendency of the metals.

TABLE 2
Chemical composition of austenitic Cr-Ni-alloys, used for fermentation processes (taken from AVESTA information 8559, AVESTA AB, S-77401 Avesta, Sweden)

Typ of steel			chemical composition in %					
AVESTA	DIN	ASTM	Cr	Ni	Mo	C	N	Cu
17-12-2.5	1.4436	316	17	11	2.7	0.05	-	-
17-14-4LN	1.4439	S31726	17	13	4.2	0.03	0.14	-
904L	1.4539	N08904	20	25	4.2	0.02	-	1.5
254SMO	-	S31254	20	18	6.2	0.02	0.02	Cu

The unusually high content of molybdenum in 254SMO typically increases the material's resistance at higher temperatures (compare pitting potential in Tab.3), whereas the Cu content confers resistance to a non-oxidizing environment. A low carbon content is essential to avoid chromium carbide formation during the process of welding, whereas nitrogen influences the mechanical characteristics of the material.

TABLE 3
Pitting potential (mV/SCE) of different types of steel in 3.5% NaCl, taken as a measure for the materials resistance at higher temperatures (from AVESTA information 8559, AVESTA AB, S-77401 Avesta, Sweden)

Type of material	Temperature 25°C	60°C	90°C
1.4436 (V4A)	550	-	-
1.4539 (904L)	1000	830	<500
254SMO	1000	1000	870

A Model Organism (Actonomycete A5-1):

Our screening program has been aimed at the isolation of halotolerant organisms which exhibit an interesting range of compounds and are easy and cheap to grow using common fermentation techniques. Such a model organism which grows on a basic synthetic medium and uses starch as the sole carbon source is strain A5-1. This currently unidentified actinomycete was isolated from mud samples of the salinas in Alicante (Spain) and exhibits a rather broad salt tolerance (Table 4).

TABLE 4
Relative growth rate of the actinomycete strain A5-1 at different salt concentrations (100 % is equivalent to a doubling time of 9.5 h)

NaCl (g/kg water)	50	100	150
relative growth rate	100 %	80 %	30 %

Strain A5-1 is able to produce a wide spectrum of compatible solutes including ectoine and trehalose. As can be judged from the remaining signals of the NMR spectrum (Fig.2) a number of unidentified compounds are also present at high concentrations. Acetonitrile was added as a reference at an amount equivalent to a cytoplasmic concentration of 100 mM. It can, therefore, be concluded that the intracellular concentration of trehalose is relatively low (approx. 200 mM) whereas the other compounds are approaching molar concentrations within the cytoplasm.

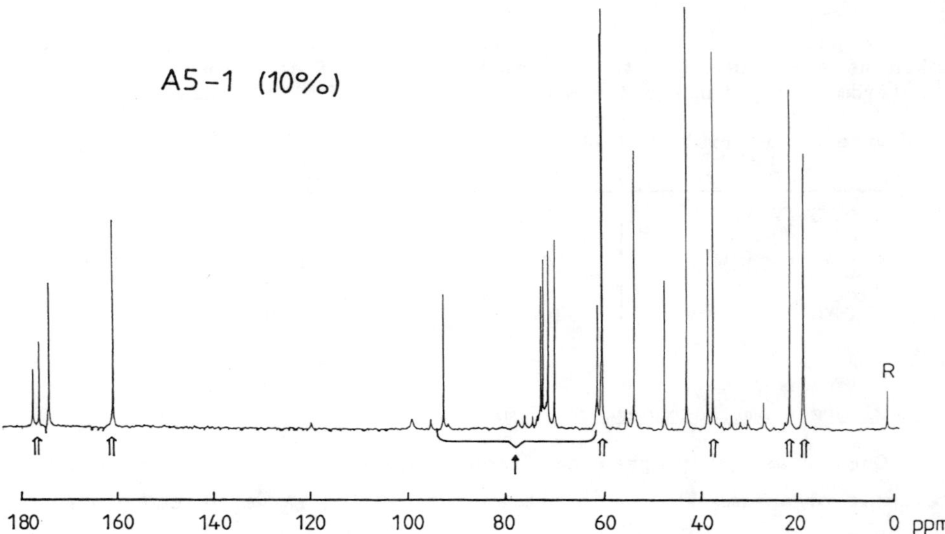

Figure 2: Natural abundance (^{1}H decoupled) ^{13}NMR spectrum of an aqueous extract from a halotolerant actinomycete, strain A5-1, grown at 10 % NaCl. Some of the signals can be allocated to trehalose (↑) and ectoine (⇑), the remainder are due to the presence of unidentified compounds. R = acetonitrile reference (10 μl/1.8 ml).

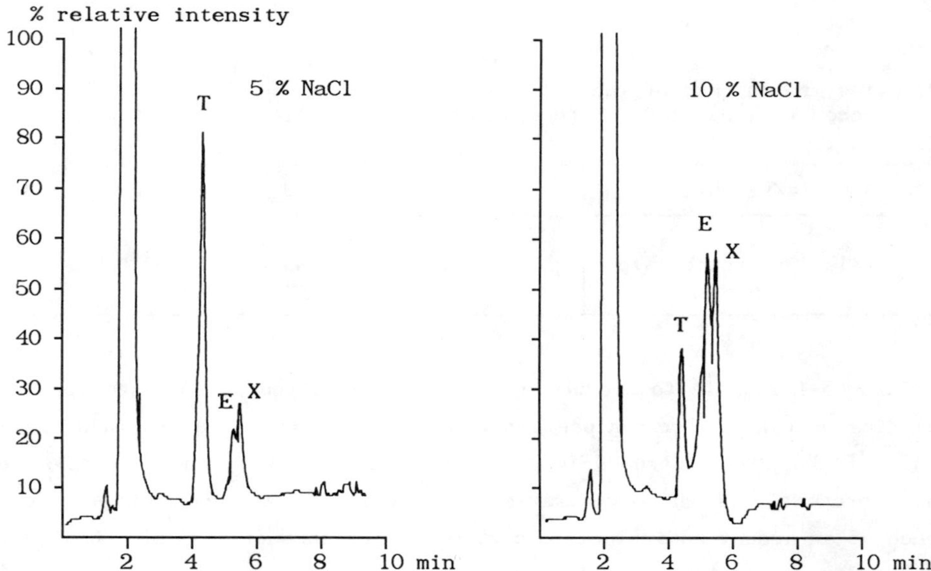

Figure 3: HPLC separation of compatible solutes from strain A5-1 grown at different salinities (Nucleosil-5NH_2; solvent: acetonitrile/methanol/ water, 3:1:1 v/v); T = trehalose, E = ectoine, X = unidentified compound.

To study possible variations of the compatible solute spectrum we have subsequently applied a chromatographic analysis of cell material grown at different salt concentrations. Despite uncertainties as to whether the whole spectrum of compounds has been detected using the HPLC method described it can be concluded from the chromatograms that at least three different types of compatible solutes are produced. Comparison of the signal intensities under different growth conditions (5 - 15 % salt) reveals a remarkable variation in the relative proportion of cytoplasmic compounds. The tendency can be described as follows: trehalose is accumulated at low salinity whereas ectoine and other compounds become the dominant compatible solutes at increased NaCl concentrations. It seems possible, therefore, to channel the bioproduction of compatible solutes towards certain products by varying the osmolality of the surrounding medium. To elevate the influence of the ionic strength of the medium we have subsequently studied the possible use of non-ionic substitutes to maintain a low water potential.

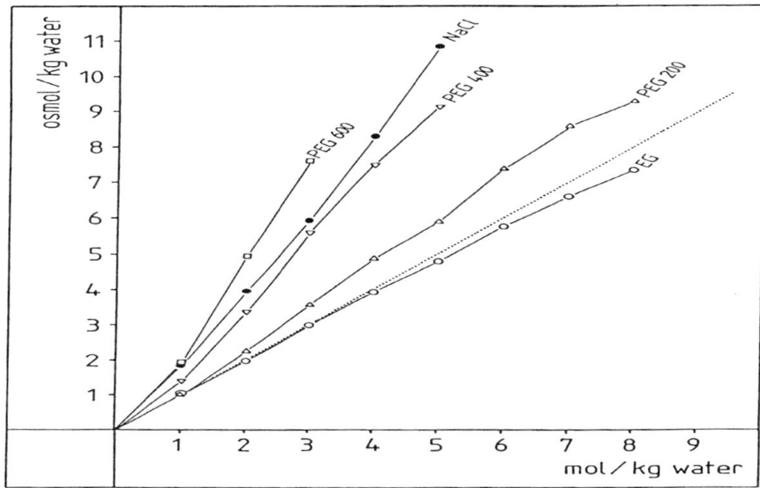

Figure. 4: Osmolality of different types of polyethylene glycol as compared to NaCl solutions;PEG = polyethylene glycol (average molecular mass of 200, 400, 600); EG = monomeric ethylene glycol

Unlike low molecular mass compounds (e.g. ethylene glycol) polymeric ethylene glycol does not conform to Van't Hoffs law of osmotic pressure in ideal solutions. Fig.4 illustrates how the osmotic effect of polyethylene glycol increases according to the molecule's average molecular mass and strongly deviates from the value for ideal solutions (dotted line). It is, therefore, apparent that comparison of different substitutes for NaCl must be drawn on an osmolal basis.

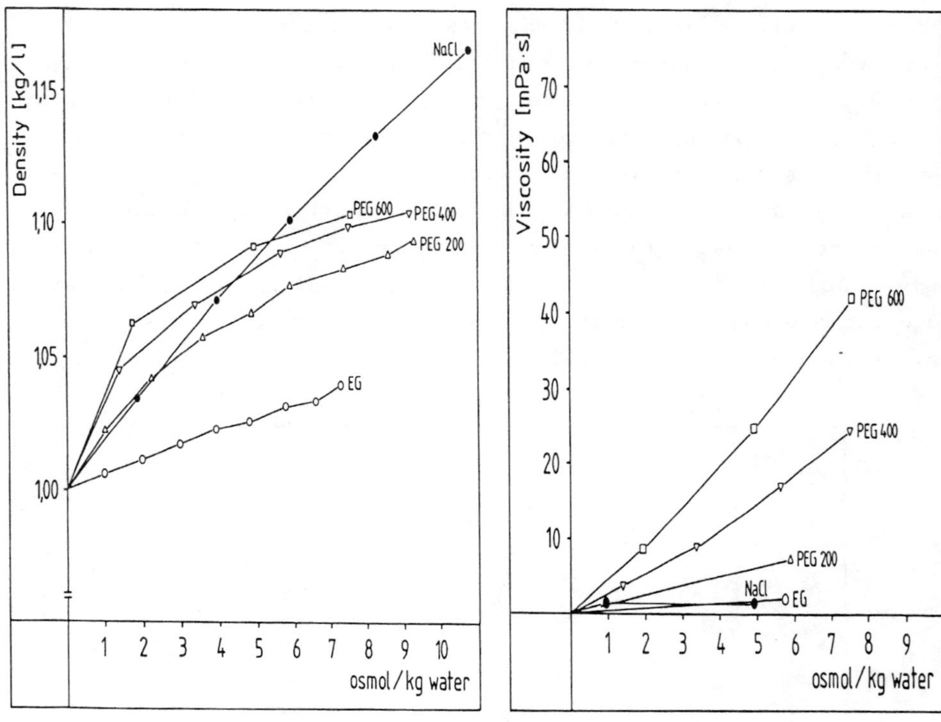

Figure 5: Density (kg/l) and viscosity (mPa x s) of different types of polyethylene glycol (average molecular mass of 200, 400, and 600) as compared to NaCl of equal osmolality; PEG = polyethylene glycol; EG = monomeric ethylene glycol

The physical parameters density and viscosity of equiosmolal solutions have been compared in Fig.5. Whereas the density of a number of polyethylene glycols is very similar to that of NaCl within the range of potential application (3-6 osmolal), viscosity strongly increases with increasing molecular mass of the polymer. Viscosity will, therefore, play an important part when choosing a substitute for NaCl in fermentation media.

DISCUSSION AND CONCLUSION

It can be concluded from published data (Tables 2 and 3), as well as from our own investigations, that Cr-Ni-alloys with a composition similar to steel type 1.4539 (904L) may be used in fermentations up to 60°C, wheras at higher temperatures an alloy of the 254SMO-type (very high molybdenum content) would be more suitable. Nonetheless we advise against sterilizing saline media <u>in</u> the fermentation vessel, as the continual subjection to high temperatures as well as elevated pressure may result in stress corrosion cracking. Technically these problems can be solved using appropriate materials. However long-term experience is still needed to prove the protective value of teflon coatings on submersed equipment like stirrer, aeration tube and electrode housings.

The upper limit of the salt tolerance of strain A5-1 (approx. 15%) may be determined by a number of factors, including the inhibitory effect of NaCl on exoenzymes. Partial replacement of salts by organic (polymeric) media components, therefore, seems to be rather promising not only from the point of view of corrosion but also with respect to widening the organism's range of osmotolerance. Recovery of the organic osmoticum, however, largely depends on the type of polymer used. The separation of organic and aqueous phases are common practice in down stream processing of enzymes [5], the high viscosity of polyethylene glycol solutions may, however, impose a severe obstacle to its use as an osmoticum in halo-fermentation.

As the relative proportion of different types of compatible solutes varies considerably with the salinity of the growth medium we have obtained a means of specifically directing biosynthesis towards the production of certain compounds. In addition other unidentified compatible solutes are to be expected which will widen the spectrum of compounds available for bioproduction. Using fermentation media containing 1 % of an organic carbon source we have so far been able to produce 10 g/l of bacterial cell mass (wet weight) and to gain approximately 0.4 g/l of compatible solutes. The economical feasibility of the fermentation process can still be improved and will certainly profit from the use of overproducing or leaky mutants.

An enzyme protective action of compatible solutes has been reported in a number of cases [7, 8] and it has also been shown that glycerol, betaine, and proline are able to fully or partly relieve salt inhibition on certain enzymes [9, 10, 11]. Though a number of theories have been put forward to explain the protective action of compatible solutes the underlying principle is by no means clear. A systematic study of this class of compounds may, therefore, provide a significant contribution towards the understanding of enzyme action and their protection by compatible solutes.

I would like to thank Dr. B.J. Tindall for proof-reading of the manuscript

REFERENCES

1. Trüper, H.G. and Galinski, E.A., Concentrated brines as habitat for microorganisms. Experientia, 1986, 42, 1182-1187.

2. Brown, A.D., Microbial Water Stress. Bact. Rev., 1976, 40, 803-846.

3. Galinski, E.A., Pfeiffer, H.P. and Trüper, H.G., 1,4,5,6-Tetrahydro-2-methyl-4-pyrimidinecarboxylic acid, a novel cyclic amino acid from halophilic phototrophic bacteria of the genus Ectothiorhodospira. Eur. J. Biochem., 1985, 149, 135-139.

4. Galinski, E.A., Salzadaption durch kompatible Solute bei halophilen phototrophen Bakterien, Ph.D.thesis, University of Bonn, April 1986.

5 Bligh, E.G. and Dyer, W.J., A rapid method of lipid extraction and purification. Can. J. Biochem. Physiol., 1959, 37, 911-917.

6. Hustedt, H., Kroner, K.H., Menge, U. and Kula, M-R., Protein recovery using two-phase systems. Trends in Biotechnol.,1985, 3, 139-144.

7. Coughlan, S.J. and Heber, U., The role of glycine betaine in the protection of spinach thylakoids against freezing stress. Planta, 1982, 156, 62-69.

8. Jolivet, Y., Larher, F. and Hamelin, J., Osmoregulation in halophytic higher plants: The protective effect of glycine betaine against the heat destabilization of membranes. Plant Sci.Letters, 1982, 25, 193-201.

9. Pollard, A. and Wyn Jones, R.G., Enzyme activities in concentrated solutions of glycine betaine and other solutes. Planta, 1979, 144, 291-298.

10. Pavlicek, K.A. and Yopp, J.H., Betaine as a compatible solute in the complete relief of salt inhibition of glucose-6-phosphate dehydrogenase from a halophilic blue-green alga. Plant Physiol., 1982, 69, 58 Suppl.

11. Luard, E.J., Activity of isocitrate dehydrogenase from three filamentous fungi in relation to osmotic and solute effects. Arch. Microbiol., 1983, 134, 233-237.

INTERNATIONAL CONFERENCE ON BIOREACTORS AND BIOTRANSFORMATIONS GLENEAGLES, SCOTLAND, UK: 9-12 NOVEMBER 1987

Paper F1

ENZYMATIC REDUCTION OF BETA-KETOESTERS USING IMMOBILISED YEASTS

M. Christen and D.H.G. Crout
Department of Chemistry
University of Warwick
Coventry, West Midlands, CV4 7AL.

ABSTRACT

In the reduction of beta-ketoesters by yeasts, immobilisation on alginate or celite gave products with optical purities greater than those obtainable using free yeasts. This factor, together with other advantages including higher stability and easier handling, improves significantly the performance of yeasts as biocatalysts in biotransformations.

1. INTRODUCTION

In recent years there has been a marked increase in interest in biotransformations with yeasts [1]. Baker's yeast reductions of beta-ketoesters have enjoyed great popularity because this yeast is readily available to the non-microbiologist [2]. However, there are some serious drawbacks associated with the use of baker's yeast. The work up often proves to be difficult and tedious and substrate concentrations have to be kept low owing to the low solubility of many substrates in water. This last factor leads to low productivity numbers (PN) [3]. In addition, reproducibility, particularly with respect to the enantiomeric purity of the products, is often low. We therefore decided to investigate the reduction of a variety of beta-ketoesters with a readily-available, stable biocatalyst, and to attempt to overcome these problems by using immobilisation techniques. Although there have been many reports on the use of immobilised baker's yeast, there have been no systematic investigations of the behaviour of such systems as biocatalysts in organic synthesis [4-6].

2. MATERIALS AND METHODS

2.1. Chemicals

Beta-ketoesters with sulphur substituents at C-4 (cf. (1)) were synthesised by a published procedure [7]. Triton X-100 was purchased from Sigma, Bacto YM broth from Difco Laboratories, and Celite 630 from Fluka AG. All other chemicals were purchased from the Aldrich Chemical Company Ltd.

2.2. Microorganisms and Cultivation Conditions

A pure strain of *Saccharomyces cerevisiae* was obtained locally and deposited in the National Collection of Yeast Cultures as NCYC 1765. *Candida guilliermondii* NCYC 973 was obtained from the same source. Typically, cultivation was carried out at 30°C in conical flasks in a rotary shaker at 100 r.p.m. in Difco YM broth. Growth was followed by determination of the optical density, at 630 nm. The cells are best immobilised in the last stage of the growth phase.

2.3. Immobilisation on Alginate

To 100 ml of culture broth containing about 0.5 g of wet cells, was added sodium alginate (0.5 g). The suspension was dropped into calcium chloride solution (100 ml, 10% w/v). The gel beads (20 g, 4-5 mm in diameter) were collected and washed with water [8-9]. The beads can be stored at 4°C in 5 mM calcium chloride for several months, if the calcium chloride solution is regularly replaced (every two weeks) with fresh sterilised solutions.

2.4. Immobilisation on Celite

To 100 ml of culture broth containing about 0.5 g of wet cells was added Celite 630 (2.5 g) and the suspension was stirred for 3 hours [10]. The celite was collected by filtration, washed with water and stored in phosphate buffer (0.1 M, pH 6.0) at 4°C. The buffer was replaced with fresh, sterilised solution every two weeks.

2.5. Reductions with Free Cells

To 100 ml of the culture broth (0.5 g of wet cells) was added sucrose (100 mg, 0.3mmol), Triton X-100 (50 mg, 0.08 mmol) and substrate (0.1 mmol) in 5 ml of cosolvent. When the reaction was judged to be complete (by t.l.c.), Celite 535 (2 g) was added. The resulting suspension was filtered and the filtrate was extracted three times with diethyl ether. The combined organic extracts were dried ($MgSO_4$), filtered and evaporated under reduced pressure. The residue was dried under high vacuum. If necessary, products were separated and purified by flash chromatography [11].

2.6. Reductions with Immobilised Cells

To alginate beads (20 g) or Celite 630 (2.5 g) in 45 ml of 5 mM $CaCl_2$ solution or 0.1 M phosphate buffer (pH 6) respectively, was added sucrose (100 mg, 0.3 mmol), Triton X-100 (50 mg, 0.08 mmol) and substrate (0.1 mmol in cosolvent (5 ml)). On completion of the reaction, the mixture was filtered and the filtrate was extracted with diethyl ether. The combined organic extracts were filtered, evaporated and dried as above (Section 2.5).

2.7. Further Experimental Details

Reactions were monitored by thin layer chromatography on silica gel using the solvent system ethyl acetate: light petroleum (b.p. 40-60°C) (1:1). Compounds were visualised by spraying the plate with phosphomolybdic acid (10% in ethanol), followed by heating at 150°C for 5 minutes. HPLC analysis was carried out on a C-18 reversed-phase column (Spherisorb S5ODS2) in acetonitrile: water (7:3). Concentrations were determined using a calibration curve constructed at 254 nm. Enantiomeric excesses were determined by ^{1}H NMR in $CDCl_3$ using the chiral shift reagent europium (D-3-trifluoroacetyl camphorate)$_3$. Alternatively, optical rotations were determined in those cases in which the rotatons of the pure enantiomers were known.

3. RESULTS AND DISCUSSIONS

3.1. Influence of Immobilization on Enantiomeric Purity

The reduction of nine different beta ketoesters was studied (Scheme 1). The results obtained are summarised in Table 1.

$R^1C(O)CH_2C(O)OR^2$ (1) $\xrightarrow{\text{biocatalyst}}$ $R^1CH(OH)CH_2C(O)OR^2$ (L - 2) + $R^1CH(OH)CH_2C(O)OR^2$ (D - 3)

Scheme 1

TABLE 1

Reduction of Beta-Ketoesters Using Free and Immobilised Yeast Cells

Substrate		*S. cerevisiae*			*C. guilliermondii*		
R^1	R^2	Free cells	Alginate	Celite	Free cells	Alginate	Celite
		(a) (b)					
a H	CH_3	75/70L	70/80L	70/78L	55/70D	60/75D	60/78D
b H	CH_2CH_3	80/80L	75/90L	75/92L	40/35D	43/50D	50/50D
c Cl	CH_3	60/70L	70/75L	75/77L	50/78D	53/80D	55/83D
d C_2H_5S	CH_3	55/70L	60/80L	65/80L	8/90D	30/90D	35/90D
e n-C_3H_7S	CH_3	49/65L	60/70L	60/73L	12/90D	48/95D	50/94D
f n-C_4H_9S	CH_3	67/70L	70/80L	60/75L	30/90D	52/93D	55/95D
g n-$C_5H_{11}S$	CH_3	30/58L	45/80L	50/78L	14/80D	55/88D	53/93D
h C_6H_5S	CH_3	42/73L	50/80L	50/75L	32/88D	58/90D	60/92D
i p-ClC_6H_4S	CH_3	40/50L	60/80L	60/82L	38/80D	60/87D	65/85D

(a) % isolated yield (b) % enantiomeric excess configuration

In all cases studied, the use of immobilized cells gave an increase in the proportion of the major enantiomer in the product by comparison with the free-cell system. If it is assumed that the partial stereospecificity of yeast reduction is a consequence of the operation of different oxido reductases [12], it can be concluded that immobilisation increases the difference in the rates of these individual enzymatic reactions. Our results contrast strongly with those of Ohno [6] who reported an increase of the D-enantiomer, even when the L-enantiomer was the major product of the biotransformation in the free-cell system.

Our conclusions are supported by the observation that it is possible to increase the optical purities of the products of yeast reductions by slow, continuous addition of the substrate, thereby taking maximum advantage of the differences in the rates of the different enzymatic reactions (which depend on V and K_M) [13-14]. Indeed, it was possible, using a laboratory reactor system in which substrate concentrations were kept low and products were continuously removed, to obtain 2b and 2i in an enantiomerically pure form using *Saccharomyces cerevisiae* immobilized in alginate. Because the hydroxyesters 2i and 3i are crystalline and can be recrystallised readily to optical purity from mixtures of the enantiomers, efforts were concentrated on substrate ester 1i.

3.2. Influence of Cosolvents

Using substrate 1i, the effect of five cosolvents (ethanol, dimethylsulphoxide, dimethylformamide, dimethoxyethanol and acetonitrile) on the reduction with *S. cerevisiae* was investigated. The results are summarised in Table 2.

TABLE 2

Effect of cosolvent on the Productivity Number (PN) in the Reduction of Beta-Ketoester 1i Using Alginate-Immobilised *S.cerevisiae*

	PN		
Cosolvent	Free cells	alginate immobilised cells	celite immobilised cells
10% EtOH	190	180	170
10% DMSO	170	120	140
10% DMF	550	500	420
10% DME	320	340	340
10% AN	570	600	590

Surprisingly, there were hardly any differences in the reaction rates between the free-cell and immobilised systems. Unfortunately, it was not possible to determine a productivity number for the reduction in the absence of cosolvent, because the ester 1i is quite insoluble in water. However, the reduction of substrate 1b, amply described in the literature [15], proceeded with a PN of about 850, depending on the amount of sucrose used, and using S.cerevisiae immobilised on alginate. Based on these results, acetonitrile (AN) was selected as the most suitable cosolvent for use with our biocatalyst. It affords perfectly clear solutions with a large range of crystalline substrates.

3.3. Reusability and Stability

Experiments were carried out on the reduction of beta-ketoester 1i, using both yeasts immobilised on celite and with acetonitrile as cosolvent. The results are shown in Table 3.

TABLE 3

Productivity Numbers in Three Consecutive Experiments with Celite-Immobilised Biocatalysts

Number of Consecutive Reductions	S. cerevisiae	C.guilliermondii
1	590	620
2	320	410
3	90	80

Both catalysts lost most of their activity after their second use. However, they could be stored at 4°C for two months with 95% retention of activity.

4. CONCLUSIONS

We have shown that the use of immobilised yeasts overcome most of the disadvantages associated with the use of free-cell reductions. It provides for easy work up, controlled reaction conditions, acceptable productivity numbers and good reproducibility. Given that yeasts generally are easy to grow, we have, in immobilised yeasts, a very useful biocatalyst with a wide substrate range, high stereospecificity and one which provides for simple handling. In addition, we have already demonstrated the use of the chiral synthons 2i and 3i in the synthesis of optically pure, polyhydroxy esters [16].

ACKNOWLEDGEMENT

We thank Glaxo Group Research Ltd. for financial support.

5. REFERENCES

1. Jones, J.B., Tetrahedron, 1986, 42, 3351-403

2. For a recent example see Brooks, D.W., Wilson, M. and Webb, M., J. Org. Chem., 1987, 52, 2244-48.

3. $$PN = \frac{\text{amount of product [mmol]}}{\text{dry weight of catalyst [kg] x time [h]}}$$

   Simon, H., Bader, J., Gunther, H., Neumann, S. and Thanos, J., Angew. Chem. Int. Ed. Engl., 1985, 24, 539-53.

4. Utaka, M., Konishi, U., Okubo, T., Tsuboi, S. and Takeda, A., Tetrahedron Lett., 1987, 28, 1447-50.

5. Lis, R., Caldwell, W.B., Rudd, I.G. and Lumma, W.C., Tetrahedron Lett., 1987, 28, 1487-90.

6. Nakamura, K., Higaki, M., Ushio, K., Oka, S. and Ohno, A., Tetrahedron Lett., 1985, 26, 4213-16.

7. Campaigne, E. and Homfeld, E., J. Heterocyclic Chem., 1979, 16, 1321-24.

8. Kierstan, M. and Buche, C.; Biotechnol. Bioeng., 1977, 19, 387-97.

9. Chotani, G.K. and Consfantinides, A., Biotechnol. Bioeng., 1984, 26, 217-20.

10. Fluka, A.G. Personal Communication.

11. Still, W.C., Kahn, M. and Mitra, A., J.Org.Chem., 1978, 43, 2923-25.

12. Chen, C.S., Zhou, B.N., Girdaukas, G., Shieh, W.R., Van Middlesworth, F., Gopalan, A.S. and Sih C.J., Bioorganic Chem., 1984, 12, 98-117.

13. Wipf, B., Kupfer, E., Bertazzi, R. and Leuenberger, H.G.W., Helv. Chim. Acta, 1983, 66, 485-88.

14. Shieh, W.R., Gopalan, A.S. and Sih, C.J., J. Am Chem. Soc. 1985, 107, 2993-94.

15. Seebach, D., Sutter, M.A., Weber, R.H. and Zuger, M.F., Org. Synth., 1985, 63, 1-9.

16. Christen, M. and Crout, D.H.G., communication in preparation.

INTERNATIONAL CONFERENCE ON BIOREACTORS AND BIOTRANSFORMATIONS
GLENEAGLES, SCOTLAND, UK: 9-12 NOVEMBER 1987

Paper F2

Language problem severe
Read all slides in great detail.

MICROBIAL TRANSFORMATION OF ISOPRENOID HYDROCARBONS AND RELATED COMPOUNDS

K. Nakajima
Petroleum Fermentation Division,
Microbe Engineering Department,
Fermentation Research Institute,
Agency of Industrial Science & Technology,
Yatabe-machi, Ibaraki,
Japan

ABSTRACT

Rhodococcus sp. BPM 1613, capable of utilizing pristane as a sole carbon and energy source was isolated from soil specimens, grown on isoprenoid hydrocarbons such as pristane, phytane, 1-pristene, norpristane, farnesane and squalene as the sole carbon source, resulting in accumulation of oxidation products in culture broth. The oxidation products of their substrates in the respective culture broth were isolated and their chemical structures were determined, indicating that these major oxidation products from pristane were pristanol and pristanic acid. Synthetic oligomers 2,4-dimethyl-1,4-nonadiene and 2-methyl-2-nonanol were submitted to microbial conversion by BPM 1613. One of their oxidation products was identified as 8-hydroxy-8-methylnonanoic acid.

The cultural conditions for the production of the pristanol and pristanic acid from pristane and microbial utilization of the isoprenoid hydrocarbon mixture derived from shale oil by the strain BPM 1613 were also discussed.

INTRODUCTION

Isoprenoid hydrocarbons such as pristane (2,6,10,14-tetramethylpentadecane), phytane (2,6,10,14-tetramethylhexadecane), 1-pristene (2,6,10,14-tetramethyl-1-pentadecene), norpristane (2,6,10-trimethylpentadecane) and farnesane (2,6,10-trimethyldodecane) have been known to be contained in shale oil obtained by dry distillation btween 280～305°C [1]. Among them, pristane and phytane have been reported to occur widely in nature, for example, in the tissues of men and animals. It is now generally admitted that branched-chain hydrocarbons are more amenable to microbial oxidation than their straight-chain counterparts [2]. There have been already several reports on microbial oxidation of pristane [3,4]. Microbial oxidation of phytane, norpristane and related compounds has been reported only by Cox _et al_. [5] According to Cox _et al_., when _Mycobacterium fortuitum_ was grown on each of these compounds, monocarboxylic acids were derived through terminal oxidation and further β-oxidation of the oxidation products, which were identified by GLC-MS.

Isoprenoid hydrocarbon squalene (2,6,10,15,19,23-hexamethyl-2,6,10,14,18,22-tetracosahexaene) with pristane has been found in large quantities in shark liver oil and occurs in small amounts in olive oil, wheat germ oil, rice bran oil and yeast [6]. Yamada _et al_. [7] reported the _trans_-geranylacetone and four carboxylic acids as squalene-derived metabolites but nothing has been reported on the microbial terminal oxidation of squalene.

Recently, the synthetic methods of alternating cooligomers which have been the base structure of diene and olefin as isoprene or 1,3-butadiene and propylene have been developed and similar compounds with isoprenoid hydrocarbons have been synthesized. Microbial conversions of natural isoprenoid hydrocarbons such as terpenes have been mentioned in several reports but reports on microbial conversion of their synthetic oligomers have not been observed.

The present report deals with the isolation and identification of the microbial oxidation products from pristane, phytane, 1-pristane, norpristane, farnesane and squalene as isoprenoid hydrocarbons and 2,4-dimethyl-1,4-nonadiene and 2-methyl-2-nonanol as synthetic oligomers by _Rhodococcus_ sp. BPM 1613.

MATERIALS AND METHODS

Rhodococcus sp. BPM 1613, a microorganism isolated from soil and able to utilize hydrocarbons such as _n_-paraffin and pristane was mainly used [8,9]. The medium employed for precultivation was prepared by dissolving $NaNO_3$ 5.0g (or NH_4NO_3 2.5g), KH_2PO_4 1.5g, Na_2HPO_4 1.5g, $MgSO_4 \cdot 7H_2O$ 0.5g, $FeSO_4 \cdot 7H_2O$ 0.01g, $CaCl_2 \cdot 2H_2O$ 0.01g and yeast extract 0.2g in distilled water, the final volume was 1 liter, and the pH was adjusted to 7.2. To the solution was added 0.5% (v/v) of mixed _n_-paraffin as the carbon source. The mixed _n_-paraffin was replaced in the medium for accumulation of the oxidation products by 1 or 3% (v/v) each of substrates and the amount of yeast extract added was mainly increased to 2.0g. In the case of yeast, pH of culture medium was adjusted to 5.5.

A 50 ml portion of the precultivation medium in a 500 ml shaking flask was sterilized and inoculated from an agar slant culture. The inoculated medium was incubated at 30°C for 2 days on a reciprocal shaker (120 rpm). This seed cell suspension was added at 8% (v/v) to shaking flasks containing 50 ml of the accumulation medium followed by incubation at 30°C for 5 days as described for the preliminary cultivation.

Pristane was purchased from Aldrich Chemical Co., Inc. Squalene was a product of Tokyo Kasei Kogyo Co., Ltd. Phytane was synthesized from phytol and farnesane was synthesized from farnesol [10]. 1-Pristene was prepared from pristanol [11]. 2,4-Dimethyl-1,4-nonadiene and 2-methyl-2-nonanol were synthesized by Taniguchi a co-researcher.

Thin layer chromatographic (TLC) analyses for purity test and monitoring of column were carried out on 0.25 mm-thick silica gel plates (Merck). For preparative separation, 2 mm-thick silica gel plates (Merck) were used. Silica gel (200 mesh) for column chromatography was a product of Wako Pure Chem. Ind., Ltd.

Gas chromatography (GLC) was performed with a Shimadzu Model GC-4APF or GC-5A equipped with a flame ionization detector. Infrared (IR), nuclear magnetic resonance (NMR) and mass spectrometry were used for physical measurment.

Major oxidation products from pristane were isolated and identified as follows; The culture broth (1 liter) from the shaking flasks inoculated with strain BPM 1613 was adjusted to pH 2 with sulfuric acid and extracted with diethyl ether. The ether solution was concentrated after dehydration over anhydrous sodium sulfate. The concentrate was dissolved in a small quantity of hexane and fractionated on a silica gel column. Comparison of IR and ^{1}H-NMR spectral data of major alcohol products suggested that pristanol (Figure 1) is produced from pristane by oxidation of the terminal methyl group. The final proof of the structure of pristanol was obtained on the basis of the demonstration that ketone chemically derived from product through the procedure was identical with an authentic specimen of phytone. One of the major products could be dissolved in an aqueous solution of 10% potassium carbonate with generation of soap bubbles. Reduction of the ester with lithium aluminium hydride as alcohol which gave the same feature of IR spectrum as that of pristanol. These results indicated that this product is pristanic acid (Figure 1). The yields of pristanol and pristanic acid were 7.0 g and 1.7 g/liter, respectively.

RESULTS

Identification of the oxidation products from pristane

Rhodococcus sp. BPM 1613 accumulated several oxidation products of pristane in the culture broth. Among them, major oxidation products were isolated by the use of silica gel column chromatography. On the basis of data of IR, ^{1}H-NMR, etc., these products were identified as pristanol (2, 6,10,14-tetramethyl-1-pentadecanol, Figure 1) produced from pristane by oxidation of a terminal methyl group and pristanic acid (2,6,10,14-tetramethylpentadecanoic acid, Figure 1) formed by further oxidation of pristanol.

In addition to the major products described above, minor oxidation products were also detected on TLC. A large scale cultivation with a jar fermenter was attempted in order to obtain sufficient amounts of these products for their characterization. Minor products were characterized pristyl pristanate (0.37 g/liter) and pristyl aldehyde (11 mg/liter) by IR spectrum, etc., respectively. Three acidic metabolites were also extracted with diethyl ether from the culture broth and purified by column chromatography and preparative thin-layer chromatography on silica gel. On the basis of instrumental analysis, these products were determined as 2,6 - dimethylnonanedioic acid , 2,6 - dimethylheptanedioic acid and 2 - methylpentanedioic acid (yield; 38 mg, 138 mg and 90 mg/liter), respectively.

Pristanol and pristanic acid of major oxidation product from pristane can be converted into phytone which is a synthetic intermediate for Vitamin E and K.

Production of pristanol and pristanic acid

The cultural conditions for the production of pristanol and pristanic acid for pristane by Rhodococcus sp. BPM 1613 were studied. Among them, it was found that sodium nitrate was the most suitable nitrogen source and the addition of yeast extract was effective (TABLE 1). Under the optimal conditions, yields of pristanol and pristanic acid were 7.0 g/liter and 1.7 g/liter, respectively.

Identification of the oxidation products from phytane, 1-pristene, norpristane and farnesane

Rhodococcus sp. BPM 1613 grows on isoprenoid hydrocarbons such as phytane, 1-pristene, norpristane and farnesane as the sole carbon source, resulting in accumulation of oxidation products in the culture broth. The oxidation products of phytane (product D and E), 1-pristene (product F),

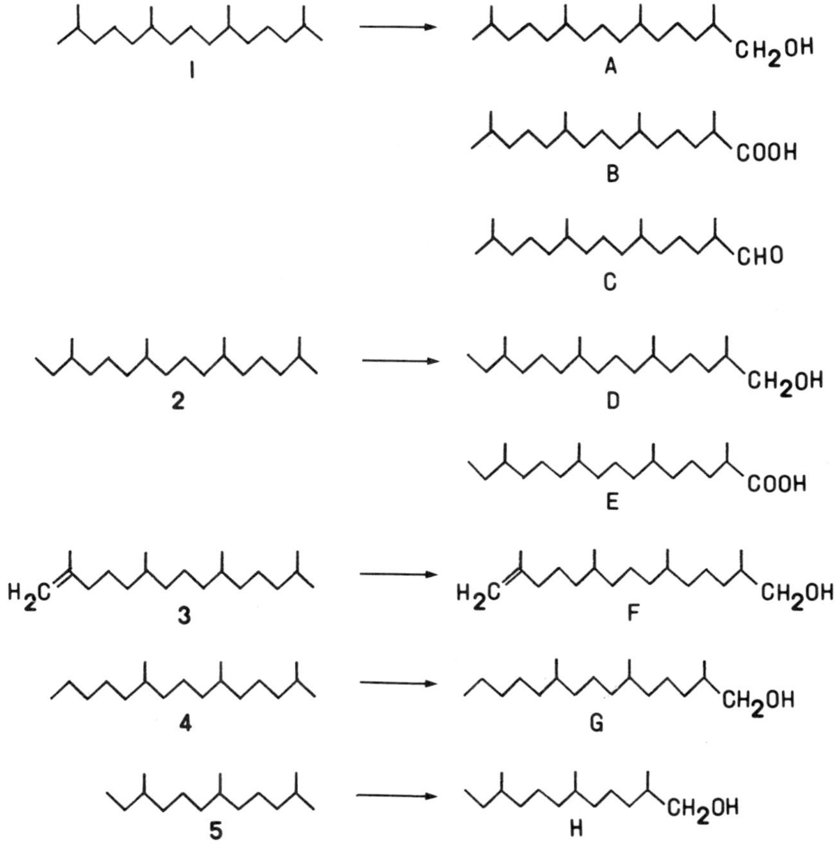

Figure 1. Transformations of isoprenoid hydrocarbons contained in shale oil by *Rhodococcus* sp. BPM 1613.
1, pristane; **2**, phytane; **3**, 1-pristene; **4**, norpristane; **5**, farnesane; **A**, pristanol; **B**, pristanic acid; **C**, pristyl aldehyde; **D**, 2,6,10,14-tetramethy-1-hexadecanol; **E**, 2,6,10,14-tetramethylhexadecanoic acid; **F**, 2,6,10,14-tetramethyl-14-pentadecen-1-ol; **G**, 2,6,10-trimethyl-1-pentadecanol; **H**, 2,6,10-trimethyl-1-dodecanol.

norpristane (product G) and farnesane (product H) in the respective culture broth were isolated. On the basis of the data of instrumental analyses such as IR, NMR and mass spectrometry, the oxidation products of phytane were identified as 2,6,10,14-tetramethyl-1-hexadecanol (Figure 1) which was formed on oxidation of the isopropyl terminus of phytane and 2,6,10,14-tetramethylhexadecanoic acid formed on further oxidation of product D. The product of 1-pristene was identified as 2,6,10,14-tetramethyl-14-pentadecen-1-ol, the product of norpristane as 2,6,10-trimethyl-

TABLE 1
Effect of yeast extract concentration on production of pristanol and pristanic acid

Conc. of Yeast extract (g/liter)	Final pH	Cells (g/liter)	Pristanol (g/liter)	Pristanic acid (g/liter)	Total (g/liter)
0.2	7.2	1.1	3.12	0.81	3.93
0.5	7.2	0.9	2.97	0.71	3.68
1.0	7.2	0.9	2.86	0.76	3.62
2.0	7.2	3.7	4.78	1.31	6.09
3.0	7.4	6.0	4.40	1.34	5.74

Cultivation was carried out at 30°C for 5 days.

1-pentadecanol, and that of farnesane as 2,6,10-trimethyl-1-dodecanol derived on oxidation of each substrates at the isopropyl terminus (Figure 1). The yields of products D, E, F, G and H were 3.1 g, 0.05 g, 0.32 g and 2.9 g/liter, respectively.

Using the same organism, microbial assimilation of the isoprenoid hydrocarbon mixture derived from dry-distilled Colorado shale oil (distillate between 280～305) from the U.S.A. and oxidation products from their isoprenoid hydrocarbon mixtures was studied. Overall , approximately 65% of the substrate had been utilized after 7 day's cultivation. The utilization percentages of the main components (phytane, pristane, 1-pristene and norpristane) were 44, 58, 58 and 83, respectively. Regarding monoalcohols among the oxidation products, pristanol from pristane was most abundant, followed by those from phytane and norpristane. One from norpristane was minimum. On the other hand, little monocarboxylic acids were determined; only pristanic acid from pristane was detected.

Identification of the oxidation product from squalene

From the culture broth of strain BPM 1613 on squalene, the product was extracted with diethyl ether, chromatographed on a silica gel column and purified by repeated recrystallizations from hexane. The chemical structure was defined by instrumental analyses such as IR, NMR and mass spectrometry as squalenedioic acid (2,6,10,15,19,23-hexamethyl-2,6,10,14,18,22-tetracosahexaene-1,24-dioic acid, Figure 2). Among nitrogen sources tested, urea was the most suitable one (yield of squalenedioic acid, 3.41 g/liter).

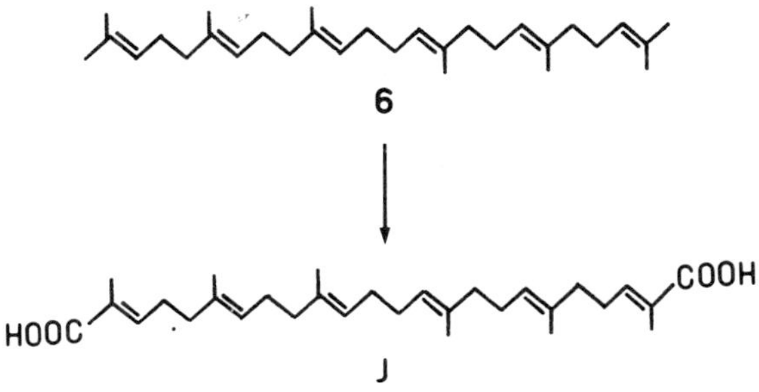

Figure 2. Transformation of squalene by Rhodococcus sp. BRM 1613. 6, squalene; J, squalenedioic acid.

The assimilability of pristane, squalene and squalane, and oxidation products from these compounds

The assimilability of pristane, squalene and squalane was tested by ten strains of standard cultures (Candida lipolytica, Can. rugosa, Can. tropicalis, Pichia farinosa, Arthrobacter simplex, Corynebacterium equi, Cor. fascians, Mycobacterium smegmatis, Nocardia asteroides and N. corallina) and four strains of isolated cultures (BPM 7713, BPI 3143 and NB 1802), including Rhodococcus sp. BPM 1613. Pristane was easily assimilated by one standard culture (N. corallina) and all strains of isolated cultures, and aqualene was utilized by the majority of tested strains, but squalane was utilized assimilated only by NB 1802. Oxidation products of pristane, squalene and squalane were also examined by six strains (Can. lipolytica, Cor. equi, M. smegmatis, Rhodococcus sp. BPM 1613, BPM 7713 and NB 1802) which accumulated oxidation products in culture broth. All strains produced monoalcohol (pristanol) and monocarboxylic acid (pristanic acid) from pristane, and four strains produced dicarboxylic acid (squalenedioic acid) from squalene. On the other hand, no strain tested accumulated oxidation producrs from squalane.

Identification of the oxidation products from 2,4-dimethyl-1,4-nonadiene and 2-methyl-2-nonanol

Alternating cooligomers (trimer) of isoprene or 1,3-butadiene and propylene or 1-butene, and their derivatives were submitted to microbial conversion by Rhodococcus sp. BPM 1613. n-Hexadecane-grown cell of BPM 1613 strain produced an oxidation products from 2,4-dimethyl-1,4-nonadiene (trimer of isoprene and propylene) and 2-methyl-2-nonanol (hydrogenated alcohol of trimer of 1,3-butadiene and propylene). The products were extracted with diethyl ether from the culture broth and isolated by column chromatography on silica gel. Their chemical structures were determined on the basis of IR, NMR and nass spectrometry. The oxidation product of 2,4-dimethyl-1,4-nonadiene was identified as 4,6-dimethyl-3,6-heptadienoic acid (Figure 3), and that of 2-methyl-2-nonanol as 8-hydroxy-8-methylnonanoic acid (Figure 3). These products are new compounds.

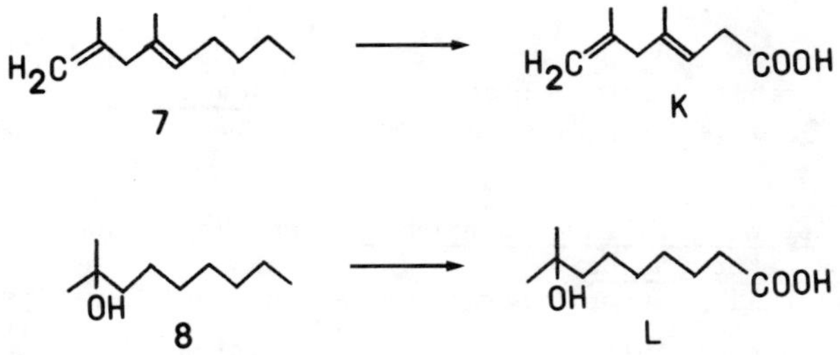

H_2C 7 → H_2C COOH K

OH 8 → OH COOH L

Figure 3. Transformations of synthetic oligomers by Rhodococcus sp. BPM 1613.
7, 2,6-dimethyl-1,4-nonadiene; **8**, 2-methyl-2-nonanol; **K**, 4,6-dimethyl-3,6-heptadienoic acid; **L**, 8-hydroxy-8-methylnonanoic acid.

DISCUSSIONS and CONCLUSIONS

Pristane oxidation by a new isolate Rhodococcus sp. BPM 1613 was examined in detail. Pristanol and pristanic acid as major monoterminal oxidation products, pristyl pristanate and pristyl aldehyde as minor monoterminal oxidation products, and three acidic intermediates in the metabolic pathway of pristane were identified. McKenna and Kallio [4] have reported the isolation and identification of 4,6,10-trimethyltridecanoic and 2-methylpentanedioic acids as the metabolites of pristane by Corynebacterium sp. Pirnik et al. [12] have reported three monocarboxylic acids and six dicarboxylic acids as the metabolites of pristane by Brevi bacterium erythrogenes. Rhodococcus sp. BPM 1613 as shown by this investigation metabolized pristane through the following two pathways.

(1) pristane→pristanic acid→β- oxidation, and

(2) pristanic acid→pristanedioic acid (ω- oxidation)→β- oxidation.

Pristanol as well as pristanic acid can be converted in to phytol which is useful for synthesis of vitamin E and K. The cultural conditions for production of pristanol and pristanic acid from pristane by Rhodococcus sp. BPM 1613 were studied. Under the optimal conditions, yields of pristanol and pristanic acid were 7.0 g/liter and 1.7 g/liter, respectively.

The same organism was used for metabolic conversion of phytane, 1-pristene, norpristane and farnesane. The oxidation products of these four substrates were identified as 2,6,10,14-tetramethyl-1-hexadecanol and 2,6,10,14-tetramethylhexadecanoic acid, 2,4,10,14-tetramethyl-14-pentadecen-1-ol, 2,6,10-trimethyl-1-pentadecanol, and 2,6,10-trimethyl-1-dodecanol, respectively. Among these oxidation products, the former three alcohols are new compounds. Microbial oxidation of phytane, norpristane and related compounds has been reported only by Cox _et al_. [4] They had _Mycobacterium fortuitum_ grown on 2,6,10,14-tetramethylheptadecane, phytane, norpristane and 2,6,10-trimethyltetradecane as the sole carbon source, isolated and identified the oxidation products as monocarboxylic acids derived on monoterninal oxidation and as further oxidized monocarboxylic acids. In the case of BPM 1613, pristane having isopropyl groups as both termini was found to be oxidized by BPM 1613 through two metabolic pathways. Phytane having isopropyl and ethyl termini is oxidized at isopropyl terminus, resulting in a monoalcohol and monocarboxylic acid, and farnesane with a terminal structure similar to that of phytane is oxidized at the isopropyl terminus to a monoalcohol. The results agree well with those obtained in the preliminary experiments in which the same organism was shown to be able to utilize 2,7-dimethyloctane with two isopropyl groups as the termini but unable to utilize 3,6-dimethyloctane with ethyl groups as two termini. 1-Pristene having isopropyl and isopropenyl terminus like phytane is oxidized at the isopropyl terminus to a monoalcohol. In the case of norpristane with isopropyl and _n_-pentyl termini, only a small quantity of the monoalcohol was formed on terminal isopropyl oxidation and accumulated without accumulation of the oxidation product at the _n_-pentyl terminus. The poor accumulation of oxidation product was not due to the inability of the organism to use norpristane, but attributable to oxidative degradation starting from the relatively lengthy _n_-pentyl cahin.

Using the same organism, microbial utilization of the isoprenoid hydrocarbon mixture which derived from shale oil (distillate between 280 ~ 305°C) and containing phytane, pristane, 1-pristene and norpristane, and their oxidation products were studied. About 65% of substrate was utilized during 7 days and the reduction of norpristane being especially remarkable. Regarding monoalcohols among the oxidation products, pristanol from pristane was most abundant, followed by those from phytane and 1-pristene. One from norpristane was minimum.

Microbial oxidation of squalene has been reported only by Yamada _et al_. [7] They isolated and identified as oxidative degradation products _trans_-geranylacetone and four monocarboxylic acid metabolites by _Arthrobacter_ sp. _Rhodococcus_ sp. BPM 1613 only accumulated diterminal oxidation product squalenedioic acid and other oxidation products had been not recognized. There has been supported that product squalenedioic acid was metabolized via β-oxidation.

The assimilability of pristane, squalene and squalane was tested by ten strains of standard cultures (_Candida lipolytica_, _Can. rugosa_, _Can. tropicalis_, _Pichia farinosa_, _Arthrobacter simplex_, _Corynebacterium equi_, _Cor. fascians_, _Mycobacterium smegmatis_, _Nocardia asteroides_ and _N. corallina_) and four strains of isolated cultures (BPM 7713, BPI 3143, NB 1802 and _Rhodococcus_ sp. 1613). Pristane was easily assimilated by one standard culture (_N. corallina_) and all strains of isolated cultures, and squalene was utilized by the majority of tested strains, but squalane was assimilated only by NB 1802. Oxidation products of pristane, squalene and squalane were also examined by six strains (_Can. lipolytica_, _Cor. equi_, _M. smegmatis_, _Rhodococcus_ sp. BPM 1613, BPM 7713 and NB 1802). All strains produced monoalcohol and monocarboxylic acid from pristane, and four strains (_Can. lipolytica_, _Cor. equi_, _Rhodococcus_ sp. BPM 1613 and NB 1802) produced dicarboxylic acid from squalene. On the other hand, no strain tested accumulated oxidation products from squalane.

_n_-Hexadecane-grown cells of BPM 1613 strain accumulated the oxidation products from 2,4-dimethyl-1,4-nonadiene and 2-methyl-2-nonanol. The product of 2,4-dimethyl-1,4-nonadiene was identified as a new compound 4,6-dimethyl-3,6-heptadienoic acid and the ethyl ether of this compound has a flavor like pineapple. The product of 2-methyl-2-nonanol was determined as a new compound, term 8-hydroxy-8-methylnonanoic acid and this compound is similar compound of growth inhibitor, myrmicacin (3-hydroxydecanoic acid), and which is able to expect to have any physiological effect.

Acknowledgments

The author wishes to thank Dr. A. Sato, Fermentation Research Institute, for his extremely valuable guidance and instruction throughout this work. The author also expresses his thanks to Dr. T. Higashihara, Fermentation Research Institute, Dr. T. Iida, the Institute of Physical and Chemical Research, and Dr. M. Taniguchi, Marzen Petrochemical Co., as co-researcher for their valuable suggestions and help.

The author thanks Dr. R. Kurane for valuable suggestions.

REFERENCES

1. Iida, T., Kitatsuji, E. and Hayashi, S., Studies of the components in NTU shale oil. III: Identification of acyclic isoprenoid hydrocarbons. Yakugaku Zasshi, 1976, 96, 796-800.

2. Davis, J.B., Petroleum Microbiology, ed Elsevier Publishing Co., New York, 1959, p. 340.

3. Jones, d.f., Microbiological oxidation of long-chain aliphatic compounds. Part II: Branched-chain alkanes. J. Chem. Soc. (C), 1968 2809-15.

4. McKenna, E.J. and Kallio, R.E., Microbial metabolism of the isoprenoid alkane pristane. Proc. Natl. Acad. Sci., 1971, 68, 1552-4.

5. Cox, R.E., Maxwell, J.R. and Mayers, R.N., Monocarboxylic acids from oxidation of acyclic isoprenoid alkanes by Mycobacterium fortuitum. Lipids, 1976, 11, 72-6.

6. The Merck Index, ed by M. Windholz and S. Budavari, 9th ed., Merck & Co., 1976, pp. 1133-4.

7. Yamada, Y., Motoi, H., Kinoshita, S., Takada, N. and Okada, H., Oxidative degradation of squalene by Arthrobacter species. Appl. Microbiol., 1975, 29, 400-4.

8. Nakajima, K., Sato, A., Misono, T., Iida, T. and Nagayasu, K., Microbial oxidation of isoprenoid hydrocarbons. Part I: Microbial oxidation of the isoprenoid alkane pristane. Agric. Biol. Chem., 1974, 38, 1859-65.

9. Nakajima, K. and Sato, A., Microbial oxidation of isoprenoid hydrocarbons. Part IV: Microbial metabolism of isoprenoid alkane pristane. Nippon Nogeikagaku Kaishi, 1983, 57, 299-305.

10. Nakajima, K., Sato, A., Takahara, Y. and Iida, T., Microbial oxidation of isoprenoid hydrocarbons. Part V: Microbial oxidation of isoprenoid alkanes, phytane, norpristane and farnesane. Agric. Biol. Chem., 1985, 49, 1993-2002.

11. Nakajima, K., Sato, A., Takahara, Y. and Iida, T., Microbial oxidation of isoprenoid hydrocarbons. Part VI: Microbial oxidation of isoprenoid hydrocarbon, 1-pristene. Agric. Biol. Chem., 1985, 49, 2763-5.

12. Pirnik, M.P., Atlas, R.M. and Bartha, R., Hydrocarbon metabolism by *Brevibacterium erythrogenes*: Normal and branched alkanes. J. Bacteriol., 1974, 119, 868-78.

INTERNATIONAL CONFERENCE ON BIOREACTORS AND BIOTRANSFORMATIONS
GLENEAGLES, SCOTLAND, UK: 9-12 NOVEMBER 1987

Paper F3

Well presented

THE APPLICATION OF ORGANIC SOLVENTS FOR THE BIOCONVERSION OF BENZENE TO CIS-BENZENEGLYCOL

W.J.J. van den Tweel[1,2], E.H. Marsman[2], M.J.A.W. Vorage[2],
J. Tramper[2] and J.A.M. de Bont[1]
Agricultural University Wageningen,
[1]Department of Microbiology and
[2]Department of Food Science, Food- and Bioengineering Group,
De Dreijen 12, 6703 BC Wageningen, The Netherlands

ABSTRACT

Benzene was oxidized to cis-benzeneglycol (CBG) by a mutant of a Pseudomonas sp. (strain Mt 92). A prolonged continuous production of CBG from benzene was achieved by supplying benzene to Mt 92 cells growing on succinate under nitrogen-limited conditions in a chemostat. During such continuous production experiments the benzene concentration should kept low to minimize the toxic effect of benzene. Incubation experiments with Mt 92 showed that n-hexadecane is a suitable solvent to circumvent benzene toxicity during the bioformation of CBG from benzene. Moreover, the addition of n-hexadecane did not significantly effect the rate of CBG formation. Since the bioconversion could easily be performed in the presence of n-hexadecane the feasibility of a recently described liquid-impelled loop reactor was investigated. Preliminary experiments showed that such bioreactor can indeed be used for the conversion of benzene to CBG.

INTRODUCTION

CBG and other cis-dihydrodiols may be useful building blocks for the production of synthetic polymers [1] and various pharmaceuticals [2]. Already in 1970 Gibson and coworkers [3] isolated a mutant of Pseudomonas putida which was able to convert benzene to CBG (Figure 1). Using this mutant ICI was able to produce CBG in kilogram quantities for the produc-

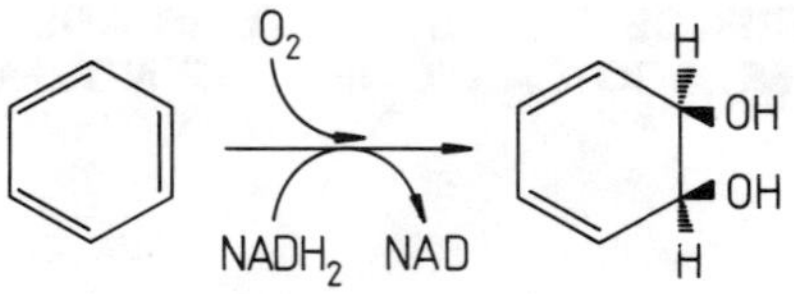

O_2 ... $NADH_2$ NAD ... H OH OH H

Figure 1. Bioconversion of benzene to CBG

tion of polymerization monomers [4]. Moreover, the broad substrate specificity of the pertinent dioxygenase also enabled the production of a range of substituted cis-dihydrodiols [5]. In spite of the huge interest in CBG only minor information is available on bioproduction processes for CBG [6,7]. Both biological and process engineering aspects that limit the exploitation of the CBG production process apparently need more detailed investigations. In the present work we have concentrated on the aspect of benzene toxicity towards the biocatalyst and it was attempted to overcome this toxicity by using n-hexadecane as a second liquid phase. The influence of many water-immiscible solvents on retention of activity of immobilized Mycobacterium cells was determined by Brink and Tramper [8] and it was found that retention of activity is usually favoured by low solvent polarity in combination with a high molecular weight. LogP, which is defined as the logarithm of the partition coefficient of a given compound in a standard octanol-water two phase system is a very useful parameter to describe a correlation between biocatalytic activity and solvent properties [9]. Solvents with a logP smaller than 2 were least suitable for biocatalysis while solvents having a logP above 4 were readily applicable.

This paper shows that n-hexadecane, which has a logP value of 8.8 is a suitable solvent to circumvent the inhibitory effect of benzene on the biocatalyst. Moreover, preliminary experiments are presented on the use of a liquid-impelled loop reactor, a new type of density-difference-mixed bioreactor [10], for the bioformation of CBG from benzene.

MATERIALS AND METHODS

Organism.

The isolation and characterization of Mt 92 which is able to convert

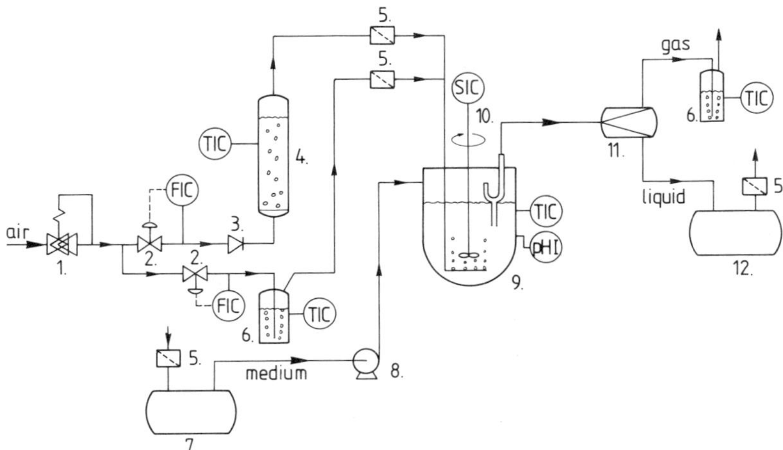

Figure 2. Schematic representation of the chemostat used for continuous CBG production. Pressure reducer, 1; mass flow controller, 2; one-way valve, 3; benzene column, 4; air filter, 5; gas washer, 6; medium tank, 7; pump, 8; fermentor vessel, 9; stirrer, 10; separator, 11; effluent tank, 12.

benzene stoichiometrically to CBG will be described elswhere [7].

Continuous bioproduction of CBG.

Continuous formation of CBG was achieved by Mt 92 cells growing on succinic acid under nitrogen-limited conditions in a chemostat (working volume 0.93 dm3, 30°C, pH 7.0, dilution rate 0.085 h^{-1}; Figure 2) [7]. Air at a rate of 60 cm^3 per minute was supplied to the fermentor through an inlet beneath the stirrer. Strain Mt 92 was routinely grown in a mineral salts medium containing in 1 dm^3 of deionized water: K_2HPO_4, 1.55 g; NaH_2PO_4, 0.85 g; NH_4Cl, 0.15 g; Na_2SO_4, 0.125 g; $MgCl_2 \cdot 6H_2O$, 0.075 g; yeast extract, 0.2 g and 0.2 ml of a trace element solution [11]. Succinic acid was added to this mineral salts medium at 2.5 g dm^{-3}. At steady state a part of the air was passed through a column containing benzene [7]. As a result of the appearance of benzene in the culture medium the benzene dioxygenase was induced and CBG started to accumulate.

Incubation experiments with induced Mt 92 cells.

The effluent culture medium of the continuous bioproduction experiment was collected in a tank which was kept at 0°C. After one day these induced cells were harvested by centrifugation (16,000 x g for 5 min at

4°C), washed with potassium phosphate buffer pH 7.0 (50 mM), resuspended in the same buffer and used for the incubation experiments. Incubations were performed at 30°C in 315 cm^3 serum bottles containing 19 cm^3 of the following nitrogen-free mineral salts medium: K_2HPO_4, 1.55 g dm^{-3}; NaH_2PO_4, 0.85 g dm^{-3}; Na_2SO_4, 0.1 g dm^{-3}; $MgCl_2 \cdot 6H_2O$, 0.075 g dm^{-3} and 0.2 cm^3 dm^{-3} of the forementioned trace element solution. Unless stated otherwise 0.5 g dm^{-3} succinic acid was added to the medium for cofactor regeneration. After addition of the specified amount of benzene the incubation was initiated by the addition of 1 cm^3 induced Mt 92 cells (total protein 14 mg). In some experiments 20 cm^3 n-hexadecane was added to the serum bottles prior to benzene addition. Samples, taken at various times from the water phase, were analysed for the concentration of CBG by HPLC.

Production of CBG in the liquid-impelled loop reactor.

Figure 6 gives a schematic representation of the employed liquid-impelled loop reactor. The height of the reactor was 60 cm, the internal diameter of the downcomer and riser was 25 and 40 mm, respectively. The total volume of the reactor was 1.25 dm^3 (1.0 dm^3 water and 0.25 dm^3 n-hexadecane). The design and principles of mixing of the liquid-impelled loop reactor are the same as applied in the airlift loop reactor. In the liquid-impelled loop reactor the density-difference mixing was created by injection of n-hexadecane (19 dm^3 h^{-1}) which has a density (0.7733 g cm^{-3} at 20°C) smaller than that of water. To supply oxygen to Mt 92 cells the recirculating n-hexadecane was aerated in an aeration flask containing 350 cm^3 n-hexadecane. The temperature of the bioreactor was kept constant at 30°C by a thermocontrolled recycling waterbath. For the CBG production experiment the forementioned nitrogen-free mineral salts medium was used to which 0.5 g dm^{-3} succinic acid was added. After adding 5 cm^3 benzene to the hexadecane phase, the experiment was started by adding induced Mt 92 cells (total protein 2.5 g). Samples were taken from the water phase and analysed for the amount of CBG.

Analyses.

Protein concentrations were measured by the method of Lowry et al. [12]. Benzene and CBG concentrations were determined by reversed phase HPLC [7].

Materials.

Benzene was purchased from Merck-Schuchard, Hohenbrunn, FRG.

n-Hexadecane was a product from J.T. Baker Chemicals, Deventer, The Netherlands. All other chemicals were commercially available analytical grade products and were used without further purification.

RESULTS AND DISCUSSION

Continuous bioproduction of CBG.

In order to investigate in more detail some intriguing aspects of the bioconversion of benzene to CBG (Figure 1), e.g. enzyme induction, cofactor regeneration, stability of the mutant, toxicity of substrate and/or product, we have studied the continuous production of CBG by Mt 92 cells growing on succinic acid in a chemostat [7]. Since production of CBG under carbon-limited conditions was only possible for a limited period as a result of revertation of the mutant to the wild type, the bioreactor (Figure 2) was operated under nitrogen-limited conditions as described in Materials and Methods. At steady state 2.8 mmol benzene h-1 was supplied to the culture (Figure 3). After a period of enzyme induction CBG started to accumulate in the medium reaching a maximum concentration of 9.6 mM. Under these conditions a very stable bioproduction process was obtained, even after 10 days the CBG concentration was maintained maximal (Figure

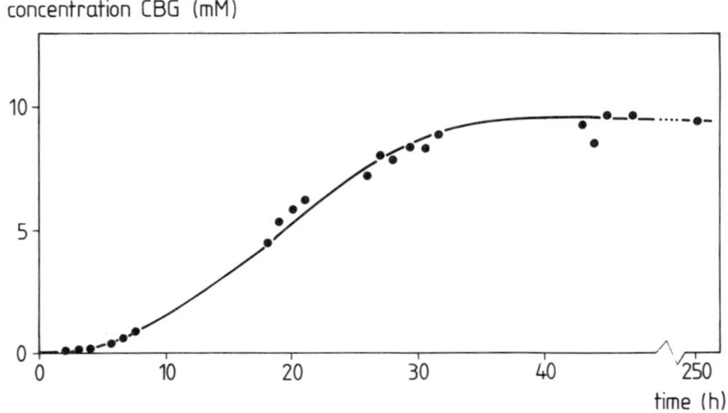

Figure 3. Continuous CBG (•) production by Mt 92 growing under nitrogen-limited conditions on succinic acid in the chemostat. At time zero (protein content 210 mg dm-3) 2.8 mmol benzene h-1 was supplied via the air stream to the fermentor.

3). During such continuous CBG production processes the benzene concentration was a very crucial parameter. Concentrations above about 2.5 mM had an inhibitory effect on both growth and induction of the benzene dioxygenase [7]. Consequently, the benzene concentrations should be carefully controlled to minimize the toxic effect of benzene.

During the continuous processes only about one mol of CBG was produced per mol of succinic acid utilized. Effluent cells which still have the ability to convert benzene to CBG however may be reused, either batchwise or continuously, to give a more economic process. Since benzene will also inhibit such bioconversion its effect on induced cells was studied in more detail.

The effect of benzene on the bioformation of CBG by induced Mt 92 cells.

To study the effect of benzene on the activity of induced Mt 92 cells incubation experiments were performed in the presence of increasing amounts of benzene (Figure 4). In order to prevent growth during these experiments the incubations were done in a nitrogen-free mineral salts medium. Under the specified conditions the addition of 10 mm^3 benzene resulted in a benzene concentration in the water phase of 0.95 mM. The best results for the bioformation were obtained using 10 and 25 mm^3 benzene (initial benzene concentrations 0.95 and 2.38 mM, respectively). After an initial CBG formation activity of about 19 nmol min^{-1} (mg protein)$^{-1}$ the rate significantly decreased, probably as a result of the depletion of reduction equivalents. In the presence of 100 mm^3 benzene

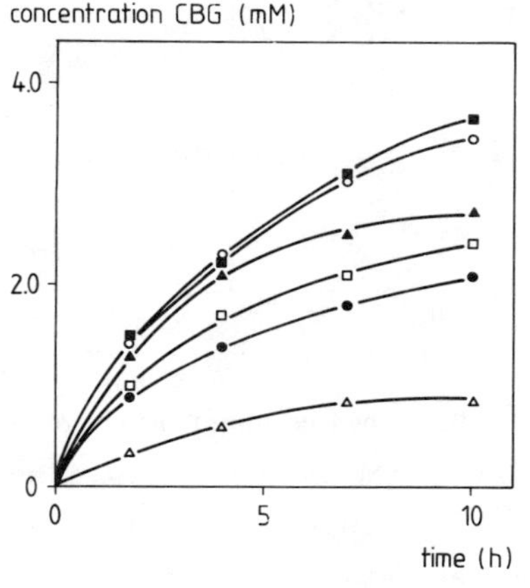

Figure 4. Effect of the amount of benzene on the formation of CBG by Mt 92. Incubations were performed as described under Materials and Methods without any hexadecane added.

■, 10; ○, 25; ▲, 50; □, 75; ●, 100; and △, 150 mm^3 benzene.

(initial concentration 9.5 mM) still a fairly good CBG production was observed. This is in contrast to growth of the wild type which was inhibited above benzene concentrations of 2.5 mM [7]. Evidently, CBG formation by induced cells is less susceptible to high benzene concentrations than is growth. To obtain an optimal bioconversion process with nongrowing cells the benzene concentration in the water phase on the one hand should be minimal wheras on the other hand a permanent supply of benzene to the cells is also required.

The use of n-hexadecane for the bioconversion of benzene to CBG by induced Mt 92 cells.

Recently, studies by Rezessy-Szabó et al. [13] have shown the potential of organic solvents with high logP values (e.g. n-hexadecane and dibutylphtalate) to circumvent benzene toxicity during growth of the wild type of Mt 92 (Pseudomonas 50). In order to investigate the potential of organic solvents to reduce the inhibitory effect of benzene on the bioformation of CBG from benzene incubation experiments were done with induced Mt 92 cells in the presence of n-hexadecane (50% v/v). In contrast to foregoing experiments high amounts of benzene in the presence of n-hexadecane did not influence the bioformation of CBG (Figure 5). Moreover, the CBG formation rate for all additions tested was about the same and almost equalled the rate which was obtained in the absence of n-hexadecane after adding 10 mm3 benzene (Figure 4). HPLC analyses showed

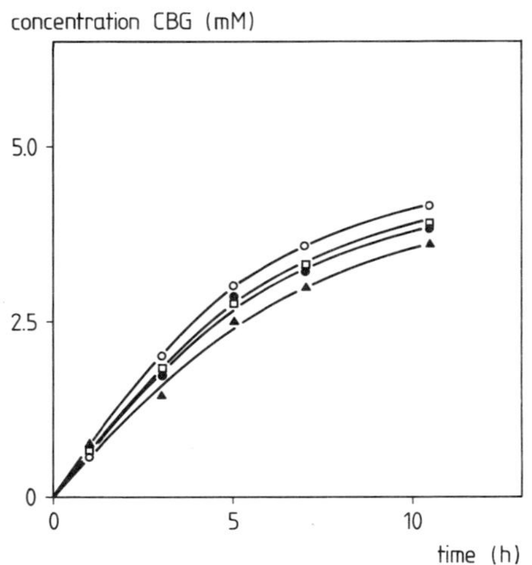

Figure 5. The use of hexadecane to decrease the toxic effect of increasing amounts of benzene on the bioconversion of benzene to CBG by Mt 92. Incubations were performed as described under Materials and Methods in the presence of hexadecane. ● , 50; □ , 100; ▲, 200; and ○ , 400 mm3 benzene.

that after addition of n-hexadecane no detectable concentrations of benzene were present in the water phase. Evidently, by reducing the benzene concentration in the water phase the inhibitory effect of benzene was circumvented. The results show furthermore that the activity of the biocatalyst was limiting the CBG formation rate rather than benzene mass transfer from the organic solvent to the water phase. Similar results have also been obtained by Furuhashi et al. [14], in that the addition of n-hexadecane reduced both the toxicity of the substrate (styrene) and the product (styreneoxide) which resulted in an enhanced production of styreneoxide from styrene by *Nocardia corralina* cells.

Bioformation of CBG in the liquid-impelled loop reactor.

Recently Tramper et al. [10] have introduced a new type of bioreactor: the liquid-impelled loop reactor. In its simplest configuration this reactor contains two phases, a water phase and a water-immiscible organic solvent phase. Density-difference mixing can be obtained by the injection of an organic solvent with a density smaller than water at the bottom of the riser. Since the use of such bioreactor might be advantageous for the bioproduction of CBG (reduction of the inhibitory effect of benzene, good mass transfer characteristics, and the possibility to enhance the supply of oxygen since oxygen is better soluble in many organic solvents than in

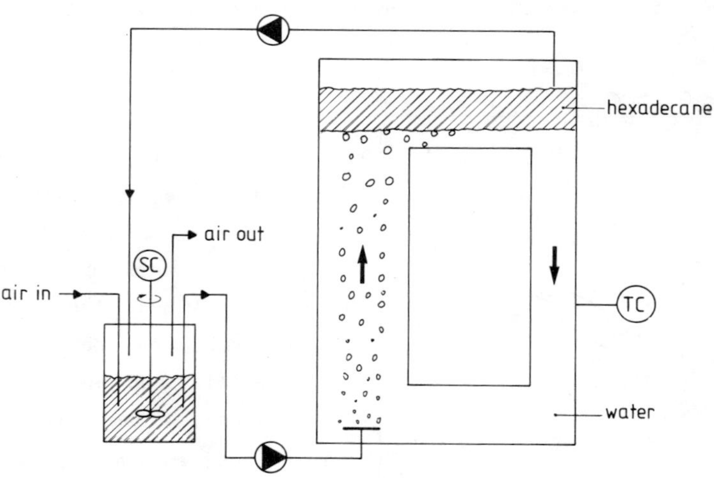

Figure 6. Schematic representation of the liquid-impelled loop reactor.

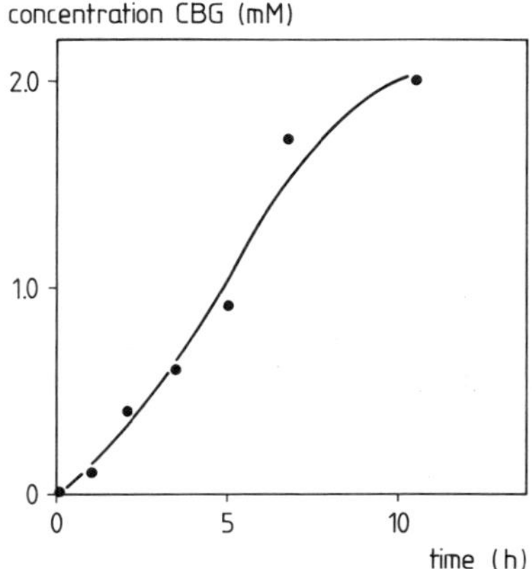

Figure 7. Formation of CBG (•) by induced Mt 92 cells in the described liquid-impelled loop reactor.

water) some preliminary experiments were performed in a n-hexadecane-impelled loop reactor. Initial experiments were performed in a bioreactor with an external loop since phase separation was easier with such a configuration [10]. Oxygen was indirectly supplied to this bioreactor by blowing air through the recirculating n-hexadecane solution as schematically shown in Figure 6. Moreover, the aeration flask can also be used to add other components to the bioreactor. Under the specified conditions the density difference between water and n-hexadecane and the recirculating flow rate of n-hexadecane were high enough to achieve a good circulation of the water phase and to allow a good phase separation. After the addition of benzene to the hexadecane phase, the experiment was started by adding induced Mt 92 cells (total protein 2.5 g) to the water phase. Immediately CBG started to accumulate in the water phase reaching a concentration of 2 mM after about 10 h (Figure 7). Currently we are investigating the use of the liquid-impelled loop reactor more fundamentally and simultaneously we are optimizing the bioformation of CBG in this new type of bioreactor.

CONCLUSIONS

Continuous bioformation of CBG from benzene by Mt 92 cells growing on succinate under nitrogen-limited conditions in a chemostat can easily be achieved. However, a low benzene concentration is a prerequisite for good performance of the bioconversion process. In order to make the continuous process economically more attractive it is also necessary to reuse the produced cells. Again a low benzene concentration is a requirement to be met for an optimal bioformation of CBG under such non-growth conditions. The addition of an organic solvent with a high logP value such as n-hexadecane can be very useful to circumvent the toxic effect of benzene on the production of CBG by either growing or non-growing cells. CBG itself also might have an inhibitory effect on the bioformation [6,15], however, because of its good solubility in water the addition of an water-immiscible organic solvent probably will not reduce this product-inhibition.

Preliminary experiments with a liquid-impelled loop reactor have shown that such type of bioreactor might be useful for the bioformation of CBG. Currently we are studying this possibility in more detail.

REFERENCES

1 Ballard, D.G.H., Courtis, A., Shirley, I.M. and Taylor, S.C., A biotech route to polyphenylene. J. Chem. Soc., Chem. Commun., 1983, pp. 954-955.

2 Ley, S.V., Sternfield, F. and Taylor, S., Microbial oxidation in synthesis: a six step preparation of (±)-pinitol from benzene. Tetrahedron Lett., 1987, 28, 225-226.

3 Gibson, D.T., Cardini, G.E., Maseles, F.C. and Kallio, R.E., Incorporation of oxygen-18 into benzene by Pseudomonas putida. Biochemistry, 1970, 9, 1631-1635.

4 Taylor, S., Enzymic synthesis of 5,6-dihydroxycyclohexa-1,3-diene. In Enzymes in Organic Synthesis. Ciba Foundation Symposium III, Pitman, London, 1985, pp. 71-75.

5 Taylor, S., Biochemical process. European Patent Application 0,076,606, 1983.

6 Jenkins, R.O., Stephens, G.M. and Dalton, H., Production of toluene cis-glycol by Pseudomonas putida in glucose fed-batch culture. Biotechnol. Bioeng., 1987, 29, 873-883.

7 van den Tweel, W.J.J., Vorage, M.J.A.W., Marsman, E.H., Koppejan, J., Tramper, J. and de Bont, J.A.M., Continuous production of cis-benzeneglycol from benzene by a mutant of a benzene-degrading Pseudomonas sp. Submitted for publication.

8 Brink, L.E.S. and Tramper, J., Optimization of organic solvent in multiphase biocatalysis. Biotechnol. Bioeng., 1985, 27, 1258-1269.

9 Laane, C., Boeren, S. and Vos, K., On optimizing organic solvents in multi-liquid-phase biocatalysis. Trends Biotechnol., 1985, 3, 251-255.

10 Tramper, J., Wolters, I. and Verlaan, P., The liquid-impelled loop reactor: a new type of density-difference mixed bioreactor. In Biocatalysis in Organic Media, Studies in Organic Chemistry 29, Elsevier, Amsterdam, 1987, pp. 311-316.

11 Vishniac, W. and Santer, M., The thiobacilli. Bacteriol. Rev., 1957, 21, 195-213.

12 Lowry, O.H., Rosebrough, N.J., Farr, A.L. and Randall, R.J., Protein measurement with Folin phenol reagent. J. Biol. Chem., 1951, 193, 265-275.

13 Rezessy-Szabó, J.M., Huijberts, G.N.M. and de Bont, J.A.M., Potential of organic solvents in cultivating microorganisms on toxic water-insoluble compounds. In Biocatalysis in Organic Media, Studies in Organic Chemistry 29, Elsevier, Amsterdam, 1987, pp. 295-302.

14 Furuhashi, K., Shintani, M. and Takagi, M., Effects of solvents on the production of epoxides by Nocardia corallina B-276. Appl. Microbiol. Biotechnol., 1986, 23, 218-223.

15 Jenkins, R.O. and Dalton, H., The use of indole as a spectrophotometric assay substrate for toluene dioxygenase. FEMS Microbiol. Lett., 1985, 30, 227-231.

INTERNATIONAL CONFERENCE ON BIOREACTORS AND BIOTRANSFORMATIONS
GLENEAGLES, SCOTLAND, UK: 9-12 NOVEMBER 1987

Paper G1

ACHIEVING SPECIFICITY WITH ALKANE MONOOXYGENASES

D. J. Leak
Centre for Biotechnology, Imperial College of Science and Technology, London SW7 2AZ

ABSTRACT

Regio- and stereospecific biological oxygen insertion reactions offer significant opportunities for the production of chiral alcohols and epoxides. However in many cases monooxygenases, particularly those acting on aliphatic/alicyclic hydrocarbons prove to have fairly low specificity. The general determinants of specificity in monooxygenases are discussed in outline and the specific example of methane monooxygenase analyzed in detail. Using alicyclic compounds as 'probes' of the active site geometry it has been possible to demonstrate highly specific biotransformations which were not predictable from consideration of the enzyme's natural substrate. It is proposed that this approach may be of general utility.

INTRODUCTION

The controlled regio- and stereospecific oxidation of alkanes and alkenes to produce alcohols or epoxides remains a challenge to chemists, although considerable progress and some elegant solutions have been attained in recent years [1-3]. However the selectivity demonstrable with some biological systems, particularly in the area of steroid bioconversions, is now legend [4]. Even here though it is evident that absolute specificity is the exception rather than the rule. Certainly for those monooxygenases more extensively characterised enzymologically (mainly those metabolising simpler substrates) low substrate specificity and multiple products have often been encountered [5-7], indicating that as a class monooxygenases do not display the level of specificity associated with many enzyme systems.

While developments in molecular biology and biochemical engineering allow us to entertain the prospects of high activity extended biocatalysis, multiple product formation will always reduce the attractiveness of this approach especially where the substrate is a relatively valuable synthetic intermediate. It is therefore informative to ask some fundamental questions about the determinants of specificity in monooxygenases in general and to investigate which of these predominate in any one enzyme or in any single biotransformation. In this way it may to possible to identify a series of biocatalysts in which the ground rules for specificity are known, from which a rational choice may be made for any desired bioconversion. To this end the bioconversion of cyclohexane and cyclohexene derivatives as active site 'probes' for the soluble [8] methane monooxygenase (MMO) from 3 methanotrophic bacteria has been studied, and more recently work has started on the selectivity of epoxidation versus hydroxylation in alkene monooxygenases. This paper will review current progress: experimental details and a more extensive presentation of results have been published elsewhere [9].

DETERMINANTS OF SPECIFICITY IN MONOOXYGENASES

Specificity in monooxygenases depends on the balance of two criteria (i) enzymic control and (ii) chemical control. The products from a particular bioconversion depend on the predominance of (i) or (ii). A general mechanism for hydrocarbon monooxygenases involves reductive activation of dioxygen followed by attack of an 'active oxygen' species on the substrate. In free solution these activated oxygen species would attack unactivated hydrocarbons to give a mixture of products [1] dependent on the nature of the active oxygen (superoxide, peroxide, hydroxyl radical or hydroxyl anion), reactivity of various carbon atoms in the substrate (primary, secondary, tertiary, saturated/unsaturated or allylic) i.e. their ability to form or stabilize any reaction intermediate, and steric effects on substrate interaction. Similarly the products from a monooxygenase in which the substrates are free to interact in any orientation (i.e. not restrained by the protein) would be governed by the same total chemical control. At the opposite extreme an enzyme in which the substrate is firmly bound in a single orientation which juxtaposes a unique carbon atom with the activated oxygen (also spatially fixed) exerts complete enzymic control. This would yield a single product, potentially hydroxylated at one of the less reactive positions and represents the ultimate in specificity desirable for a commercial bioconversion.

The nature of the active oxygen should not be ignored in these considerations. The majority of monooxygenases active on intrinsically unreactive hydrocarbons (e.g. alkanes, alicyclics) are of the cytochrome P-450 [6] or non-heme iron [5,7] types. These produce highly reactive 'active oxygen' intermediates capable of oxidising the least reactive of C-H bonds. However the high reactivity of the oxygen means that, to all intents and purposes, the oxygen is not available in free solution. Instead the mechanism must involve substrate interaction with a metal bound oxygen species and thus, even here some element of enzymic control will be evident. An additional level of selectivity is also evident with the flavin monooxygenases [10] (i.e. those with a flavin prosthetic group) which only oxidise reactive substrates e.g. amines and derivatized (amino or hydroxy) aromatics. The 'active oxygen' generated as a flavin adduct is insufficiently reactive to hydroxylate aliphatic hydrocarbons. In this respect it is interesting to note that squalene epoxidase is thought to be a flavoprotein [11].

METHANE MONOOXYGENASE

The soluble MMO from *Methylococcus capsulatus* (Bath) is comprised of 3 components. Component C is the NADH: acceptor reductase [12] which transfers electrons to component A the oxygenase which interacts with the hydrocarbon substrate [13]. Component B is involved in coupling substrate oxidation to electron transfer [14]. Component A is a non-heme iron protein containing 2mol/mol iron. Although the soluble monooxygenases from the other 2 methanotrophs *Methylosinus trichosporium* OB3b and *Methylosinus sporium* 5 have not been as extensively characterized there are some evident similarities [15].

These enzymes display a remarkably broad substrate range considering their natural substrates [16]. In many cases multiple products have been observed, with aliphatic hydrocarbons yielding primary and secondary alcohols, olefins yielding epoxides and alcohols, and aromatic substrates yielding ring and side chain hydroxylated products. Although some indications of enzymic control of specificity can be gleaned from previous work no systematic study has previously been performed to address this question. Two opposing views for the mechanism of hydroxylation of aliphatic carbons have also been presented, one suggesting a concerted insertion reaction [17] while the second proposed a hydride abstraction mechanism [18].

Products from Alicyclic Compounds

The use of a series of cyclohexane and cyclohexene derivatives as substrates enabled differentiation of regioselectivity, stereoselectivity and steric constraints on hydroxylation because of the semi-rigid nature of the substrates. As broadly similar findings were obtained with all three monooxygenases those pertinent to MMO from *M. capsulatus* will be discussed primarily, with references to product differences where appropriate.

Alkylcyclohexanes

The major product from methylcyclohexane was *cis*-3-methylcyclohexanol with 4-*cis* and 4-*trans* isomers produced in lower but roughly equal amounts. The 3-*trans* isomer was produced in the lowest yield of this series with small amounts of cyclohexylmethanol evident in all cases. In a series of monosubstituted cyclohexanes from methyl to n-propyl and isopropylcyclohexane a trend from predominantly 3-*cis* to 4-*trans* products was observed as well as increased stereospecificity at the 3 and 4 positions. The latter feature was particularly prevalent with MMO from the two *Methylosinus* species (table 1) indicating a more restricted binding site in these cases, confining hydroxylation predominantly to the equatorial (more planar) positions. t-Butylcyclohexane was not hydroxylated by any of these enzymes indicating that the bulky t-butyl group restricts access to the active site. This implies firstly that a substrate the size of methylcyclohexane must be completely buried in the enzyme and secondly that the binding site has limited vertical dimensions (taking the horizontal plane to be that of the cyclohexane ring when bound).

TABLE 1

Products from the hydroxylation of isopropylcyclohexane by soluble MMO from *M.capsulatus*, *Mts. trichosporium* and *Mts. sporium*

	cis-3 + *trans*-4	*trans*-3 + *cis*-4
Mts. trichosporium and *Mts. sporium*	90%	10%
M. capsulatus	60%	40%

The trend from 3-*cis* to 4-*trans* cyclohexanols with increasingly bulky side chain suggests that the substrate binding site is sufficiently wide to allow some lateral movement of methylcyclohexane such that the 3 position can interact with the active site (rotation to place the 2 position next to the active site is evidently not possible) but this is restricted with larger side chains. The changes in stereochemistry of hydroxylation also reveal some interesting features. With methylcyclohexane hydroxylation of the 3 position was predominantly *cis* (equatorial) whereas the 4 position yielded equimolar *cis* (axial) and *trans* (equatorial). This suggests that when bound in the 'rotated' orientation the substrate is effectively wedged. However when bound 'end on' it has considerable freedom either in binding orientation or as a reaction intermediate. The mechanistic implications of this are discussed elsewhere [9]. Suffice it to say that this is consistent with a metal bound oxygen species attacking a planar intermediate formed by hydrogen or hydride abstraction. With larger alkyl side chains this freedom is evidently lost and the major product becomes the sterically more favourable (in chemical terms) diequatorial derivative. Thus it is evident from this straightforward analysis that a previously unrecognised regio- and stereospecificity can be imposed on MMO simply by increasing the dimensions of a side chain.

Cyclohexene and Alkylcyclohexenes

The major product (86-94%) from cyclohexene was the allylic alcohol 2-cyclohexenol. While the predominance of this to the exclusion of other ring alcohols can be reconciled with a mechanism involving hydride or hydrogen abstraction (i.e. this would be the chemically predicted product) the low yield of cyclohexene oxide (table 2) was surprising given the known propensity for epoxidation with these monooxygenases. Indeed *cis*-2-butene which (superficially) has structural similarities to cyclohexene, yields the epoxide as major product. A clue to these differences arose from consideration of the structure of the epoxides formed in each case. In cyclohexene oxide the ring is forced into a boat conformation. Apart from the unfavourable torsional strain that this involves cyclohexene oxide may simply be too bulky to be easily accomodated in the active site, given the evidence for limited vertical dimensions from the work with alkylcyclohexanes. In support of this it was shown that cyclopentene oxide, a product which is not required to undergo the same conformational change during formation, was the major product (74%) from cyclopentene.

TABLE 2

Products (% of total) from the bioconversion of cyclohexene with soluble MMO from 3 methanotrophs.

	M.capsulatus	*Mts. trichosporium*	*Mts. sporium*
2-cyclohexenol	94	87	86
2-cyclohexenone	1	2	10
Cyclohexene oxide	4	11	0
Cyclohexanone	2	0	2

It should be evident at this point that enzymic control can be exerted in two ways, firstly on substrate binding and secondly on product formation. These two levels of control should be additive in the sense that hydroxylation of alkyl substituted cyclohexenes may be more specific than the equivalent alkylcyclohexane. This was confirmed with the hydroxylation of 3-methylcyclohexene where the steric influence of the methyl group combined with allylic activation by the olefin resulted in 99% regiospecificity of hydroxylation at the unsubstituted allylic position with all three organisms (some further oxidation to the ketone was evident in all cases). An additional surprising feature considering the results obtained with methylcyclohexane was the high level of stereospecificity observed particularly with MMO from M.capsulatus [19] (table3).

TABLE 3

Products (% of total) from the bioconversion of 3-methylcyclohexene with soluble MMO from 3 methanotrophs.

	M.capsulatus	Mts. trichosporium	Mts. sporium
cis-4-methyl-2-cyclohexenol	90	73	74
trans-4-methyl-2-cyclohexenol	7	15	19
3-methylcyclohexene oxide	1	1	1
4-methyl-2-cyclohexenone	2	11	5

This suggests some specific enzyme interaction with the double bond which restricts the freedom of this substrate in the binding site, or restricts the permutations on substrate binding, and yields the chemically less favourable isomer (although in the case of disubstituted cyclohexenes the barriers are less clear cut)

FUTURE POTENTIAL

A number of alkane monooxygenases from gram positive and gram negative bacteria have been characterized as non-specific [20,21,] but apart from *Pseudomonas oleovorans* [22] have not been systematically investigated in this respect. It is evident from this work with MMO that highly specific biotransformations can be realised with such enzymes but that these are not apparent from consideration of their natural substrates. By adopting this straightforward analytical approach possibilities exist for establishing a battery of enzymes with different but known specificities which would allow a rational choice for any selected biotransformation.

As outlined earlier the alkane monooxygenases are predominantly of the cytochrome P-450 or non-heme iron type and produce non-discriminatory 'active oxygen'. A new type of monooxygenase is evident from work on alkene utilizing bacteria in this and other laboratories [23]. This enzyme is capable of highly specific chiral epoxidation of terminal alkenes [24] but is apparently unable to hydroxylate alkanes. While epoxidation of terminal alkenes is also a property of the P-450 and non-heme iron monooxygenases the inability to hydroxylate alkanes is difficult to reconcile with this type of enzyme unless the alkanes are somehow excluded from the active site. Work is underway is this laboratory to determine whether these are indeed fundamentally different monooxygenases and to establish their potential as biocatalysts.

Acknowledgements I wish to thank Chris Murricane, Steve Ellison and Peter Baker for their contributions to the work and/or ideas incorporated in this paper.

REFERENCES

1. Groves, J.T., Mechanisms of metal-catalyzed oxygen insertion. In Metal Ion Activation of Dioxygen, ed. T.G. Spiro, John Wiley & Sons, New York, 1980 pp 125-162
2. Crabtree, R.H., The organometallic chemistry of alkanes. Chem. Rev., 1985 85 245-269
3. Finn, M.G. and Sharpless, K.B., In Asymmetric synthesis, ed. J.D. Morrison, Academic press, New York 1985 pp 247
4. Smith, L.L. Steroids. In Biotechnology vol 6a Ed. K. Kieslich. Verlag-chemie, Weinheim 1984 pp 31-78
5. Katopodis, A.G.,Wimalasena, K., Lee, J., and May, S.W. Mechanistic studies on non-heme iron monooxygenase catalysis: epoxidation, aldehyde formation, and demethylation by the w-hydroxylation system of Pseudomonas oleovorans. J. Am. Chem. Soc. 1984 106 7928-7935
6. Sligar, S.G. and Murray, R.I. Cytochrome P-450cam and other bacterial P-450 enzymes. In Cytochrome P-450 ed P.R. Ortiz de Montellano, Plenum Press, New York 1986 pp 426-504
7. Dalton, H. and Leak, D.J. Mechanistic studies on the mode of action of methane monooxygenase. In Gas Enzymology, eds. H. Degn, R.P. Cox and H. Toftlund, D. Reidel Publishing Co. Dordrecht.
8. Stanley, S.H., Prior,S.D., Leak, D.J. and Dalton, H. Copper stress underlies the fundamental change in intracellular location of methane monooxygenase in methane oxidizing organisms: Studies in batch and continuous cultures. Biotech. Lett. 1983 5 pp 487-492
9. Murricane, C., Leak, D.J. and Baker, P.B. Selectivity of product formation with methane monooxygenase from 3 methanotrophic bacteria: Mechanistic and structural constraints. Enz. Microb. Technol. (submitted for publication)
10. Ingraham, L.K. and Meyer, D.L. Biochemistry of Dioxygen, Plenum Press, New York 1985 pp 157-174
11. Ono,T., Takahashi, K., Odani, S., Konno, H. and Imai, Y. Purification of squalene epoxidase from rat liver microsomes. Biochem. Biophys. Res. Commun. 1980 96 522-528
12. Lund, J. and Dalton, H. Further characterization of the FAD and Fe_2S_2 redox centres of component C, the NADH acceptor reductase of soluble MMO of Methylococcus capsulatus (Bath) Eur. J. Biochem. 1985 147 291-296
13. Woodland, M.P. and Dalton, H. Purification and properties of component A of the methane monooxygenase from Methylococcus capsulatus (Bath) J. Biol. Chem. 1984 259, 53-60
14. Green, J. and Dalton, H. Protein B of soluble methane monooxygenase from Methylococus capsulatus (Bath). J. Biol. Chem. 1985 260, 15,795-15,801

15. Stirling, D.I. and Dalton, H. Properties of the methane monooxygenase from extracts of *Methylosinus trichosporium* OB3b and evidence for its similarity to the enzyme from *Methylococcus capsulatus* (Bath). *Eur. J. Biochem* 1979 **96** 205-212
16. Dalton, H. Oxidation of hydrocarbons by methane monooxygenase from a variety of microbes. *Adv. Appl. Microbiol.* 1980 **26** 71-87
17. Dalton, H., Golding, B.J., Waters, B.W., Higgins, R. and Taylor, J.A. Oxidations of cyclopropane, methylcyclopropane, and arenes with the monooxygenase system from *Methylococcus capsulatus*. *J. Chem. Soc. Chem. Comm.* 1981 **189** 482-483
18. Jezequel, S.G. and Higgins, I.J. Mechanistic aspects of biotransformations by the monooxygenase system of *Methylosinus trichosporium* OB3b. *J. Chem. Tech. Biotechnol.* 1983 **33B** 139-144
19. Leak, D.J., Ellison, S.L.R., Murricane, C. and Baker, P.B. A 90% pure hydroxylation product from methane monooxygenase bioconversion: steric and electronic effects on product distribution. *Biocatalysis* 1987 (submitted for publication)
20. Perry, J.J. Microbial cooxidations involving hydrocarbons. *Microbiol. Rev.* 1979 **43** 59-72
21. Patel, R.N., Hou, C.T., Laskin, A.I. Felix, A. and Derelanko, P. Epoxidation of n-alkenes by organisms grown on gaseous alkanes. *J. Appl. Biochem* 1983 **5** 121-131
22. Van Ravenswaay Claasen, J.C. and van der Linden, A.C. Substrate specificity of the paraffin hydroxylase of *Pseudomonas aeruginosa* Ant. van. Leeuw. 1971 **37** 339-352
23. de Bont, J.A.M., Attwood, M.M., Primrose, S.B. and Harder, W. Epoxidation of short chain alkenes in Mycobacterium E20: the involvement of a specific monooxygenase *FEMS. Microbiol. Lett.* 1979 **6** 183-188
24. Habets-Crutzen, A.Q.H., Carlier, S.J.N., de Bont, J.A.M., Wistuba, D., Schurig.V, Hartmans, S. and Tramper, J. Stereospecific formation of 1,2-epoxypropane, 1,2-epoxybutane and 1-chloro-2,3-epoxypropane by alkene-utilizing bacteria. *Enz. Microb. Technol.* 1985 **7** 17-21

INTERNATIONAL CONFERENCE ON BIOREACTORS AND BIOTRANSFORMATIONS
GLENEAGLES, SCOTLAND, UK: 9-12 NOVEMBER 1987

Paper G2

BIOTRANSFORMATIONS OF DI- AND TRICHLORINATED BENZENES

Gosse Schraa and Jan Roelof van der Meer
Department of Microbiology,
Agricultural University,
6703 CT Wageningen,
Netherlands

ABSTRACT

Transformation of 1,2-dichlorobenzene by the indigenous microbial population in an aerobic soil percolation column occurred after a lag period of 80 days and resulted in an effluent concentration of 0.1 µg/l. Inoculation of similar columns with cells of Pseudomonas testosteroni strain P51, capable of using 1,2-, 1,3-, and 1,4-dichlorobenzene and 1,2,4-trichlorobenzene as sole carbon and energy sources, caused a transformation of the compounds without a lag period. Each compound had a minimum concentration(threshold value) below which no transformation took place. The threshold value for 1,2-dichlorobenzene was 5 - 10 µg/l and for 1,2,4-trichlorobenzene 20 µg/l.

INTRODUCTION

The widespread use of aromatic compounds forms an increasing threat to the quality of terrestrial and aquatic environments. Apart from high concentrations of these chemicals in certain wastestreams from industries, low concentrations are also found in soils and surface-water. In addition, leaching processes in soil are responsible for numerous cases of groundwater contamination.

Transformation via biological reactions may be a significant process to remove these xenobiotic compounds from our environment. A large number of these compounds has indeed been found to be transformed by microorganisms [1]. In many cases, these transformations have led to the production of harmless inorganic endproducts(mineralization). However, other compounds have been observed not to be transformed at all or only partially. This persistence of a compound may be attributed to physical, chemical or environmental causes [2] such as: inaccessibility of the substrate, structural characteristics of the molecule to be degraded, concentration of the compound, absence of essential growth factors, absence of the right electron acceptor, unfavorable temperature, water activity and pH, and the inability of a microbial community to metabolize the compound due to some physiological inadequacy.

Of particular interest among the environmental factors affecting the transformations of aromatic compounds is the presence or absence of molecular oxygen. For a long time, dioxygen was thought to play a crucial role in the transformations of aromatic hydrocarbons, which contain no oxygen in their molecular structure. Molecular oxygen has been found to serve two functions during the transformations: a) as terminal electron acceptor, and b) as direct oxidant of the molecule. Benzene [1,3,4,5], toluene [1,4,5,6], xylene [1,7,8,9], ethylbenzene [1,4,7], monochlorobenzene [1,5,10], dichlorobenzene [9,11,12,13,14] and trichlorobenzene [1,5,14,15], except for 1,3,5-trichlorobenzene, are all degraded under aerobic conditions.
Recently, transformations of some of the above mentioned aromatic hydrocarbons have also been reported under anaerobic conditions: benzene and toluene under methanogenic conditions [16], _m_-xylene and toluene with nitrate as electron acceptor [17], and 1,2,4-trichlorobenzene in the presence of hydrogen [18].

Another aspect of the transformations of xenobiotic compounds is the long lag period which is often required before transformation is observed. This has been shown in the laboratory under "ideal" conditions in batch cultures [12], in soil percolation columns [9], and in percolation columns seeded with activated sludge [5]. Two explanations may be given for this lag period. The first one is the absence of enough microorganisms to transform the xenobiotic compound. The second one is the occurrence of adaptation [19,20]: microorganisms will not be able to metabolize a specific compound until the necessary genetic information for this transformation is present in one microorganism or community of microorganisms. The mechanism may be explained by a natural mutation or by transfer of genes within the community. Inoculation with microbial strains, capable of degrading specific xenobiotic compounds, may minimize this lag period.

The concentration of the organic compound may also be a significant factor affecting its susceptibility to microbial degradation. Xenobiotic compounds may persist in some environments as a result of their low concentration [21]. At these low concentrations (μg/l to ng/l range) there may not be sufficient energy available from consumption of the compound to support microbial growth.

In this paper data will be presented on the behavior of di- and trichlorinated benzenes in soil percolation columns, simulating saturated aquifers. This behavior was studied in a column with just the indigenous microbial population present and in columns which were in addition inoculated with a specific bacterial strain.

MATERIALS AND METHODS

Biotransformation of the chlorinated benzenes was examined in soil percolation columns made of hard PVC with a length of 25 cm and a diameter of 5.5 cm. The columns, with sampling ports at various heights, were filled with sandy sediments from the river Rhine near Wageningen, Netherlands. They were percolated continuously in an upflow mode under saturated conditions. The medium closely resembled the salts composition of Rhine water, was saturated with oxygen and contained di- and trichlorinated benzenes in concentration ranges of 5 - 1000 μg/l, depending on the experiment. The flow rate of the medium was around 10 ml/h and the experiments were performed at 20°C. A more detailed description of the column system is given in [9].

Pseudomonas testosteroni strain P51, capable of utilizing 1,2-, 1,3-, and 1,4-dichlorobenzene (DCB) and 1,2,4-trichlorobenzene (TCB) as sole carbon and energy sources, was used for the inoculation experiments in the soil percolation columns. This bacterium has been isolated from a mixture of different soil samples after 9 months of enrichment in the presence of 1,2,4-TCB as carbon and energy source [14]. In each inoculation experiment about 10^9 cells of *Pseudomonas testosteroni* strain P51 at a time were injected about halfway in the column. Strain P51 had been pregrown on 1,2,4-TCB in a batch culture.

Analysis of the three isomers of both dichlorobenzene and trichlorobenzene in the watersamples was performed by hexane extraction followed by capillary gas chromatography on a unit equiped with an electron capture detector [14].

RESULTS AND DISCUSSION

A mixture of 1,2-, 1,3-, and 1,4-dichlorobenzene and 1,2,3-, 1,2,4-, and 1,3,5-trichlorobenzene, ranging in concentration from 8 - 25 μg/l, was led through an aerobic soil percolation column, filled with river Rhine sediment, for a period of over 500 days. Upon breakthrough, only 1,2-DCB was removed (Figure 1).

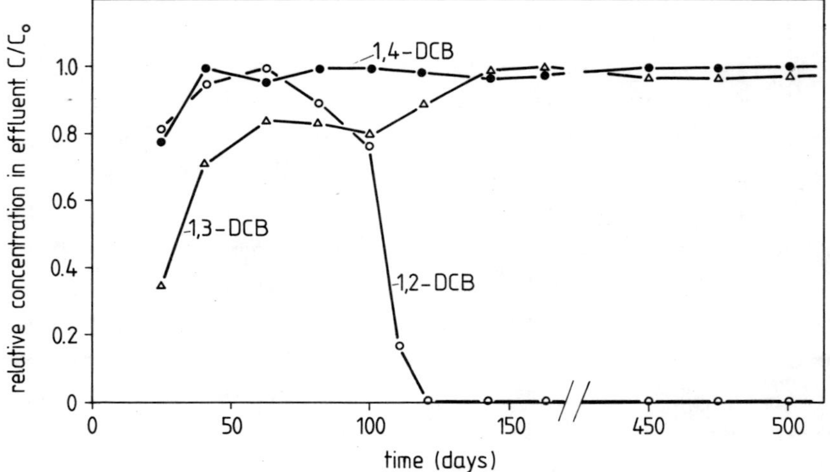

FIGURE 1. Breakthrough of the dichlorobenzene isomers in the aerobic soil percolation column and the disappearance of 1,2-DCB. Influent concentration: 1,2-DCB = 25 μg/l, 1,3-DCB = 15 μg/l, 1,4-DCB = 12 μg/l. C/C_o is the ratio of the effluent to influent concentration.

The disappearance started after a lag period of about 80 days and resulted in an effluent concentration in the order of 0.1 μg/l. Similar to 1,3- and 1,4-dichlorobenzene, nothing happened with the trichlorobenzene isomers during this period. Despite the potential of the indigenous microbial population in the sediment to degrade these compounds (a long exposure to chlorinated benzenes in river Rhine water), only one dichlorobenzene isomer was transformed after a long lag period. Long lag periods before degradation occurred have also been found with an aerobic enrichment culture where 1,4-DCB was present as carbon and energy source [12] and with an anaerobic soil percolation column with a mixture of trichlorinated benzenes. In the enrichment culture it took about 2 months before degradation of 1,4-DCB was observed and another 8 months to isolate 2 different microorganisms capable of growing on 1,4-DCB. One of these organisms, *Alcaligenes* sp. strain A175, was found to mineralize 1,4-DCB via dioxygenases in the initial steps, while ring opening proceeded via *ortho* cleavage. In ongoing experiments we have demonstrated that 1,2,3-, 1,2,4-, and 1,3,5-TCB can be transformed under anaerobic conditions in the presence of sulfate as electron acceptor and also under methanogenic conditions. It took 3 to 5 months before transformations occurred in anaerobic soil percolation columns. Via reductive dechlorination, all 3 isomers were stoichiometrically converted to monochlorobenzene [Bosma et al., in preparation].

Inoculation of bacteria which possess a specific degradation capacity may shorten this lag period. The potential of inoculating bacteria in a specific system is demonstrated in the following experiments.

In aerobic soil percolation columns, fed with a mixture of di- and trichlorinated benzenes, cells of *Pseudomonas testosteroni* strain P51 were inoculated upon complete breakthrough of the chlorinated benzenes. The columns were operated under similar conditions as before and no transformations took place before inoculation. In the first experiment with the high influent concentrations (Table 1) an immediate removal of the DCB isomers and of 1,2,4-TCB was observed beyond the point of inoculation. In time, steady effluent concentrations were reached (Table 1) and transformations were found to occur throughout the entire column. The experiment was successful for the whole test period of 80 days.

In the second experiment with lower influent concentrations, 1,2-DCB and 1,2,4-TCB were also immediately removed. However, we did not observe any transformation of 1,3- and 1,4-DCB. In another aspect this experiment was also less successful than the previous one. After 60 days a steady increase in the effluent concentrations of 1,2-DCB and 1,2,4-TCB took place. For some unknown reason the system had become to lose its transformation capacity. Interesting in this experiment was that initially the same effluent concentrations were reached for 1,2-DCB and

TABLE 1
Influent and effluent concentrations[a] of the chlorinated benzenes in soil percolation columns inoculated with cells of Pseudomonas testosteroni strain P51

Compound	Influent concentration			Effluent concentration
	Exp.1	Exp.2	Exp.3	
1,2-DCB	1000	200	20	5 - 10
1,3-DCB	1000	150	15	50[b]
1,4-DCB	600	80	10	30[b]
1,2,3-TCB	300	30	5	n.t.[c]
1,2,4-TCB	350	80	10	20[d]
1,3,5-TCB	-	40	5	n.t.

[a] all in µg/l
[b] only transformation observed in experiment 1
[c] n.t. = no transformation observed
[d] only transformation observed in experiment 1 and 2

1,2,4-TCB, 5 - 10 µg/l and 20 µg/l resp. as in experiment 1. The existence of threshold values was confirmed in the third experiment in which, except for 1,2-DCB, the influent concentrations were kept below the effluent concentrations of the first two experiments. Upon inoculation, only 1,2-DCB disappeared down to an effluent concentration of again 5 - 10 µg/l (Figure 2). None of the other compounds was removed. Repeated inoculations with Pseudomonas testosteroni strain P51 did not change these findings. The reason for these threshold concentrations cannot solely be explained by characteristics of strain P51. In batch cultures strain P51 has been shown to degrade the DCB isomers far below the effluent concentrations found in the inoculated column experiments, viz. 0.1 - 0.5 µg/l. In addition, the indigenous microbial population in the soil column became capable of removing 1,2-DCB down to a concentration of 0.1 µg/l under conditions similar to those for the inoculated columns. The explanation for these threshold concentrations lies probably with factors like adsorption of the chlorinated benzenes to the sediment particles, the presence of organic matter as extra carbon source, and metabolic characteristics (incl. kinetics) of strain P51 under the

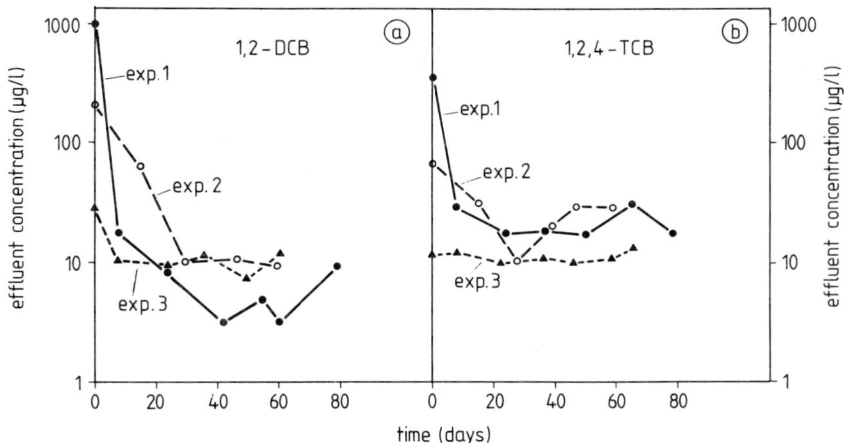

FIGURE 2. Threshold concentrations observed for the transformation of 1,2-DCB (a) and 1,2,4-TCB (b) by Pseudomonas testosteroni strain P51 in the inoculated soil columns. Inoculation took place at day 0.

prevailing conditions. It is interesting to notice that the threshold concentrations of 1,2,4-TCB were similar in the batch cultures and in the inoculated soil columns, namely around 20 µg/l.

An other finding in these experiments, which may have a practical implication, is the fact that a temporary omission of the chlorinated benzenes from the medium for a period of 4 weeks following experiment 1, resulted in the loss of the degradation capabilities of the soil column. No removal of any of the chlorinated benzenes was observed for over 50 days after a renewed addition of the chlorinated benzene mixture. Whether the cells of strain P51 were washed out of the column or that they lost their degradative capacity because of the absence of the specific substrate is hard to say.

Experiments are underway to study characteristics of Pseudomonas testosteroni strain P51 with and without the selective pressure of the chlorinated benzenes which it can utilize as sole carbon and energy source.

CONCLUSIONS

Although bacteria have been isolated which use chlorinated benzenes as sole carbon and energy sources, often long lag periods are involved before transformations are observed. To shorten this lag period soil percolation columns have been successfully inoculated with bacteria to bring about these transformations. The same may count for bioreactors which are used for the purification of wastewaters or contaminated groundwater. Research is needed to provide insight in the conversion efficiency of the system but also in the limitations of such a process. The observation of certain threshold values and the loss of the transformation capabilities of the system in our experiments are an indication of limitations. Special attention has to be paid to the behavior of the inoculated microorganisms, sub-optimal growth and transformation conditions, minimum effluent concentrations, fluctuating influent concentrations, production of intermediates, and transformations with complex mixtures of xenobiotic compounds.

REFERENCES

1. Kobayashi, H. and Rittmann, B.E., Microbial removal of hazardous organic compounds. Environ. Sci. Technol., 1982, 16, 170A-183A.

2. Alexander, M. Biodegradation of chemicals of environmental concern. Science, 1981, 211, 132-138.

3. Gibson, D.T., Koch, J.R. and Kallio, R.E., Oxidative degradation of aromatic hydrocarbons by microorganisms. I. Ezymatic formation of catechol from benzene. Biochemistry,1968, 7, 2653-2658.

4. Gibson, D.T. and Subramanian, V., Microbial degradation of aromatic hydrocarbons. In Microbial degradation of organic compounds, ed. D.T. Gibson, Marcel Dekker, Inc., New York, 1984, pp. 181-252.

5. McCarty, P.L., Rittmann, B.E. and Bouwer, E.J., Microbiological processes affecting chemical transformations. In Groundwater Pollution Microbiology, eds. G. Bitton and C.P. Gerba, J. Wiley & Sons, New York, 1984, pp. 89-115.

6. Claus, D. and Walker, N., The decomposition of toluene by soil bacteria. J. Gen. Microbiol., 1964, 36, 107-122.

7. Gibson, D.T., Mahadevan, V. and Davey, J.F., Bacterial metabolism of para and meta xylene: oxidation of the aromatic ring. J. Bacteriol. 1974, 119, 930-936.

8. Schraa, G., Bethe, B.M., van Neerven, A.R.W., van den Tweel, W.J.J., van der Wende, E. and Zehnder, A.J.B., Degradation of 1,2-dimethylbenzene by Corynebacterium strain C125. Antonie van Leeuwenhoek, Accepted for publication.

9. Kuhn, E.P., Colberg, P.J., Schnoor, J.L., Wanner, O., Zehnder, A.J.B. and Schwarzenbach, R.P., Microbial transformations of substituted benzenes during infiltration of river water to groundwater: Laboratory column studies. Environ. Sci. Technol., 1985, 19, 961-968.

10. Reineke, W. and Knackmuss, H.-J., Microbial metabolism of haloaromatics: isolation and properties of a chlorobenzene-degrading bacterium. Appl. Environ. Microbiol., 1984, 47, 395-402.

11. De Bont, J.A.M., Vorage, M.J.A.W., Hartmans, S. and van den Tweel, W.J.J., Microbial degradation of 1,3-dichlorobenzene. Appl. Environ. Microbiol., 1986, 52, 677-680.

12. Schraa, G., Boone, M.L., Jetten, M.S.M., van Neerven, A.R.W., Colberg, P.J. and Zehnder, A.J.B., Degradation of 1,4-dichlorobenzene by Alcaligenes sp. strain A175. Appl. Environ. Microbiol., 1986, 52, 1374-1381.

13. Spain, J.C. and Nishino, S.F. Degradation of 1,4-dichlorobenzene by a Pseudomonas sp. Appl. Environ. Microbiol., 1987, 53, 1010-1019.

14. Van der Meer, J.R., Roelofsen, W., Schraa, G. and Zehnder, A.J.B., Degradation of low concentrations of dichlorobenzenes and 1,2,4-trichlorobenzene by Pseudomonas testosteroni P51 in nonsterile soil columns. FEMS Microbial Ecology, in press.

15. Marinucci, A.C. and Bartha, R., Biodegradation of 1,2,3- and 1,2,4-trichlorobenzene in soil and in liquid enrichment culture. Appl. Environ. Microbiol., 1979, 38, 811-817.

16. Grbić-Galić, D. and Vogel, T.M., Transformation of toluene and benzene by mixed methanogenic cultures. Appl. Environ. Microbiol., 1987, 53, 254-260.

17. Zeyer, J., Kuhn, E.P. and Schwarzenbach, R.P., Rapid microbial mineralization of toluene and 1,3-dimethylbenzene in the absence of molecular oxygen. Appl. Environ. Microbiol., 1986, 52, 944-947.

18. Tsuchiya, T. and Yamaha, T., Reductive dechlorination of 1,2,4-trichlorobenzene by Staphylococcus epidermis, isolated from intestinal contents of rats. Agric. Biol. Chem., 1984, 48, 1545-1550.

19. Ghosal, D., You, I.-S., Chatterjee, D.K. and Chakrabarty, A.M., Microbial degradation of halogenated compounds. Science, 1985, **228**, 135-142.

20. Grady Jr., C.P.L., Biodegradation: its measurement and microbiological basis. Biotech. Bioeng., 1985, **27**, 660-674.

21. Boethling, R.S. and Alexander, M., Effect of concentration of organic chemicals on their biodegradation by natural microbial communities. Appl. Environ. Microbiol., 1979, **37**, 1211-1216.

INTERNATIONAL CONFERENCE ON BIOREACTORS AND BIOTRANSFORMATIONS GLENEAGLES, SCOTLAND, UK: 9-12 NOVEMBER 1987

Paper G3

BIOELECTROCHEMICAL TRANSFORMATIONS CATALYSED BY *Clostridium sporogenes*.

Robert W.Lovitt [1], Eurig W.James[2], Douglas B.Kell[2] and J.Gareth Morris[2].

[1]Department of Chemical Engineering, University College Swansea, Singleton Park, Swansea SA2 8PP, West Glamorgan, U.K.

[2]Department of Botany and Microbiology, University College of Wales, Aberystwyth, Dyfed SY23 3DA. U.K.

Abstract

One of the major limitations to the use of oxido-reductase enzymes in performing biotransformations is the regeneration of cofactors. One approach has been to use redox mediators indirectly or directly to couple oxido-reductase enzymes to electrochemical regeneration of cofactors. We have used methyl viologen (MV) as a mediator to drive a number of reductive reactions in the proteolytic anaerobe *Clostridium sporogenes*.

Studies have been performed with permeabilised cells. We have investigated the MV-coupled reactions catalysed by 2-oxoacid synthase and MV-NAD reductase using a dropping mercury electrode and a stirred mercury pool electrode to measure and electrochemically to reduce MV. During the course of this work we have developed polarographic assays (based on MV oxidation) for these enzymes and have subsequently characterised the reactions.

2-oxoacid synthase reactions involve the reductive carboxylation of acyl CoA derivatives to form 2-oxoacids. Using acetyl phosphate we have succesfully synthesised pyruvate. It is envisaged that these reactions could be used to synthesise 2-oxoacids, hydroxy-acids and amino-acids from 2-oxoacid derivatives using NADH regenerated via the substantial MV-NAD reductase activity, together with the aminotransferases and glutamate dehydrogenase, that this organism possesses.

INTRODUCTION

The clostridia are a diverse group of spore-forming bacteria that are differentiated on the basis of their ability to ferment different carbon sources. The biochemistry and physiology of the saccharolytic clostridia have been most widely studied, and have recently been re-examined in detail for the ability to produce solvents and organic acids from carbohydrates [1-3]. Similarly the proteolytic clostridia, those which ferment amino acids, have also been the subject of renewed interest, although attention has focussed upon potentially useful enzymes that these organisms produce. Apart from extracellular hydrolases, most notable are enoate [4], nitroaryl [5], linoleate [6], 2-oxoacid [7], proline [8] and glycine reductases [9] (table 1). Another reduction reaction involved in the synthesis of amino acids is that of ferredoxin-linked reductive carboxylation. These reactions are used by certain bacteria for the synthesis of amino acids and in some cases to drive a reductive tricarboxylic acid cycle [10-12].

TABLE 1
Potentially useful reductive enzymes in the clostridia

Reduction reaction	Electron donor	Organism	Ref
aldehyde/ketone dehydrogenase			
steroids	?	C. paraputrificum	[13]
		C. bifermentans	[14]
methyl ketones	NADPH	C. thermohydrosulfuricum	[15]
ketones	?	C. pasteurianum	[17]
	?	C. tyrobutyricum	[17]
2-oxoacid synthase			
Fatty acids	ferredoxin	C. sporogenes	[18]
acetate	"	C. kluyveri	[19]
Linoleic reductase			
linoleic acid	?	C. sporogenes	[6]
Enoate reductase			
cinnamic acid	NADH	C. sporogenes	[20]
crotonic acid	"	C. tyrobutyricum	[21]
2-oxoacid reductase			
phenylpyruvic acid	NADH	C. sporogenes	[20]
Nitroaryl reductase			
chloramphenicol	ferredoxin/flavodoxin	C. acetobutylicum	[22]
metronidazole	"		[23]
p-nitrobenzoate	"		[5]
2 nitrobenzene	"		[5]
Proline reductase			
proline	NADH/MV	clostridia	[8,24]
Glycine reductase			
glycine	ferredoxin	clostridia and other anaerobes	[9]
Lipoamide dehydrogenase			
NAD/lipoamide	lipoamide/NADH	C. kluyveri	[25]
MV-NAD(P) reductase			
NAD(P)/ferredoxin	ferredoxin/NAD(P)H	clostridia and other anaerobes	[23]

Clostridium sporogenes performs the Stickland reaction in which pairs of amino acids are fermented, one amino acid acting an electron donor (e.g. valine,leucine isoleucine) whilst the other acts as an electron acceptor (e.g. proline or glycine). Phenylalanine can act either as donor or acceptor. We have investigated the biochemistry and physiology of these reactions [17,24,27-29]. In this respect C. sporogenes has been shown to synthesise many of its required amino acids from deaminated and decarbonylated fatty acid analogues; the relevant enzymes have been shown to be highly active and to show broad specificity in the case of oxo-glutarate aminotranferase, 2-oxoacid synthase, acyl-CoA phosphotranferase and acyl kinase reactions [28].

The main carbon and electron flow pathways of Clostridium sporogenes have been investigated by several workers [e.g. 8,9, 27-29] and are shown in figure 1. It should be noted that the reduction of acyl phosphate derivatives proceeds by a series of reductive steps, the derivatives being first reductively carboxylated to 2-oxoacids and further reduced to hydroxyacids and/or amino acids.

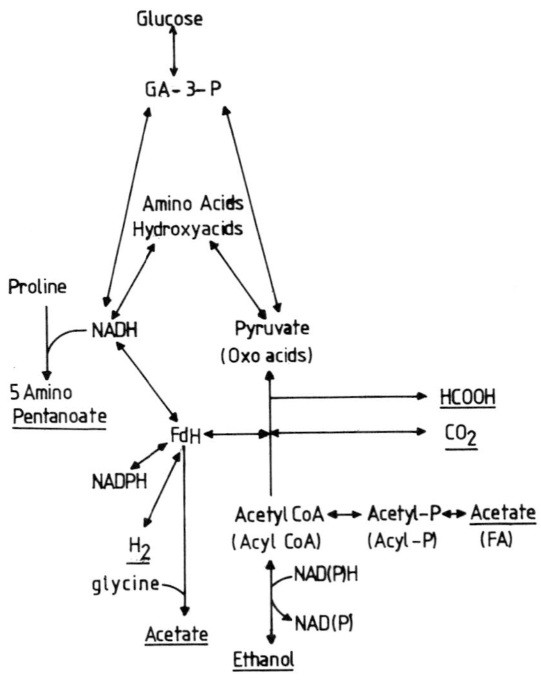

FIGURE1

Carbon and electron flow pathways of C. sporogenes

To harness these reductive reactions for commercial biotransformation processes, a number of approaches are possible: (1) direct incubation of the substrate in the presence of whole cells. (2) incubation of substrate with permeabilised cells, or (3) incubation of substrates with cell-free extracts or purified enzymes.

The major limitation of approach (1) is the access of the substrate to the intracelluar enzymes, whilst the more significant limitation on the use of oxidoreductase enzymes in (2) and (3) is the problem of cofactor recycling. To overcome the problem of cofactor recycling two approaches are possible: (a) recycling using another dehydrogenase reaction, e.g. formate dehydrogenase [30], though it is not possible to reduce ferredoxin by this means, or (b) recycling using direct or indirect electrochemical means as pioneered by Simon and coworkers [4,21,26] . The direct and reversible electrochemical reduction of ferredoxin and other iron-sulphur proteins is also possible [31]; however at present, these proteins are expensive to produce and have slow reduction kinetics at electrodes, due to their low diffusion coefficients.

In this paper we report on our studies of two reductive enzymes in permeabilised cells of C. sporogenes viz. 2-oxoacid synthase and MV-linked NAD reductase. The first is used to synthesise pyruvic acid and other 2-oxo acids from acyl phosphate derivatives, whilst the second is used to recycle NAD using MV as a mediator between an electrode and the enzyme.

EXPERIMENTAL

Microbiological and Biochemical methods

Clostridium sporogenes NCIB 8053 was grown and maintained on a defined medium as outlined elsewhere [18]. Cells were harvested from overnight cultures (12-14 h) in late exponential phase (OD 0.9-1.0), washed once in anaerobic 0.1 M potassium phosphate (pH 7.0), resuspended in the same buffer and stored anaerobically in sealed vials. The organisms were then permeabilised by the addition of 20 µl of 10% v/v toluene in ethanol per ml of cell suspension [18]

The preparation of Acyl phosphates

Apart from Acetyl phosphate (obtained from Sigma), these compounds were synthesised, by the hydrolysis of fatty acid anhydrides in the presence of potassium phosphate, and purified, according to the method of Stadtman [32].

Electrochemical methods

The determination of Methyl Viologen (MV)-linked enzyme activities was performed using a PAR 174A Potentiostat and model 303 dropping mercury electrode cell (in DC sampling mode), by measuring polarographically at

-875 mV (vs Ag/AgCl (3M KCl)) the rate of MV oxidation in a 3-electrode cell completed with a Pt wire counter elctrode. It was demonstrated using these methods that it is possible, after calibration, accurately and continuously to measure the amount of oxidised MV in the system, and thus the activities of 2-oxoacid synthase and MV-NAD reductase.

The assays were performed in a 5 ml volume using permeabilised cells at 20°C. For 2-oxoacid synthase the reaction mixture contained 100mM potassium phosphate pH 7.0, 10 mM Acyl phosphate, 50 μM CoA, 10 mM $KHCO_3$ and 5 U of phosphotransferase (Sigma, optional). The reaction was initiated by the addition of bicarbonate. For MV-NAD reductase the reaction mixture contained 100mM potassium phosphate pH 7.0, 1 mM reduced MV, (prereduced in the preparative cell, see below) and 0.5mM NAD. The reaction was initiated by the addition of NAD.

The preparative reactor cell is illustrated Figure 2. The cell is a glass vessel with a mercury pool electrode at the base, the surface of which is stirred by a glass-coated magnetic follower coupled to a magnetic stirrer. The mercury pool is the working electrode and is the reactive surface at which MV is reduced. The reference electrode, Ag/AgCl (3M KCl) is present in the cell whilst the Pt auxiliary electrode is connected to the cell via a salt bridge so as to isolate it from the highly reducing environment of the reactor cell. The cell is kept anaerobic by constant gassing and additions are made though the gas outlet port. The cell was jacketed and was operated at 20°C with a total reaction volume of 5 to 10 ml.

For preparative work the reaction medium consisted of 200 mM potassium phosphate pH 7.0, 1 to 25 mM Acetyl-PO_4, 50 μM CoA (optional) and 5U phosphotranferase (optional). The system was purged and gassed with CO_2 before cells and other additions are made and the reaction was carried out under a CO_2 headspace. The reaction in dilute cell suspensions (<.1 mg dry wt ml^{-1}) is dependent upon the addition of CoA; however at high cell concentrations (>2 mg dry wt ml^{-1}) the reaction is dependent only upon added Acetyl-PO_4.

RESULTS

Characterisation of the reductive carboxylation reactions of *C. sporogenes*

The reductive carboxylation reactions of *C. sporogenes* were first demonstrated by showing that certain fatty acids could substitute for amino acids required for anabolism and hence for growth. It was shown that acetate, 2-methyl propionate, 2-methyl butyrate and 3-methyl butyrate could replace serine, valine, leucine and isoleucine respectively [18].

The requirements of the reductive carboxylation reaction were investigated using MV as a reducing agent and are shown in Table 2. As can be seen, the reaction was shown to require CO_2, CoA and Acetyl phosphate. Exogenous phosphotranferase stimulated the reaction but was not required.

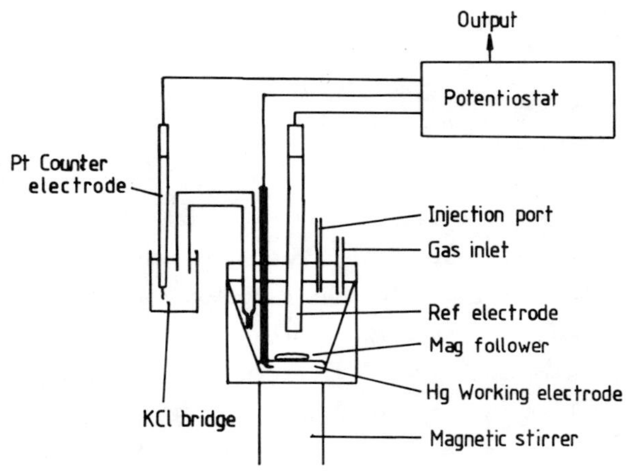

FIGURE 2

Diagram of preparative reactor cell

TABLE 2.

The characteristics of the reductive carboxylation reaction

	Activity μmol. (min mg dry wt)$^{-1}$
Complete system	0.172
- Bicarbonate	0.031
- CoA	0.026
- Phosphotranferase (PTA)	0.133
- Acetyl Phosphate	0.027

The complete system, reactor volume 5ml, contained 200 mM potassium phosphate, (pH 7.0), 10 mM Acetyl phosphate, 1mM MV, 20 mM $KHCO_3$, 50 μM CoA, 5U PTA, and Cells 1-2 mg (dry wt. ml)$^{-1}$

The pH optimum for the reaction was in the range 6.7 to 7.0, and tris and "Good"-type buffers were more inhibitory to the reaction than was phosphate buffer. The CoA requirement for the reaction was small and the reaction was unaffected even when the added CoA was reduced to 30 μ M. The K_m for bicarbonate was approximately 6 mM at pH 7.0.

To test the specificity of 2-oxoacid synthase the activity of permeabilised C. sporogenes cells towards other acyl phosphates was assessed. The observed activity of the cells to acyl phosphates is shown in Table 3. In this assay system no addition of PTA was made, as this commercial enzyme is specific for acetyl phosphate [33]. The activities for these compounds were lower than those observed with for acetyl phosphate, although significant activity was detectable.

TABLE 3

2-oxoacid synthase activity towards various acyl phosphates

substrate	Activity μ mol. (min mg dry wt)$^{-1}$
Acetyl phosphate	0.104
Butyryl phosphate	0.044
2-methyl propionyl phosphate	0.034
Pentyl phosphate	0.042
3-methyl butyryl phosphate	0.060
Hexyl phosphate	0.030

Conditions of Assay: temperature 21°C, 100 mM potassium phosphate pH 7.0, 10 mM Acyl phosphate, 20mM $KHCO_3$ and cells as described in the legend to Table 2.

2. Reductive carboxylation in a preparative cell: the synthesis of pyruvate from acetyl phosphate

Now that the reductive carboxylation reaction had been characterised and optimised, the reaction was investigated on the preparative scale. A 5-10 ml reactor (Figure 2) was used. A typical experiment involved, firstly, the reduction of the MV in the reactor. After the addition of cells, CoA and PTA, the reaction was started by the addition of Acetyl phosphate. Figure 3a shows a typical current-time profile. The current passed in the reactor is proportional to the activity of the MV reductase processes involved and thus its time integral, in a highly coupled reaction, to the amount of substrate added.

In experiments with high cell concentrations where the reaction no longer required additions of CoA or PTA to obtain good activity, it is presumed that the concentration of cell-derived CoA was sufficient to satisfy the requirement for the reaction. Figure 3b shows a current-time profile for a reaction involving a high cell concentration. By integration of the reaction profiles the amount of charge used can be calculated and is equivalent (within 5%) to the amount of Acetyl phosphate added (assuming a two electron reduction of acetyl phosphate).

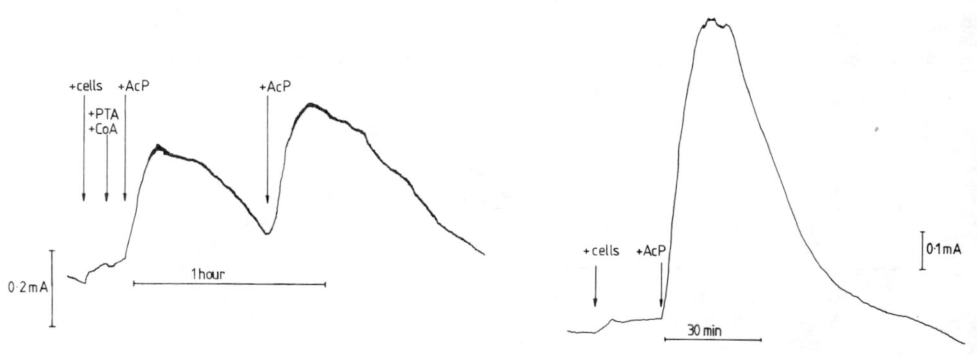

Fig 3a

Fig 3b

FIGURE 3

Current-time profiles observed on the addition of acetyl phosphate.

Fig 3a shows 2 additions of acetyl phosphate, of 5 and 15 μ mol respectively at low cell density (0.15 mg dry wt ml^{-1}). Figure 3b shows an experiment ot high cell density (1.5 mg dry wt ml^{-1}); 15 μ mol of Acetyl phosphate were added at the point indicated, note that exogenous phosphotranferase and CoA are not required.

To confirm the rate of pyruvate synthesis and to investigate the coulombic efficiency of the reaction, an experiment was performed to compare the amount of charge passed with the amount of pyruvate formed. This was done by sampling the reaction periodically and measuring the pyruvate concentration, using the standard enzymatic method based upon lactate dehydrogenase [34]. The results of this experiment are shown in Table 4 and Figure 4. The time course of pyruvate production was approximately linear and 7 μ mol of pyruvate were formed in 2.8 h. As mentioned above the amount of charge passed was proportional to the amount of acetyl phosphate; however as revealed by this experiment, the amount of pyruvate formed from the acetyl phosphate added showed only a 10-28% yield depending upon the method of calculation (Table 4). The reaction appeared to be stable for at least 24 h.

TABLE 4

Coloumbic efficiency of pyruvate formation from acetyl phosphate

time (h)	Pyruvate [mM]	Pyruvate µMole in the reactor	charge (C)	charge -endog (C)	Pyruvate equivalent charge (C)	% conversion -endog	% conversion total
0	0	0	0	0	0	0	0
0.33	-	-	1.74	1.14	-	-	-
0.66	0.256	1.7	3.36	2.28	0.328	14.3	9.7
1.00	0.401	2.8	4.98	3.42	0.540	15.7	10.8
1.33	0.502	3.5	6.42	4.38	0.675	15.2	10.5
1.66	0.550	3.8	7.62	5.06	0.733	14.4	9.6
2.00	0.610	4.2	8.82	5.82	0.829	14.2	9.3
2.80	1.01	7.0	11.97	7.97	1.350	16.9	11.2
11.5	9.00	63.0	55.71	44.19	12.15	27.6	21.8

Conditions. The transformation was performed at 22°C in a reaction volume of 7 ml. The reaction took place in saturating carbon dioxide, 200 mM potassium phosphate pH 7.0, 16 mg dry wt cells. 5 Units PTA, 50 mM Acetyl phosphate, 50 µ mol CoA. The assumption is made that the endogenous rate is constant throughout the experiment an assumption which minimises the calculated yield. 1 µ mol equivalent of electrons is 0.096487 C.

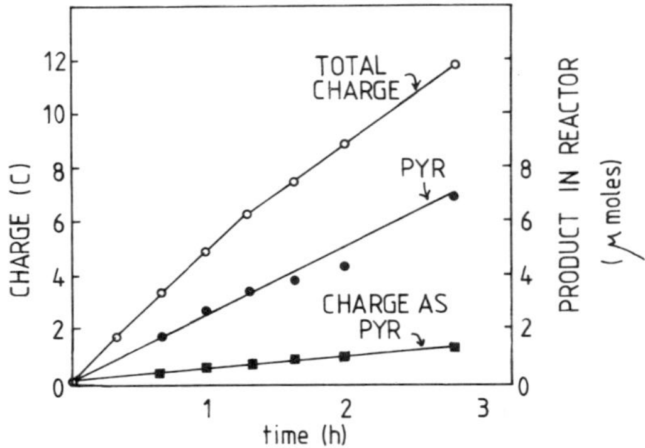

FIGURE 4

Time course of the reductive carboxylation reaction using acetyl phosphate as substrate.

The figure shows the amount of pyruvate produced, the total charge passed during the reaction and the amount of charge incorporated into the pyruvate formed (see also table 4).

3. MV-NAD reductase

MV-NAD reductase is of potential utility in systems where the regeneration of NADH is required. This enzyme system is highly active in many clostridia including C. sporogenes [26]. The enzyme is generally present at an activity of 1 μ mole min^{-1} mg dry wt^{-1} although on occasions up to 2 μ mole min^{-1} mg dry wt^{-1} were measured. The enzyme is expressed thoughout the growth cycle of the organism. The characterisation of this enzyme is now underway. The first and most important characteristic to be assessed was its stability. This was investigated by incubating the toluenised preparations at various temperatures, and assaying the activity at 20°C. The results are presented in figure 5. Incubation at 50°C rapidly inactivated the enzyme; however at 37°C, 21°C and 1°C the enzyme remained stable for many hours. After an initial short period where the activity was stimulated, the activity was then slowly lost. The approximate half-lives for the enzyme were between 65 h (37°C) and 175 h (21°C). Indeed considerable activity remained after 2 weeks. We are now endeavoring to employ this reaction to drive NADH-linked oxidoreductase enzymes.

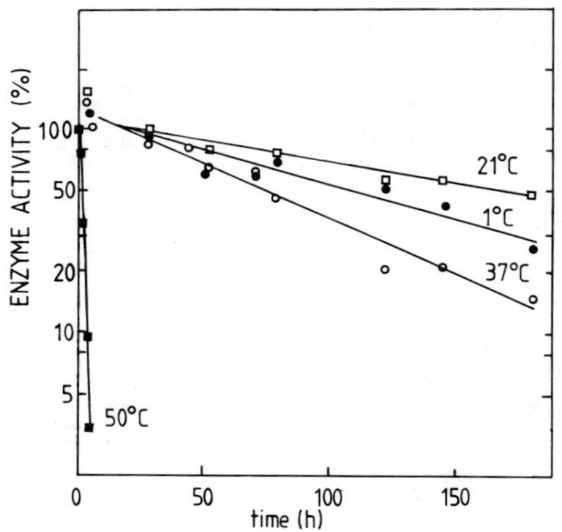

FIGURE 5
The effect of temperature on the stability of MV-NAD reductase of C. sporogenes

DISCUSSION

Pyruvate synthase and 2-oxoacid synthesis

We report here our studies of two reactions performed by C. sporogenes. Using the 2-oxoacid synthase, MV-NAD reductase and additional enzymes present in C. sporogenes it is possible to drive these reactions electrochemically to produce a series of oxo-, and thus hydroxy- and amino-acids. Several important aspects of the 2-oxoacid synthase remain to be investigated; in particular its temperature optimum and stability need to be determined and this work is currently being undertaken.

At present we have no explanation for the apparently low coloumbic yield of pyruvate formation. There are a number of possible routes whereby the Acetyl phosphate may be reduced. Acetyl CoA may be reduced to acetaldehyde by aldehyde dehydrogenase; this would require the presence of NADH, but if this reaction were taking place, acetaldehyde would most probably be further reduced to ethanol using the alcohol dehydrogenase known to be present in this organism (figure 1) [18]. If this were the case, the charge passed would be greater than the two electron equivalents calculated from the current-time profiles (figure 3a and 3b). This argument, and the fact that little or no NAD is present, would tend to militate against such a possibility. The loss of the dependence of the reaction on added CoA when high cell concentrations are used would suggest that under these conditions, other cell-derived cofactors such as NAD(P), FAD and FMN are present in concentrations sufficient to allow significant side reactions to occur. In tranformations using low cell concentrations (and which are CoA-dependent), the pyruvate yield is improved to 30-35% and this may be due to the decreased rate of side reactions. The remaining alternative to explain the low efficiency is the possibility that the reaction is forming pyruvate which is then converted to other products such as acetolactate. Investigations are currently taking place to obtain accurate material balances within the reactor.

MV-NAD reductase and its use in cofactor recycling

The MV-NAD reductase present in C. sporogenes is a suitable enzyme for the recycling of NAD to NADH and is stable at physiological temperatures. An alternative enzyme, pig heart lipoamide dehydrogenase, is obtainable commercially and also shows good stability [35]. However, MV-NAD reductase is obtained from a microbial source in a simple permeabilised cell preparation and in conjunction with the 2-oxoacid synthase and other enzymes allows a variety of multistep transformations using only one organism.

Reactor design and mediated reactions

The limits to the rates of bioelectrochemical reductions are in fact strongly governed by the electrochemical reaction at the working electrode of the reactor since it is easy to saturate the overall reaction rate by addition of high concentrations of permeabilised cells. For efficient conversion and scale up, the relative area of the working electrode must be increased.

ACKNOWLEDGEMENTS

We are indebted to the SERC Biotechnology Directorate and ICI Bioproducts Business for generous financial support.

REFERENCES

[1] Zeikus J.G. (1980). Chemicals and fuel production by anaerobic bacteria. Ann. Rev. Microbiol. 34, 423-356

[2] Zeikus J.G. (1985). Biology of spore-forming anaerobes. In The Biology of industrial microorganisms, pg 79-114. Ed A.L. Demain and N.A. Solomon. Benjamin Cummings, Menlo Park, Calfornia, USA.

[3] Morris J.G. (1982). Anaerobic fermentations- some new possibilities. Biochem. Soc. Symp. 48, 147-172

[4] Simon H., Gunther H., Bader J. and Neumann S. (1985). Chiral products from non-pyridine nucleotide-dependent reductases and methods for NAD(P)H regeneration. In "Enzymes in Organic synthesis", pp 97-111. (CIBA foundation syposium No. 111) Pitman. London.

[5] Angermaier L. and Simon H. (1983). On nitroaryl reductase activities in several Clostridia. Z. Physiol. Chem. 354, 1653-1663

[6] Verhulst A., Semjen., Meerts U., Janssen G., Parmentier G., Asselburghs S., Van Hespen H. and Eyssen H. (1985). Biohydrogenation of linoleic acid by Clostridium sporogenes, Clostridium bifermentans Clostridium sordelli and Bacteriodes sp. FEMS Microbial Ecol. 31, 255-259

[7] Giesel H. and Simon H. (1983). On the occurence of enoate reductase and 2-oxo-carboxylate reductase in Clostridia and some observations on the amino acid fermentation by Peptostreptococcus anaerobius. Arch. Microbiol. 135, 51-57.

[8] Seto B. and Stadtman T.C. (1976). Purification and and properties of proline reductase from Clostridium sticklandii. J. Biol. Chem. 251, 2435-2439.

[9] Tanaka H. and Stadtman T.C. (1979) Selenium-dependent clostridial glycine reductase: purification and chacterization of the two membrane-associated protein components. J. Biol. Chem. 254, 447-452.

[10] Buchanan B.B. (1972) Ferredoxin-linked carboxylation reactions. In The Enzymes, vol IV, 3rd edition, pp 193-216. Edited by P.D Boyer. Academic Press. New York.

[11] Fuchs G., Stupperich E. and Eden G. (1980). Autotrophic CO_2 fixation in Chlorobium liminicola. Evidence for the operation of a reductive carboxylation cycle in growing cells. Arch. Microbiol. 128, 64-71

[12] Shiba H., Kawasumi K., Igarashi Y., Kadama T. and Minoda Y. (1985). The CO_2 assimulation via the reductive tricarboxylic acid cycle in an obligately autotrophic, aerobic hydrogen utilising bacterium, Hydrogenobacter thermophilus. Arch. Microbiol. 141, 198-203

[13] Harnick M., Aharonowitz. Y., Lamed R. and Kushman Y. (1983). Tetra- and hexahydro derivatives of aldosterone and 18 hydroxycorticosterone by chemical and microbial reductions. J. Steroid Biochem. 19, 1441-1450.

[14] Archer R.H., Maddox I.S. and Chong R. (1981). 7α-dehydroxylation of cholic acid by Clostridium bifermentans Eur. J. Appl. Microbiol. Biotechnol. 12, 46-52

[15] Lamed R., Keinan E. and Zeikus J.G.(1981). Potential applications of an alcohol/ketone oxidoreductase from thermophilic bacteria. Enzyme Microbial Technol. 3, 144-148.

[16] Butt S. and Roberts S.M. (1987). Opportunities for using enzymes in organic synthesis. Chem. Br.23, 127-134

[17] Holt, R.A., Thompson A.N. and Morris J.G. unpublished observations

[18] Lovitt, R.W. Morris J.G. and Kell, D.B. 1987. The growth and nutrition of Clostridium sporgenes NCIB 8053 in defined media. J. Appl. Bacteriol. 62, 71-80.

[19] Andrew I.G. and Morris J.G. (1965). The biosynthesis of alanine by Clostridium kluyveri. Biochim. Biophys. Acta. 97, 176-179.

[20] Simon H. and Gunther H. (1983). Chiral synthons by biohydrogenation or electro-enzymatic reduction In Studies in Organic Sythesis, Eds Yoshida et al, Vol 13, pp 207-227 . Elservier, Amsterdam

[21] Simon H., Bader J., Gunther H., Neumann S and Thanos J. (1985) Chiral compounds synthesized by biocatlytic reductions. Angew. Chem. Int. Ed. Engl. 24, 539-553.

[22] O'Brien R.W. and Morris J.G. (1971). The ferredoxin-dependent reduction of chloramphenicol by Clostridium acetobutylicum. J. Gen. Microbiol. 67, 265-271.

[23] Angermaier L and Simon H. (1983). On the the reduction of aliphatic and aromatic compounds by Clostridia, the role of ferredoxin and its stabilization. Z. Physiol. Chem. 364 961-975.
See also O'Brien R.W. and Morris J.G. (1972) Effect of metronidazole on hydrogen production of Clostridium acetobutylicum. Arch. Mikrobiol. 84, 225-233.

[24] James E.W., Kell. D.B., Lovitt R.W., and Morris J.G. (1988) submitted to Bioelectrochem. Bioenerg.

[25] Kaplan F., Setlow P. and Kaplan N.O. (1969). Purification and properties of DPNH/TPNH diaphorase from Clostridium kluyveri. Arch. Biochem. Biophys. 132, 91-98.

[26] Bader J., Gunther H., Nagata S., Schuetz H-J., Link M-L. and Simon H. (1984). Unconventional and effective methods for regeneration of NAD(P)H in microrganisms and crude extracts of cells. J. Biotechnol. 1, 95-109.

[27] Lovitt, R.W. Kell, D.B. and Morris J.G. (1987). The physiology of Clostridium sporogenes NCIB 8053 growing in defined media. J. Appl. Bacteriol. 62, 81-92

[28] Lovitt R.W., James E.W., Kell D.B. and Morris J.G. (1988). Characterisation of the carbon and electron flow pathways involved in the Stickland reaction of C. sporogenes, NCIB 8053. submitted to J. Appl. Bacteriol.

[29] Lovitt, R.W. Kell, D.B. and Morris J.G. (1986) Proline reduction by Clostridium sporogenes is coupled to vectorial proton ejection. FEMS Microbiology Letters 36, 269-273

[30] Wichmann R., Wandrey C., Buckmann A.F. and Kula M.R. (1981). Continuous enzymatic transformation in a membrane reactor with simultaneous NAD(H) regeneration. Biotechnol. Bioeng. 23, 2789-2802.

[31] Armstrong F.A., Hill H.A.O. and Walton N.J. (1986). Reactions of electron transfer proteins at electrodes. Q. Rev. Biophys. 18, 261-322

[32] Stadtman E.R. (1957). Preparation and assay of acetyl phosphate. Meth. Enzymol. 3 228-231

[33] Stadtman E.R. (1955). Phosphotransacetylase from Clostridium kluyveri Meth. Enzymol. 1 596-599

[34] Bergmeyer H.U. (1965) Methods of enzymatic analysis. Academic Press Inc. New York.

[35] Gunther H and Simon H. (1987). The use of pig heart dihydrolipoamide dehydrogenase (diaphorase) for the regeneration of NADH or NAD. Appl. Microbiol. Biotechnol. 26 9-12

INTERNATIONAL CONFERENCE ON BIOREACTORS AND BIOTRANSFORMATIONS
GLENEAGLES, SCOTLAND, UK: 9-12 NOVEMBER 1987

Paper G4

BIOCONVERSIONS IN PERMEABILIZED CELLS

H.R.Felix
AGRO Research,
Sandoz Ltd.,
4002 Basel,
Switzerland

ABSTRACT

Whole cells made permeable by chemical or physical treatment allow the conversion of substrates to useful, expensive products. Permeabilized cells are helpful if substrate or product or both cannot enter the cell. As many commercially interesting bioconversions require cofactor-dependent enzymes, regeneration of expensive cofactors is necessary. This is possible in permeabilized cells. Thus, permeabilized cells are useful in bioconversions, where cofactors are required and where substrates or products or both are not taken up or released by intact cells.

INTRODUCTION

The plasma membrane and the cell wall have to protect the living plasma from the environment. They create and preserve good conditions for cell metabolism. They play an important role in the regulation of substrate exchange between the cell and the surrounding medium. By active transport certain substances can be accumulated within the cell. On the other hand, the penetration of molecules can be limited or totally prevented, e.g. phosphorylated compounds cannot enter the intact cell. A series of methods has been developed to partially or totally overcome this barrier [1]. Cells can be permeabilized, without lysis of cells or destruction of enzyme systems. The morphology of the cells remains intact, low-molecular-weight molecules can freely enter and leave the cell. After permeabilization with organic solvents the cells most often are no longer viable, which can be useful, as energy is no longer wasted for the synthesis of cell mass. The viability depends largely on the concentration of the agent used and the time of its application.

Enzymes are capable of catalyzing complex and specific reactions efficiently and under mild conditions. Of these enzymes currently exploited industrially almost all are hydrolases because of their ready availability at low cost and their relative simplicity of action. They do not require low molecular mass cofactors or coenzymes for catalytic activity. Cofactors and coenzymes are expensive,thus regeneration is a necessity. Regeneration is possible in permeabilized cells as the necessary substrates can freely enter the cells. Often regeneration and bioconversion is carried out by the same organism.

It is a matter of specific circumstances whether treatment with organic solvents is defined as permeabilization or extraction. In some cases solubilization of substrates or products with organic solvents could be combined with membrane permeabilization. Product removal from a reactor (extractive bioconversions) is not considered here. In the future the integration of bioconversion, permeabilization and downstream processing will certainly open new possibilities.

PERMEABILIZATION OF CELLS

For bioconversions, cells (microorganisms, plants) are very often immobilized. External and internal mass tranfer limitations influence the rate of substrate conversion by immobilized biocatalysts in artificial and natural systems. The cell permeability coefficient is important in the context of this paper and can be increased by treating cells with compounds that change membrane permeability [1] . Depending on the membrane permeability characteristics of educt and product of a enzymatic conversion, permeabilization is necessary.

For bioconcersions, microorganisms were permeabilized with toluene [2-5] , benzene/n-heptane [6] , benzene/n-hexane/sorbitane monolaurate [7] , ethanol [8] and a bile extract [9] . Microorganisms can also be rendered permeable by different ways of drying: air-drying [4,10,11] , freeze-drying [12] , acetone-drying [3,4,6] . Sonication combined with an activation at 37°C [13] , pH treatment for activation [14,15] also make cells permeable. The choice of a special permeabilization method depends on the composition of the cell wall and the cell membrane. Among other things the difference between gram-positive and gram-negative bacteria has to be taken into account.

Plant cells are most frequently treated with dimethylsulfoxide (DMSO), known to extract sterols from eucaryotic membranes. In some bioconversions a reversible permeabilization would be interesting. Several methods exist such as phenethyl alcohol treatment of microorganism [16,17] , DMSO treatment of plant cells [18] . Plant cells entrapped in agarose or alginate, were permeabilized with DMSO, cell viability was retained. A specific concentration of DMSO and application time is necessary. Catharantus roseus cells intermittantly treated with DMSO released the produced ajmalicine isomers. DMSO also permeabilizes cells of the plant Cinchona ledgeriana and allows the release of alkaloids [19] . The cells however appear to be permanently damaged by the levels of DMSO necessary to produce release. In this respect DMSO is unsuitable for the long term harvesting of alkaloids from these C.ledgeriana cells. Clearly the suitability of DMSO for permeabilization varies with the culture. Part of the reason for this might be related to the conditions necessary for product release. In C.roseus a 30 min treatment with 5% DMSO produced nearly complete release of alkaloid [18] . In C.ledgeriana such a treatment was ineff-tive and much higher DMSO levels were necessary. This may reflect heterogenity of the cell population, but may also be due to the alkaloids being stored primarily in the vacuole, which requires slightly higher levels of DMSO for permeabilization than does the plasmalemma [20] . Perhaps in the case of C.roseus there may have been a rapid exchange of alkaloid between the vacuole and the cytoplasm which allowed a substantial release of alkaloid following permeabilization of the plasmalemma alone. Other possible explanations for the differences in behaviour between C.roseus and C.ledgeriana can also be proposed. It has for example been shown that there is considerable cell-to-cell variation in the accumulation of alkaloids in C.roseus cultures and possibly those cells in which alkaloid is concentrated are particular sensitive to DMSO. This may thus allow extensive release of alkaloids with little effect on overall viability in this species. It must also be mentioned that C.roseus cells were immobilized [18] , whereas C.ledgeriana were not [19] . Whatever the exact explanation for the different behaviour of various species, it is apparent that the DMSO permeabilization method is not universally applicable. The choice of an appropriate method depends on the organism and the composition of the cell wall and cell membrane.

Immobilized cells exist in an environment of reduced water activity resulting in distinct alterations of metabolic behaviour. In combination with limited oxygen supply, some mysteries reported for immobilized cells can be explained The immobilization itself can lead to altered membrane permeability. After immobilization in alginate Mucuna pruriens cells release 90% of the product dihydroxyphenylalanine [21] . A controlled release of oxalate from Amaranthus tricolor after immobilization and permeabilization with chitosan was observed [22] . Permeabilized C.roseus cells immobilized in alginate did not carry out any reaction dependent on phosphorylated cofactors [23] . This is most likely due to Ca^{2+} present in the incubation mixture (required for stabilization of the alginate gel), which prevents the reaction by ionic binding to the negatively charged phosphorylated coenzymes. Thus for certain reactions a neutral carrier such as agarose or a polyurethane foam is adequate.

A simple method, employing high-voltage electric discharge (electroporation), was developed to introduce phosphorylated nucleosides into the cytoplasm of viable cells. The same method has been applied for the release of secondary metabolites from plant cells [24] . It is, however, not clear, whether this method can also be used for large reactors.

REGENERATION OF COFACTORS OR COENZYMES

Of the about 2000 different enzymes to which a specific number has been assigned, about one third requires one of the four adenine coenzymes - NAD, NADP, CoA, and ATP - to participate in the enzymatic reactions. Many of the enzymes likely to be useful for the production of commodity chemicals and fine chemicals require an energy-rich cofactor. Without an efficient regeneration system any such process would be rendered impossible for economic reasons. Of those enzymes exploited industrially almost all are hydrolases because of their ready availability at low cost and their relative simplicity of action.

There are several methods of recycling oxidized and reduced forms of NAD(P) : chemically (autoxidizable electron acceptors), electrochemically (electrodes), enzymatically (isolated enzymes, permeabilized cells), photochemically. The enzymatic regeneration is the preferred one at present. Immobilized, permeabilized cells of the plant Catharantus roseus reduce the alkaloid cathenamine to ajmalicine. The NADPH required in the transformation is regenerated by isocitrate through the isocitrate dehydrogenase found in the same cell (Fig.1) [23] .

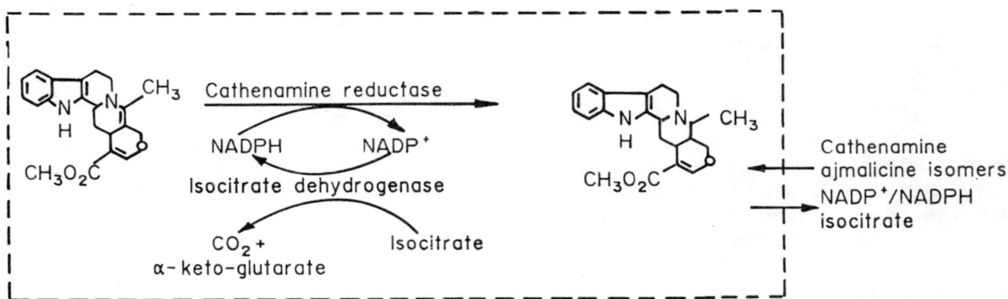

Figure 1. Conversion of cathenamine to ajmalicine isomers using permeabilized cells of the plant C.roseus and NADPH as coenzyme. NADPH is built up with isocitrate dehydrogenase from $NADP^+$. (---) Cell wall with permeable cell membrane, (⇌) free diffusion of substrates, products and cofactors [23].

An interesting method of $NADP^+$ reduction was found in the system of a methanogen using formic acid or hydrogen as electron donor [25] . Work has to be carried out under anaerobic conditions, which is certainly a disadvantage of the method, the advantage of cheap precursors does not outlevel that.

Extensive studies have been carried out on bioreactors, combining ATP-consuming and ATP-generating systems with immobilized biocatalysts to produce complex compounds in microorganisms. Acetate kinase using acetylphosphate to regenerate

ATP has been used in different bioreactor sytems [2,4,5] . Immobilized microbial cells can be utilized as a tool to regenerate ATP. Yeast cells phosphorylate adenosine, AMP, ADP to yield ATP using glycolytic energy [3,4,7,8,10,12,26] . A stable ATP-generating system was prepared by entrapping S.cerevisiae cells in polyacrylamide gel [3] .
Dextran-bound ADP was also used [4,5]. However, glutathione-producing activity of these gels was rather low, probably for the following reasons. First, the cosubstrate activity of dextran-bound ATP is lower than that of ATP itself. Second, steric interferences restricted the availability of ATP. Third, in the reaction system employed, the dextran-bound ATP is involved in the three sequential enzyme reactions catalyzed by acetate kinase and the constituents of the glutathione-forming enzyme system. Probably a single form of dextran-bound ATP does not show high activity in all three enzyme reactions.

APPLICATION OF PERMEABILIZED CELLS

Although permeabilized cells free in solution show higher enzymatic activities than do immobilized, permeabilized cells [2] , only the latter were investigated to any greater extent, because their operational stability is much larger. Bioconversions were carried out in microorganisms and plant cells. Industrial application seems possible.
Microorganisms. The ATP-generating systems were combined with ATP-consuming systems in the same microbial cell or in other cells.

CMP and choline were added to the ATP generating system (glycolysis) of Saccharomyces cerevisiae, CDP-choline formation was observed [7,10,12] .

S.cerevisiae cells entrapped in polyacrylamide gel were used for continuous production of glutathione from L-glutamate, L-cysteine and L-glycine using the glycolytic system in the same organism as energy source (Fig.2) [26] .

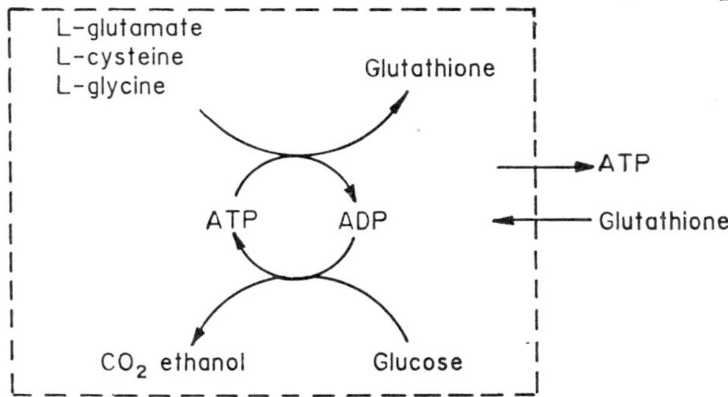

Figure 2 . Bioreactor (yeast cells) containing an ATP-regenerating system(glycolysis) and a glutathione synthesizing system. Yeast cells became permeable to glutathione after immobilization in polyacrylamide gel [26].

Because of its simplicity this sytems of *S.cerevisiae* [26] was found to be superior to the coimmobilized system of *E.coli* (synthesizing glutathione) and *S.cerevisiae* (generating ATP) [3,4] . The highest activity was found in an immobilized *E.coli* system using acetylphosphate as phosphate donor for ATP-generation [2] . The inhibitory effect of phosphate on glutathione synthesis seems to be the barrier to further development of the glutathione production system utilizing the glycolytic pathway as a tool for ATP generation. No such problems occur with *Candida tropicalis*, which produces extracellular glutathione during growth in filamentous form caused by adding 2.5% (v/v) ethanol [8] . Myo-inositol added at a physiological concentration (5µg/ml) prevented the ethanol-induced production of extracellular glutathione probably by stabilizing membranes. It is probable that the release of glutathione to the culture medium is the consequence of the effects of ethanol on the structure and function of cell membranes.

A co-immobilized system was also constructed for the production of CoA from pantothenate and cysteine. *Brevibacterium ammoniagenes* (synthesizing CoA) and *S.cerevisiae* (generating ATP) showed a higher activity of CoA synthesis using glucose as energy source than immobilized *B.ammoniagenes* alone, supplemented with ATP [7,11] .

Nicotinamide adenine dinucleotide phosphate ($NADP^+$) was produced by the phosphorylation of NAD^+ with *B.ammoniagenes* having NAD kinase activity and *S.cerevisiae* having glycolytic activity [3,4] .

ATP regenerated by the glycolytic pathway in acetone-dried *S.cerevisiae* can also be supplied to air-dried *S.cerevisiae* producing S-adenosyl-L-methionine [4] .

A series of other reactions was performed in permeabilized cells: L-malic acid synthesis from potassium fumarate in *Brevibacterium flavum* [9] , L-alanine synthesis from aspartic acid using pyridoxalphosphate as cofactor in this decarboxylation reaction [13] and L-alanine from ammonium fumarate [14,15] in *Pseudomonas dacunhae*.

Enzymatic and fermentative bioconversions, even in cases where water-insoluble, lipophilic compounds are substrates, usually have been carried out in aqueous systems, because enzymes and microbial cells were believed to be unstable in organic solvents. Immobilization often gives stability to biocatalysts against denaturation by organic solvents. Thus, conversion of water-insoluble compounds, such as steroids, have been carried out with immobilized microbial cells in organic solvent systems [27].

Although many organic compounds have been excluded from enzymatic reactions by the low solubility in aqueous solution, the examples mentioned [27] may encourage transformations of various hydrophobic compounds by using immobilized cells in organic solvents, as it has been common practice with isolated enzymes for quite some time [e.g. 28] . One limitation of these methods exist. Enzymes dependent on membrane systems, e.g. mixed function oxygenases carrying out hydroxylations, are excluded, because organic solvents destroy the essential membrane system.

Plant cells. In plant cells the advantageous utilization of immobilized biocatalysts may be impeded, as the produced metabolites of interest are often stored within the cell. The permeability of the cell membrane can be changed upon immobilization, either as a side effect of the immobilization method or by design. Products which can move within a plant across membranes either by diffusion or by active transport mechanism are likely to be exported from cells in culture at least under certain conditions.

Immobilization affects the partitioning of products between cells and the culture medium. In Catharantus roseus, free cells retain all of their alkaloid products within the vacuole while immobilized cells release up to 90% of the alkaloid produced into the medium. In cells of Mucuna pruriens immobilization in alginate leads to the release into the medium of L-dihydroxyphenylalanine (L-DOPA) formed from exogeneously supplied L-tyrosine while free cells of the same line make L-DOPA by de novo synthesis and store the product intracelluarly [21]. The modified metabolic behaviour of immobilized cells compared with free cells can be traced to the chemical and physical properties of the support material and the effect this has on the microenvironment. Charged groups on the polymer bind water and so effectively reduce water activity within the vicinity of cells in the gel leading to stress, increased maintenance metabolism and reduced growth.

Membrane properties can be manipulated externally, also. As mentioned earlier an intermittant release of alkaloids from C.roseus is possible [18] . Treatment of other plants with the same method resulted in cell death [19]. Relatively high DMSO concentrations were necessary to release intracellularly stored quinoline alkaloids from Cinchona ledgeriana [19] . There is a dependence between concentration of permeabilizing agent and degree of permeabilization. Various cell lines respond differently. Only tolerant cell lines may be permeabilized with preserved viability.

Except for certain specialist biotransformations, it seems unlikely that true immobilized plant cells can compete effectively in this area with a complex biocatalyst such as immobilized enzymes or microbial cells.

CONCLUSIONS

Permeabilized cells can be a good tool for bioconversions especially in microorganisms.As permeabilization of cells is often achieved with organic solvents, the latter can serve at the same time for the solubilization of hydrophobic substrates and products. In many cases authors showed the feasibility of using permeabilized cells.

REFERENCES

1. Felix, H., Permeabilized cells. Anal.Biochem., 1982, **120**, 211-234.

2. Murata, K., Tani, K., Kato, J. and Chibata, I., Glutathione production coupled with ATP regeneration system. Europ.J.Appl.Microbiol.Biotechnol., 1980, 10, 11-21.

3. Murata, K., Tani, K., Kato, J. and Chibata, I., Glycolytic pathway as an ATP generating system and its application to the production of glutathione and NADP. Enzyme Microb. Technol., 1981, 3, 233-242.

4. Murata, K., Tani, K., Kato, J. and Chibata, I., Continuous production of glutathione using immobilized microbial cells containing ATP generation system. Biochimie, 1980, 62, 347-352.

5. Murata, K., Tani, K., Kato, J. and Chibata, I., Application of immobilized ATP in the production of glutathione by a multienzyme system. J.Appl.Biochem., 1979, 1, 283-290.

6. Omata, T., Tanaka, A., Yamane, T. and Fukui, S., Immobilization of microbial cells and enzymes with hydrophobic photo-crosslinkable resin prepolymers. Europ.J.Appl.Microbiol.Biotechnol., 1979, 6, 207-215.

7. Samejima, H., Kimura, K., Ado, Y., Suzuki, Y. and Tadokoro, T., Regeneration of ATP by immobilized microbial cells and its utilization for the synthesis of nucleotides. In Enzyme Engineering, eds Brown, Manecke, Wingard, Plenum Press, New York, 1978, 4, 237-244.

8. Yamada, Y., Tani, Y. and Kamihara, T., Production of extracellular glutathione by Candida tropicalis Pk 233. J.Gen.Microbiol., 1984, 130, 3275-3278.

9. Takata, I., Yamamoto, K., Tosa, T. and Chibata, I., Immobilization of Brevibacterium flavum with carrageenan and its application for continuous production of L-malic acid. Enzyme Microb.Technol., 1980, 2, 30-36.

10. Kimura, A., Tatsutomi, Y., Mizushima, N., Tanaka, A., Matsuno, R. and Fukuda, H., Immobilization of glycolysis system of yeasts and production of cytidine diphosphate choline. Europ.J.Appl.Microbiol.Biotechnol., 1978, 5, 13-16.

11. Ogata, K., Shimizu, S. and Tani, Y., Studies on the metabolism of pantothenic acid in microorganisms. Part I. Distribution of CoA-accumulating activity in microorganisms and isolation of reaction products. Agr.Biol. Chem., 1971, 36, 84-92.

12. Ado, Y., Suzuki, Y., Tadokoro, T., Kimura, K. and Samejima, H., Regeneration of ATP by immobilized microbial cells and its utilization for the synthesis of ATP and CDP-choline. J.Solid-Phase Biochem., 1979, 4, 43-55.

13. Yamamoto, K., Tosa, T. and Chibata, I., Continuous

production of L-alanine using Pseudomonas dacunhae immobilized with carrageenan. Biotechnol.Bioeng., 1980, 22, 2045-2054.

14. Takamatsu, S., Tosa, T. and Chibata, I., Production of L-alanine from ammonium fumarate using two microbial cells immobilized with κ-carrageenan. J.Chem.Eng.Japan, 1985, 18, 66-70.

15. Takamatsu, S., Tosa, T. and Chibata, I., Industrial production of L-alanine from ammonium fumarate using immobilized microbial cells of two kinds. J.Chem.Eng.Japan, 1986, 19, 31-36.

16. Treick, R.W. and Konetzka, W.A., Physiological state of Escherichia coli and the inhibition of deoxyribonucleic acid synthesis by phenethylalcohol. J.Bacteriol., 1964, 88, 1580-1584.

17. Silver, S. and Wendt, L., Mechanism of action of phenethyl alcohol : breakdown of cellular permeability barrier. J.Bacteriol., 1967, 93, 560-566.

18. Brodelius, P. and Nilsson, K., Permeabilization of immobilized plant cells, resulting in release of intracellularly stored products with preserved viability, Europ.J.Appl. Microbiol.Biotechnol., 1983, 17, 275-280.

19. Parr, A.J., Robins, R.J. and Rhodes, M.J.C., Permeabilization of Cinchona ledgeriana cells by dimethylsulphoxide. Effects on alkaloid release and long-term membrane integrity. Plant Cell Rep., 1984, 3, 262-265.

20. Delmer, D.P., Dimethylsulfoxide as potential tool for analysis of compartmentation in living plant cells. Plant Physiol., 1979, 64, 623-629.

21. Wichers, H.J., Malingré, T.M. and Huizing, H.J., The effect of some environmental factors on the production of L-DOPA by alginate-entrapped cells of Mucuna pruriens. Planta, 1983, 158, 482-486.

22. Knorr, D. and Teutonico, R.A., Chitosan immobilization and permeabilization of Amaranthus tricolor cells. J.Agric. Food Chem., 1986, 34, 96-97.

23. Felix, H., Brodelius, P. and Mosbach, K., Enzyme activities of the primary and secondary metabolism of simultaneously permeabilized and immobilized plant cells. Anal. Biochem., 1981, 116, 462-470.

24. Brodelius, P., Shillito, R. and Potrykus, I., Verfahren zur Gewinnung von Pflanzeninhaltstoffen. Swiss Patent Application, 1986, 04063/86-3.

25. Eguchi, S.Y., Nakata, H., Nishio, N. and Nagai, S., $NADP^+$

reduction by a methanogen using HCOOH or H_2 as electron donor. Appl.Microbiol.Biotechnol., 1984, 20, 213-217.

26. Murata, K., Tani, K., Kato, J. and Chibata, I., Glutathione production by immobilized *Saccharomyces cerevisiae*, cells containing an ATP regeneration system. Europ.J.Appl. Microbiol. Biotechnol., 1981, 11, 72-77.

27. Fukui, S. and Tanaka, A., Immobilized microbial cells. Ann.Rev.Microbiol., 1982, 36, 145-172.

28. Kazandjian, R.Z., Dordick, J.S. and Klibanov, A.M., Enzymatic analyses in organic solvents. Biotechnol.Bioeng., 1986, 28, 417-421.

INTERNATIONAL CONFERENCE ON BIOREACTORS AND BIOTRANSFORMATIONS
GLENEAGLES, SCOTLAND, UK: 9-12 NOVEMBER 1987

Paper H1

THE CENTRIFUGAL FIELD BIOREACTOR - A NEW REACTOR TO OBTAIN HIGHER PRODUCTIVITIES DURING AEROBIC FERMENTATIONS, ESPECIALLY IN HIGH VISCOUS PSEUDOPLASTIC FERMENTATION BROTHS.

H. Voit , A. Mersmann
Lehrstuhl B für Verfahrenstechnik
Technical University of Munich
West-Germany

ABSTRACT

In order to increase productivity during high viscous, especially pseudoplastic fermentations the centrifugal field bioreactor was developed.

Aeration in centrifugal fields allow much higher oxygen transfer rates than in gravitational field reactors.

The paper deals with fluid dynamics and oxygen transfer rate in a centrifugal field bioreactor. Theory is compared with measured oxygen transfer rates during aerobic growth of yeast. Different viscosities were adjusted by adding Xanthan gum to the substrate.

Theoretical calculations allow a good prediction of real values.

INTRODUCTION

Productivity,P, during aerobic fermentation in common reactors usually is limited by the available oxygen transfer rate, OTR, if either cell density and growth rate are high (E-Coli, staphylococcus carnosus) or viscosity is high. Especially oxygen transfer during biosynthesis of polysacharide like Xanthan, Pullulan, Scleroglucan ... is a problem which is not yet sufficiently solved for high viscosity and extremely pseudoplastic flow behaviour.

Figure 1 for example shows theoretically possible productivities during continuous growth of E-Coli and xanthomonas campestris compared with technically realized values in common reactors.

In both cases OTR values are up to two orders of magnitude below the maximum productivities given by maximum cell density X_{max} and maximum growth rate μ_{max}.

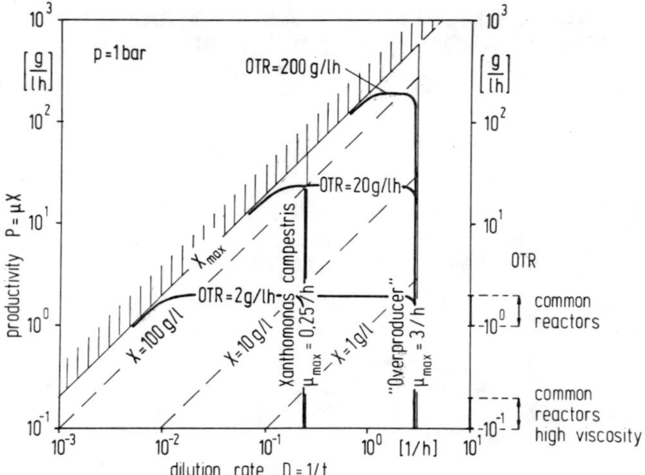

Fig. 1: Possible and oxygen limitated productivities in common reactors during aerobic growth for different organism.

In order to increase productivity during high viscous, especially pseudoplastic fermentations the centrifugal field bioreactor was developed at the Technical University of Munich.

Aeration in centrifugal fields allow much higher oxygen transfer rates than in gravitational field reactor because of

- smaller primary and secondary bubbles and higher volumetric mass transfer coefficients k_La
- higher oxygen solubility
- and lower effective viscosity in pseudoplastic fermentation browth according to

$$\eta = K^{\frac{1}{n}} \left(\frac{\Delta\rho g z d_B}{6} \right)^{1-\frac{1}{n}}$$

1. CENTRIFUGAL FIELD BIOREACTOR

Fig. 2 shows the principle of the centrifugal field bioreactor. Air passes through a sterile filter (1) and passes through a multi-way rotating device (2) into the rotor (3). Dispersion into very fine bubbles takes place at a cylindrical sieveplate or sintered ring (4).

Bubbles rise radially towards the centre and cause intensive gas-liquid mass transfer. Air leaves the rotor again over the multi-way rotating device. A mohno-pump (5) allows an external liquid circulation for indication and control of data like pH, p_{O2}, temperature, substrate concentration and cell density relevant to the process.

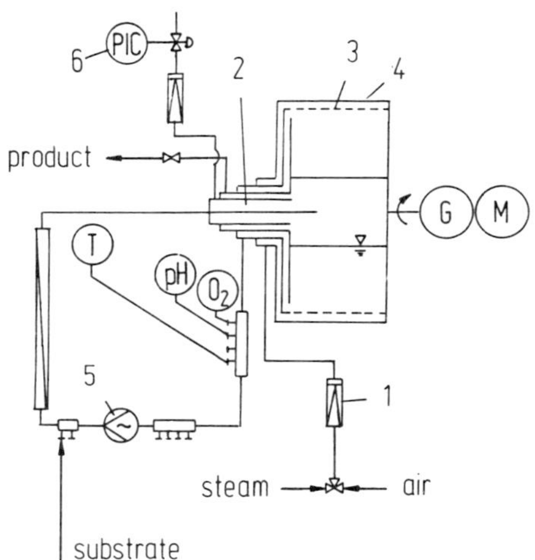

Fig. 2: Centrifugal field bioreactor

During continuous fermentation liquid volume in the centrifuge is controlled by a pressure valve (6). Precise knowledge of fluid dynamics including formation, break up and coalescence of bubbles is necessary for complete predictive calculation of oxygen transfer and productivity in a reactor as shown above.

2. PRODUCTIVITY AND FLUIDDYNAMICS IN A CENTRIFUGAL FIELD BIOREACTOR

Productivity is coupled with oxygen transfer rate and the yield coefficient Y

$$P = Y \cdot OTR \quad [1].$$

During growth for example yield coefficient $Y \cong 1$ lies almost about one [8]. Oxygen transfer rate is obtained from the product of liquid side mass transfer coefficient k_L, assuming limitation only in the liquid phase, volumetric interphase a and concentration difference Δc:

$$OTR = k_L a \cdot \Delta c$$

Since oxygen limited fermentation concentration difference is almost equal to saturation c* according to Henry's law

$$\Delta c = c^* = y \frac{p}{He} \frac{\tilde{M}_{O_2}}{\tilde{M}_{H_2O}} \cdot \rho_{H_2O}$$

The concentration difference, Δc, is proportional to p. Pressure, p, in a centrifugal field depends mainly on angular velocity ω, minimum radius r_i and actual radius r:

$$p = p_\infty + \frac{\rho}{2} r^2 \omega^2 \left(1 - \left(\frac{r_i}{r}\right)^2\right)$$

2.1 Effective shear gradient and effective viscosity in non newtonian fermentation browth, aerated in centrifugal fields

The effective shear gradient $\dot{\gamma} = \tau/\eta$ can be derived from an equilibrium of the shearing force $F_\tau = \tau \cdot S$ at the bubble surface $S = \pi d_B^3$ and the force due to buoyancy in a centrifugal field $F_A = (\pi/6)\Delta\rho g z d_B^2$, multiple acceleration due to gravity represented by the acceleration number $z = r\omega^2/g$.

In non-Newtonian liquids which obey the Ostwald de Waele rule, effective viscosity is obtained from a simple power law, $\eta = K \cdot \dot{\gamma}^{n-1}$, with flow index, n, and consistency factor, K.

Effective viscosity at the surface of a bubble moving radially in a centrifugal field finally results in

$$\eta = K^{\frac{1}{n}} \left(\frac{\Delta\rho g z d_B}{6}\right)^{1-\frac{1}{n}} \qquad [3].$$

In each case effective viscosity cannot fall below the Newtonian liquid viscosity, η_0. Fig. 3 for example shows the ground viscosity, η_0, for yeast suspensions in water.

2.2 Relative velocity of bubbles in centrifugal fields

Fig. 4 shows a diagram which allows evaluation at relative velocity of disperse particles in gravitational and centrifugal fields. This diagram is based on a diagram published by Mersmann [2] and found for Newtonian liquids in a gravitational field. With the effective viscosity introduced above this diagram can also be used for power law liquids and it is also applicable to centrifugal fields if the acceleration number is taken into account.

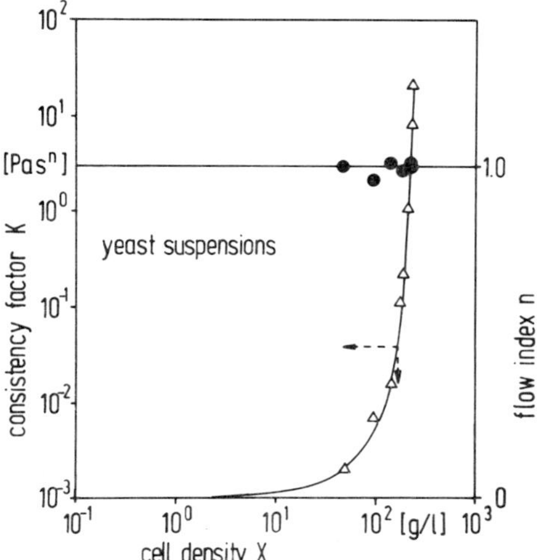

Fig. 3: Consistency factor and flow index of aqueous yeast suspensions

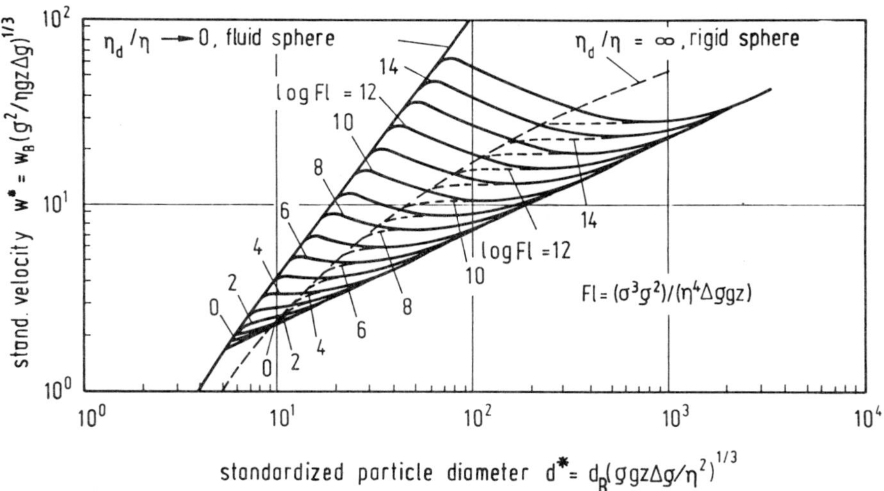

Fig. 4: Standardized relative velocity versus standardized particle diameter

2.3 Calculation of primary bubble volumes within gravitational and centrifugal fields

An extensive investigation of bubble formation in Newtonian and non-Newtonian liquids within gravitational and centrifugal fields resulted in a new standardized diagram [3]. It is based on a stationary force balance during bubble formation using the effective viscosity introduced before (Fig. 5)

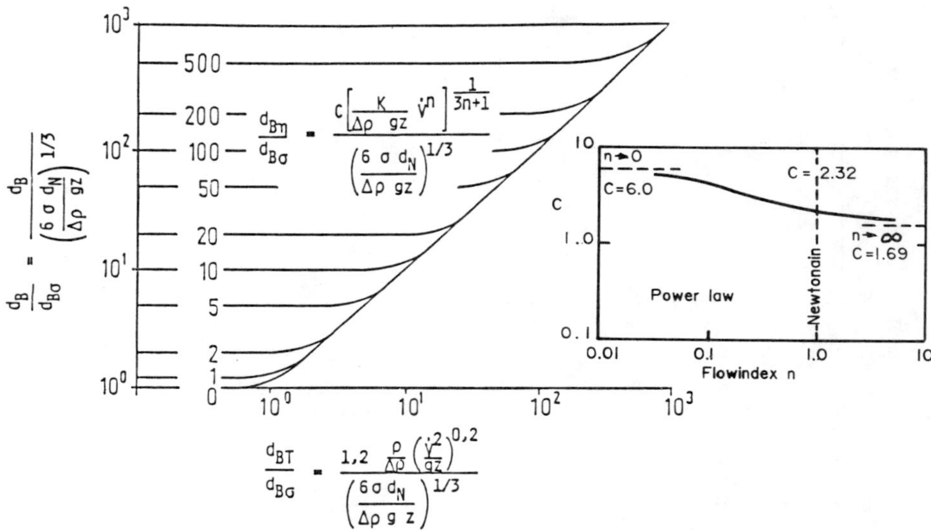

Fig. 5: Standardized particle diameter versus standardized velocity in centrifugal fields

Standardized bubble volume, $d_B/d_{B\sigma}$, is directly found with the explicitly calculable values $d_{BT}/d_{B\sigma}$ as abcissa value and the parameter $d_{B\eta}/d_{B\sigma}$. The values $d_{B\eta}$, $d_{B\sigma}$ and d_{BT} here mark those bubble volumes that result from a simplified equilibrium of viscosity, surface or inertia force and buoyancy forces.

Using this diagram as a minimum gas-flow rate, $\dot{V}$, through a single nozzle of the dispersing device it must be taken

$$\dot{V}_{min} = 1.11 \sqrt{\frac{d_N^5\,\Delta\rho gz}{\rho_d}}$$

according to minimum Froude number in analogy to sieve plates [5].

2.4 Secondary and mean bubble diameter in coalescing systems

Formation of secondary bubbles at sieve plates within gravitational fields was extensively investigated by [4].

Summarizing, secondary bubble diameter is not far away from equilibrium diameter obtained from a force balance at a single stationary moving spherical bubble

$$d_{GGW} = \sqrt{\frac{6\sigma}{\Delta\rho g}} \ .$$

This bubble diameter can also be used as mean diameter d_{32} in coalescing systems.

In an analogous way in centrifugal fields mean diameter should be taken as

$$d_{32} = \sqrt{\frac{6\sigma}{\Delta\rho g z}} \ .$$

2.5 Holdup in centrifugal fields

For predicting holdup in bubble columns there are a lot of different equations which cannot be extrapolated over their range of validity. These equations very often depend on column diameter and therefore they are not applicable to other geometrical configurations.

For predicting holdup in centrifugal fields a new holdup equation was found:

$$\psi = \frac{1}{2 + \frac{w_B}{w}}$$

This equation fits two extreme situations:

- At low superficial gas velocity holdup can be determined by the velocity ratio $\psi = w/w_B$
- At high superficial gas velocity holdup is nearly constant $\psi = 0.5$ before phase inversion occurs

This simple equation is in good agreement with values obtained in bubble columns.

Holdup in homogeneous non-coalescing systems without liquid circulation can be calculated in analogy to swarm behaviour known for drop-columns [5]. In this case a maximum superficial gas velocity, w_{max}, exist before flooding or foaming occurs:

$$w_{max} = \psi_{opt} \, (1-\psi_{opt})^{(3.6+\frac{1}{n})} \cdot w_B$$

This maximum superficial gas velocity depends on optimum holdup, ψ_{opt}, which equals $\psi_{opt} = 0.18$ in Newtonian liquids. In power law liquids

ψ_{opt} depends on flow index or Ostwald de Waele index n [6].

$$\psi_{opt} = \frac{1}{4.6+\frac{1}{n}}$$

2.5 Mass transfer coefficient

Mass transfer coefficient should be calculated assuming transfer limitation only within liquid phase:

$$Sh = 2 + f\cdot 2/\sqrt{\Pi}\,\sqrt{Pe} \qquad [7]$$

Reynolds, Peclet and Sherwood numbers are formed with bubble rising velocity, w_B, and primary bubble diameter or mean bubble diameter by using effective viscosity as introduced before.

Diffusion coefficient, D_L, was taken from pure media without poly-sacharid content.

Factor, f, is dependent on Reynolds number, see Mersmann [7]. Approximately, f can be calculated from the following equation:

$$f = 0.58 + 0.42 \left(\frac{Re}{Re + 300} \right)^{0.3}$$

2.6 Available oxygen transfer rate in centrifugal field bioreactors

Intensive investigation of fluid dynamic now allows prediction of oxygen transfer rate in centrifugal field bioreactors.

Fig. 6 shows the result of a calculation based on the above assumptions for a power law aqueous Xanthan solution ($K = 6.3 Pas^{0.18}$). The available oxygen transfer rate depends on specific aeration power ε the gas phase dissipates during aeration. An isothermal expansion is assumed.

$$\varepsilon = \frac{2\cdot\rho_N\cdot T\cdot\mathcal{R}}{\rho R} w_N\cdot\ln\left(1 + \frac{\rho g z R}{2p_\infty}\left(1 - \left(\frac{r_i}{R}\right)^2\right)\right)\cdot\frac{1}{\left(1-\left(\frac{r_i}{R}\right)^2\right)}$$

Homogeneous aeration allows higher oxygen transfer rates at low power consumptions. This effect decreases during scale up as then oxygen transfer rate becomes equal to aeration rate.

Available oxygen transfer rate is directly proportional to acceleration number. But in order to realize homogeneous aeration an optimum layout of the dispersing device is necessary.

Heterogeneous aeration does not require much constructive ideas. Available oxygen transfer rate increases with acceleration number, too

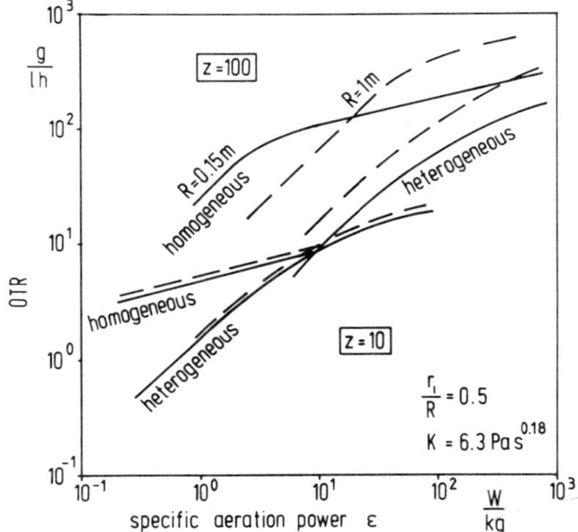

Fig. 6: Available oxygen transfer rate in a centrifugal field bioreactor versus specific aeration power

but this requires higher specific power consumption than in the case of homogeneous aeration. In comparison to oxygen transfer rates in common reactors absolute values in centrifugal field bioreactors are up to two orders of magnitude higher. This is especially valid for a different high viscous power law fermentation broth.

3. OXYGEN TRANSFER RATE IN CENTRIFUGAL FIELD BIOREACTOR DURING GROWTH OF BAKER'S YEAST

During growth of baker's yeast oxygen consumption and carbon dioxide production were measured by gas analysis.

Oxygen transfer rate was determined by a mass balance:

$$OTR = \frac{\dot{V}_N \tilde{M}_{O_2}}{V_m \, V} \left(y_o - \frac{y(1-y_o)}{(1-y-y_{CO_2})} \right)$$

Different viscosities and flow behaviour could be adjusted by adding Xanthan to the substrate.

As a dispersing device a sieve cylinder (d_N = 0.4 mm, t = 9 mm) with a wide nozzle distance was used to avoid bubble coalescence during formation and allow exact bubble size prediction.

Figure 7 shows some results obtained for liquids with different

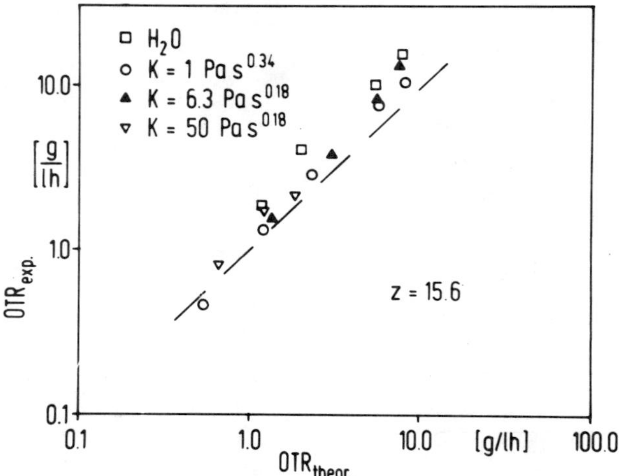

Fig. 7: Oxygen transfer rate for growth of yeast versus calculated values at different viscosities.

liquid viscosities. The results show, that even at low acceleration numbers oxygen transfer rates in aqueous newtonian and extreme power law liquids like Xanthan are almost equal. This can be explained by the fact, that effective viscosity in power law liquids can be reduced in centrifugal fields down to a small basic viscosity.

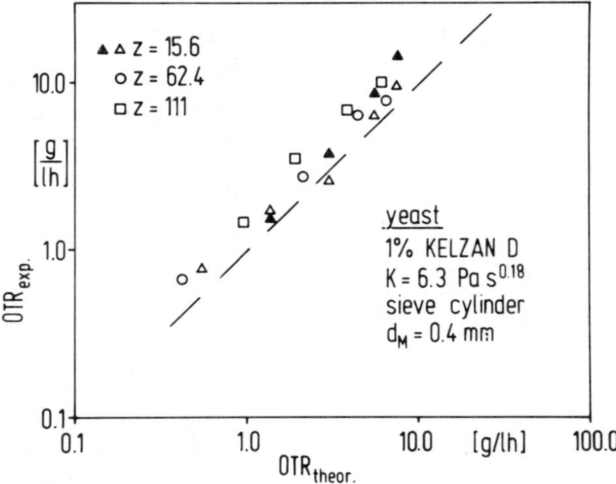

Fig. 8: Oxygen transfer rate during growth of yeast versus calculated values at differant acceleration numbers, z.

Theoretical calculation allows a good prediction of real values. Figure 8 shows some results obtained for different acceleration numbers. This figure shows that the influence of acceleration number can be predicted correctly by theory.

SYMBOLS

a	$[m^2/m^3]$	volumetric interphase
c	[g/l]	oxygen concentration in the liquid
c^*	[g/l]	oxygen saturation
d_B	[m]	bubble diameter
d_N	[m]	nozzle diameter
D_L	$[m^2/s]$	oxygen diffusivity in the liquid
F	[N]	force
$g=9.81$	$[m/s^2]$	acceleration due to gravity
He	$[m^2/s]$	Henry coefficient
k_L	[m/s]	liquid side mass transfer coefficient
K	$[Pa\ s^n]$	consistency factor
$\tilde{M}_{O2}=32$	[g/mol]	oxygen mole mass
$\tilde{M}_{H2O}=18$	[g/mol]	water mole mass
n	[-]	flow index
OTR	[g/lh]	oxygen transfer rate
P	[g/lh]	productivity
p	$[N/m^2]$	pressure
r	[m]	radius
r_i	[m]	inner radius
R	[m]	outer radius
S	$[m^2]$	bubble surface
t	[m]	nozzle distance
V	$[m^3]$	volume
$\dot{V}_N$	$[m^2/s]$	total gas flow rate at standarized conditions
$\dot{V}$	$[m^3/s]$	gas flow rate through a single nozzle
$V_m=22.4$	[l/mol]	ideal gas mole volume
w	[m/s]	superficial gas velocity
w_B	[m/s]	bubble rising velocity
y_o	[-]	mole fraction of oxygen in air
y	[-]	mole fraction after passing the reactor

y_{CO2}		mole fraction of carbon dioxide after passing the reactor
Y	$[\frac{\text{g product}}{\text{g } O_2 \text{ consumption}}]$	yield coefficient
z	$= \frac{r\omega^2}{g}$	acceleration number
$\dot{\gamma}$	[1/s]	shear rate
ψ	[-]	holdup
$\eta = K\dot{\gamma}^{n-1}$	[Pas]	dynamic viscosity power law liquids
η_o	[Pas]	ground viscosity of water or cell suspension
ρ	[kg/m³]	liquid density
ρ_d	[kg/m³]	gas density
ω	[1/s]	angular velocity

$$Re = \frac{w_B \, d_B \, \rho}{\eta}$$ Reynolds number

$$Pe = \frac{w_B \, d_B}{D_L}$$ Peclet number

$$Sh = \frac{k_L \, d_B}{D_L}$$ Sherwood number

REFERENCES

[1] Schügerl, K., Reaction engineering fundamentals relating to the design and operation of Bioreactors. Int. Chem. Engn. 28 (1986) (2), pp. 204-230

[2] Mersmann, A., Beyer von Morgenstern, I. and Deixler, A., Deformation, Stabilität und Geschwindigkeit fluider Partikel. Chem.-Ing.-Tech. 55 (1983) (11), pp. 865-867

[3] Voit, H., Zeppenfeld, R. and Mersmann, A., Calculation of primary bubble volume in gravitational and centrifugal fields. Chem. Eng. Technol. 10 (1987), pp. 99-103

[4] Klug, P., Diss. TU Clausthal 1983

[5] Mersmann, A., Auslegung und Maßstabsvergrößerung von Blasen- und Tropfensäulen. Chem.-Ing.-Tech. 49 (1977) 679, 691

[6] Brea, F.M., Edwards, M.F. and Wilinson, W.L., The flow of non-newtonian slurries through fixed and fluidized beds. Chem. Eng. Science 31, (1976) pp. 329-336

[7] Mersmann, A., Voit, H. and Zeppenfeld, R., Brauchen wir Stoffaustauschmaschinen ? Chem.-Ing.-Tech. 58 (1986) (2), pp. 87-96

[8] Bailey, J.E. and Ollis, D.F., Biochemical Engineering Fundamentals, 2nd Edition. McGraw-Hill Book Company, New York, 1986

INTERNATIONAL CONFERENCE ON BIOREACTORS AND BIOTRANSFORMATIONS
GLENEAGLES, SCOTLAND, UK: 9-12 NOVEMBER 1987

Reasonable but nervous presentation

Paper H2

THE BIOTECHNOLOGICAL POTENTIAL OF MICROBIAL HOLLOW-FIBRE BIOREACTORS

E.A. Linton, C.J. Knowles, A.W. Bunch and G. Higton
Biological Laboratory, University of Kent at Canterbury,
Canterbury, Kent CT2 7NJ, U.K.

ABSTRACT

The potential advantages of microbial hollow-fibre bioreactors include the high density to which cells may be grown, the simultaneous separation of product and biomass and the possibility of biocatalyst regeneration. A major disadvantage of hollow-fibre cell cultures is the difficulty of monitoring and controlling their growth and metabolism. A facultative anaerobe, *Streptococcus faecalis* var. *zymogenes*, and a strict aerobe, *Pseudomonas testosteronii*, were immobilized within the matrix of anisotropic hollow fibres. A means of continuously monitoring the growth of the cultures *in situ*, by electronic measurement of the transmembrane pressure flux and correlation of this parameter with total reactor protein is described.

INTRODUCTION

The development of novel bioreactors will be of crucial importance to the commercial viability of many biotechnological processes. Hollow-fibre bioreactors have been used for several years with enzymes as catalysts (1), for the cultivation of mammalian cells (2), and for the production of plant cell metabolites (3). Some basic, small-scale engineering principles have been established (4). More recently the technique has been applied to microbiological processes, including the production of lactic acid (5), conversion of L-histidine to urocanic acid (6) and biosynthesis of β-galactosidase (7).

The cultivation and immobilization of microorganisms within hollow fibres offers a number of inherent advantages over other immobilization methods. These include the high density to which cells may be grown, elimination of washout and simultaneous separation of product and biocatalyst. Other advantages include the possibility of biocatalyst regeneration, reduced capital and operating costs, the amenability of the system to scale-up and the achievement of growth and immobilization in a single step. A major disadvantage of hollow-fibre cell cultures is the difficulty of monitoring and controlling their growth and metabolism. Other disadvantages are the low mass transfer rates encountered at high cell densities and the problems associated with excessive growth, leading to blinding and rupture of the internal ultrafiltration membrane. In addition, little is known about the effect of containment of micro-organisms on their physiology, long term viability and productivity of simple or complex biotransformations.

In this paper we demonstrate immobilization of the lactic acid bacterium *Streptococcus faecalis* var. *zymogenes* and the strict aerobe *Pseudomonas testosteronii* within the matrix of anisotropic hollow fibres. The end-product profiles of the lactic acid bacteria have been examined as they are indicative of the physiological conditions under which they are grown, as are variations in parameters such as pH and dissolved oxygen tensions which cause characteristic alterations in major fermentation end-products. A novel system for the monitoring of growth *in situ* has been developed. Methods of enhancing low mass transfer rates of nutrients and dissolved gases, particularly oxygen, are currently being investigated.

METHODS

Organisms and Growth Conditions

Streptococcus faecalis var. *zymogenes* was grown on tryptone ($10gL^{-1}$), yeast extract ($2gL^{-1}$) buffered to pH 6.5 with K_2HPO_4 ($5gL^{-1}$) and HCl, to which glucose (10 or 28 mM) was added. Shake flask cultures were grown in 100 ml volumes in 250 ml conical flasks. Incubation was at 35°C in an orbital shaker (200 r.p.m.). The bacterium was maintained on the above medium containing glucose ($5gL^{-1}$) and purified agar (2% w/v).

Pseudomonas testosteronii was grown on minimal medium comprising M-9 salts (8) and glutamate ($2gL^{-1}$) at 30°C. The organism was maintained on nutrient agar slopes.

Hollow-Fibre Reactor

The overall scheme of the microbial hollow-fibre reactor system is illustrated in Figure 1. The reactor consisted of a cylinder of stainless steel (150mm length, 11mm internal diameter). Anisotropic hollow fibres

(PM 100; 1.84mm o.d., 1.1mm i.d., Romicon; Massachusetts), manufactured from polysulphone, containing an internal ultrafiltration membrane with a nominal molecular weight cut-off of 100,000 were used as the immobilization matrix. Nine fibres of 170 mm length were potted in Rapid Araldite (Ciba-Geigy Plastics, Cambridge). The fibre matrices (excluding the lumen space) occupied a total volume of 2ml, resulting in a matrix to shell-space ratio of 1:7.

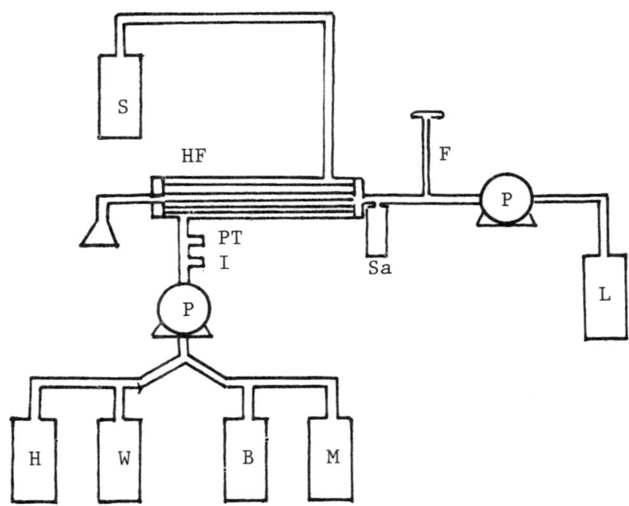

Figure 1. Schematic diagram of the microbial hollow-fibre experimental system. S, shell-side outlet; P, pump; F, flow meter; L, lumen outlet; Sa, sampling port; I, inoculation port; H, hypochlorite; W, water; B, buffer; M, growth medium; HF, hollow-fibre reactor; PT, pressure transducer.

Before inoculation the fibres were treated with acid, alkali and hypochlorite (9). Reactor components, with the exception of the hollow fibre bundle, were individually autoclaved, assembled aseptically, then the whole apparatus sanitised with sodium hypochlorite (200 ppm available chloride) and rinsed thoroughly overnight with sterile distilled water. A mid-exponential phase bacterial culture (20ml) was introduced into the shell-space and the pressure on the lumen side of the ultrafiltration membrane reduced by peristaltic pumping for 1h. The shell-space was then rinsed for 1h with K_2HPO_4/KH_2PO_4 buffer (pH 6.5, 100mM) to remove residual free-living bacteria from the reactor into the shell-space outlet vessel.

The shell-space outlet was closed and growth medium delivered to the culture at 10mlh^{-1}. The reactor was operated in the transverse mode i.e. medium was pumped into the shell-space, passed through the matrix and across the ultrafiltration membrane into the fibre lumen. The shell-side hydrostatic pressure was monitored using an electronic pressure transducer linked to a BBC microcomputer. The lumen pressure remained in equilibrium with the atmosphere throughout reactor operation.

Analytical Methods

The pressure transducer was obtained from Farnell Electronic Components Ltd. (Leeds). All other components in the pressure monitoring system were from RS Components Ltd. (Corby, Northants). The detection system was sensitive to 10^{-5} p.s.i. Thick walled silicone tubing was used to connect the transducer to a sterile hypodermic needle which was inserted into the reactor in the position indicated in Figure 1.

The concentrations of acetate and ethanol in the cell-free lumen outlet stream were determined by gas-liquid chromatography.

L-(+)-lactate was assayed using the L-(+)-lactate dehydrogenase method (10).

Glucose concentrations were measured by the glucose oxidase enzyme method (Boehringer Mannheim Test Combination Kit).

Microbial biomass in the shell-space, on the surface of fibres and within the hollow fibre matrix were individually determined using the modified Biuret method for whole cell protein determination (11).

Cell packing characteristics were studied using scanning electron microscopy. Fibre sections (1cm) were fixed, dehydrated, critical point dried, gold coated and then examined with a StereoScan S600 Electron Microscope (Cambridge Instruments).

RESULTS

Two microorganisms, the facultative anaerobe *Streptococcus faecalis* var. *zymogenes* and the strict aerobe *Pseudomonas testosteronii* have been immobilized within the matrix of anisotropic hollow fibres. The lactic acid bacterium reached a high density within the fibre matrix, growth occuring at and around the ultrafiltration membrane and within the interstices of the fibre wall (Figure 2).

In contrast, growth of the pseudomonad within the fibre matrix occurred at lower densities, being associated only with the surfaces of the chamber walls (Figure 3).

Figure 2. Streptococcus faecalis var. zymogenes growing within the matrix of a PM100 hollow-fibre; magnification x1000.

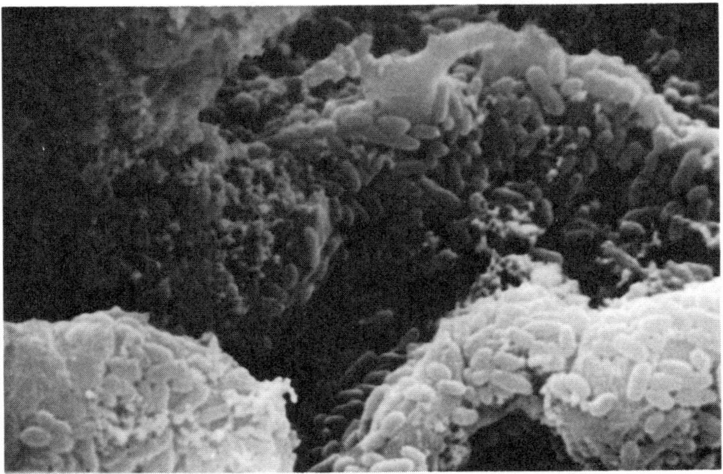

Figure 3. Pseudomonas testosteronii growing within the matrix of a PM100 hollow-fibre; magnification x 6800.

Anaerobic fermentation of glucose by S. faecalis var. zymogenes leads to production of lactate, whereas acetate is the major end-product of aerobic metabolism. Figure 4 describes the time course of a S. faecalis var. zymogenes fermentation in a hollow-fibre reactor under conditions of glucose limitation.

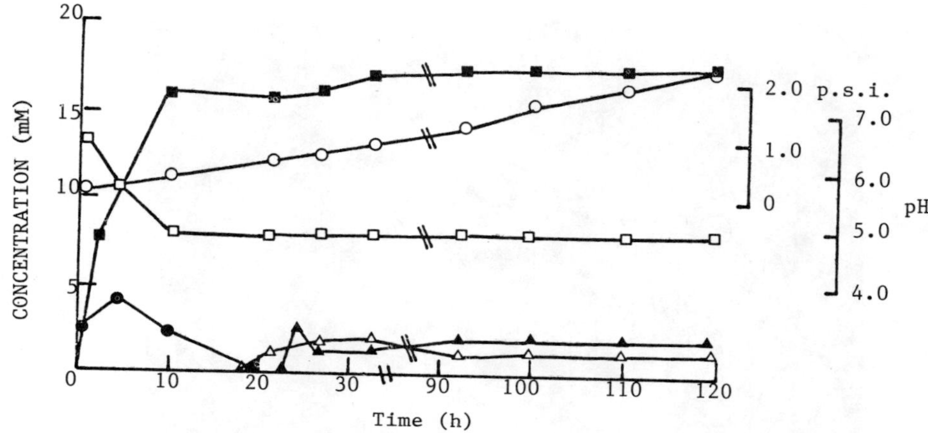

Figure 4. Growth of Streptococcus faecalis var. zymogenes in a hollow-fibre reactor under glucose limited conditions.
Glucose, ●; Acetate, ▲; Ethanol, △; pH, □; Lactate, ■; p.s.i., O.

The pH of the outflow medium fell to 4.9 reflecting acid production from the limiting amount of glucose present, the concentration of which fell rapidly to an undetectable level. No attempt was made to influence the dissolved oxygen tension of the incoming growth medium. Oxygen levels in the reactor were rapidly depleted by aerobic metabolism of glucose by the bacterial culture which produced small but significant amounts of acetate throughout the fermentation. However, the major end-product was lactate, indicating that anaerobic conditions prevailed in the greater part of the reactor. Ethanol also appeared as a fermentation product. This has not previously been reported as a product of glucose metabolism by batch cultures of this bacterium. The shell-side hydrostatic pressure flux (ΔP) was monitored throughout the fermentation. As a result of bacterial growth within the fibre matrix, the effective barrier between the shell-side and lumen phases increased and partial blinding of the membrane was observed. This raised the hydrostatic pressure necessary to force aqueous medium across the membrane at a constant fluid flow rate. During the course of the fermentation the pressure flux increased as the biomass increased in the reactor. The final ΔP value of 2.20 p.s.i. corresponds to a total reactor protein content of 104.2 mg, of which 41.7 mg were immobilized in the fibre matrix and 62.5 mg were associated with the fibre surface or were in free culture in the reactor shell-space.

Figure 5 illustrates the course of a culture growing under conditions of glucose excess.

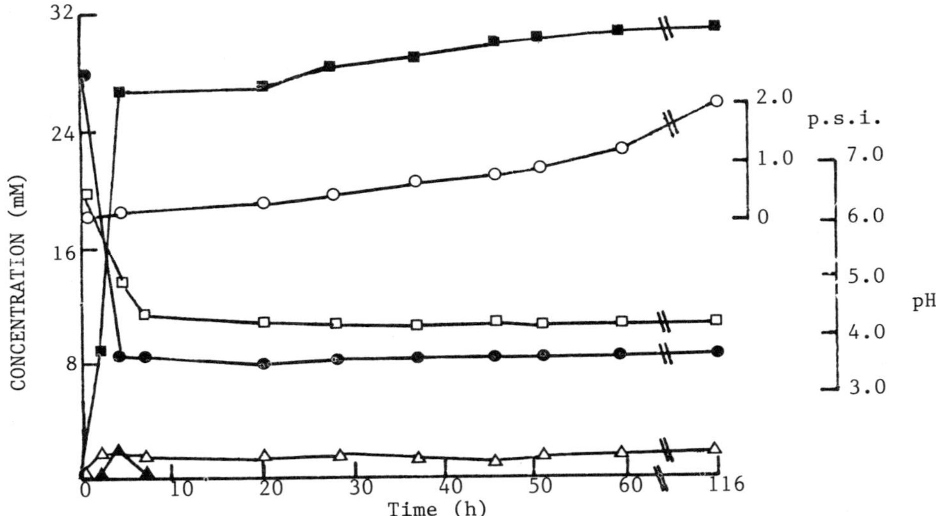

Figure 5. Growth of Streptococcus faecalis var. zymogenes in a hollow-fibre bioreactor under conditions of glucose excess. Glucose, ●; Acetate, ▲; pH, □ ; Lactate, ■ ; p.s.i., O; Ethanol, Δ.

The culture pH fell to a lower value than under glucose limitation, reflecting a greater degree of acid production. Lactate was again observed as the major fermentation end-product indicating the mainly anaerobic conditions within the reactor. Acetate appeared transiently, and ethanol was produced in small amounts throughout the fermentation. The pressure flux increased to a final value of 2.00 p.s.i., following a similar pattern to that observed under glucose limitation. This corresponded to a total reactor protein content of 87.9 mg, of which 33.4 mg were immobilized in the fibre matrix and 54.5 mg were associated with the fibre surface or were in free culture in the reactor shell-space.

The flux in shell-side hydrostatic pressure (ΔP) can be correlated with the total biomass within a reactor. Individual S. faecalis var.

zymogenes fermentations were allowed to proceed for a variety of times and a range of final shell-side hydrostatic pressures generated. Total reactor protein was assayed and plotted as a function of ΔP (Figure 6). The relationship is linear over the range examined, indicating that ΔP may be used as an index of total biomass within the reactor.

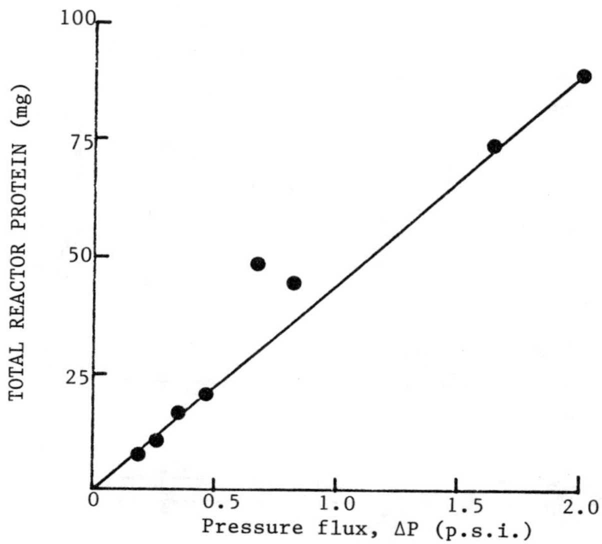

Figure 6. Total reactor protein as a function of the flux in shell-side hydrostatic pressure.

DISCUSSION

Streptococcus faecalis var. zymogenes and Pseudomonas testosteronii were immobilized within the matrix of anisotropic hollow fibres. However, the extent to which growth occurred differed; the lactic acid bacterium reached extremely high densities throughout the fibre matrix whereas growth of the pseudomonad was restricted to the surfaces of chambers within the anisotropic matrix structure. This presumably reflects the capacity of lactic acid bacteria to proliferate in the regions of the reactor where oxygen has been depleted, whereas pseudomonads cannot grow in such an environment. Methods to increase the supply of dissolved oxygen to microbial cultures within the hollow fibres are currently being investigated.

Figure 6 indicates that the trans-membrane pressure difference (ΔP) is a satisfactory index of the total protein present in the reactor over the range examined. Use of the pressure transducer with an electronic output opens up the possibility of coupling to computer control, permitting switching from a "growth" medium to a "maintenance" medium once the desired level of biomass is achieved and prior to membrane blinding and/or rupture. This will overcome the major obstacle to reactor breakdown due to excessive microbial growth. Similarly, an organism may be cultivated on an inexpensive growth medium, then presented with a more expensive substrate for biotransformation once a dense culture has been established. The capacity for biocatalyst regeneration can also be exploited using this approach.

If the biotechnological potential of hollow fibre reactors is to be fully realised it is essential that the effects of long term containment on the physiology of microbial populations is understood. Physiological problems such as those associated with mass transfer, particularly oxygenation, and the maintenance and extension of viability and biotransformation potential are currently being addressed.

REFERENCES

1. Chambers, R.P., Cohen, W. and Baricos, W.H. Physical immobilization of enzymes by hollow-fibre membranes. Methods in Enzymology, 1976, 44, 291-317.

2. Ku, K., Kuo, M.J., Delente, J., Wildi, B.S. and Feder, J. Development of a hollow-fibre system for large scale culture of mammalian cells. Biotechnol. Bioeng., 1981, 23, 79-95.

3. Shuler, M.L., Hallsby, G.A., Pyne, J.W. and Cho, T. Bioreactors for immobilized plant cell cultures. Ann. N.Y. Acad. Sci., 1986, 469, 270-278.

4. Webster, I.A. and Shuler, M.L. Whole cell hollow-fibre reactor: Efffectiveness factors. Biotechnol. Bioeng., 1979, 21, 1725-1748.

5. Vick-Roy, T.B., Blanch, H.W. and Wilke, C.R. Lactic acid production by Lactobacillus delbruckii in a hollow-fibre fermenter. Biotechnol. Lett., 1982, 4. 483-488.

6. Kan, J.K. and Shuler, M.L. Urocanic acid production using whole cells immobilized in a hollow-fibre reactor. Biotechnol. Bioeng., 1978, 20, 217-230.

7. Inloes, D.A., Smith, W.J., Taylor, D.P., Cohen, S.N., Michaels, A.S. and Robertson, C.R. Hollow-fibre membrane bioreactors using immobilized Escherichia coli for protein synthesis. Biotechnol. Bioeng., 1983, 25, 2653-2681.

8. Miller, J.H. Experiments in molecular genetics. Cold Spring Harbor Laboratory, Cold Spring Harbor, N.Y., 1972, 431-433.

9. Devereux, N. and Hoare, M. Membrane separation of protein precipitates: Studies with cross flow in hollow fibres. Biotechnol. Bioeng., 1986, 28, 422-431.

10. Gutmann, I. and Wahlefeld, A.W. L-(+)-lactate. Determination with lactate dehydrogenase and NAD. In Methods of Enzymatic Analysis (2nd edition), ed. H.U. Bergmeyer,1974, Academic Press, New York, 1974, 3, 1464-1468.

11. Gornall, A.C., Bardawill, C.J. and David, M.M. Determination of serum proteins by means of the biuret reaction. J. Biol. Chem., 1949, 177, 751-766.

INTERNATIONAL CONFERENCE ON BIOREACTORS AND BIOTRANSFORMATIONS
GLENEAGLES, SCOTLAND, UK: 9-12 NOVEMBER 1987

Paper H3

ENZYMATIC HYDROLYSIS OF 7 PHENYLACETAMIDODEACETOXYCEPHALOSPORANIC ACID TO 7-AMINODEACETOXYCEPHALOSPORANIC ACID WITH IMMOBILIZED PENICILLIN AMIDASE (EC 3.5.1.11) AT LABORATORY AND PILOT PLANT SCALES USING A NOVEL TYPE OF ENZYMATIC REACTOR

Joaquim Pereira Cardoso*, Manuel Bento Correia da Costa+,
Joaquim Perdigão Queiroga+ and Maria da Conceição Brandão+

*+ Cipan-Companhia Industrial Produtora de Antibióticos, SA
2580 Alenquer - PORTUGAL

* Laboratório de Engenharia Bioquímica,
Instituto Superior Técnico
1096 Lisboa Codex - PORTUGAL

ABSTRACT

This paper describes the studies done on the enzymatic conversion of 7 phenylacetamidodeacetoxycephalosporanic acid (Cephalosporin G) to 7 aminodeacetoxycephalosporanic acid (7-ADCA) both at laboratory and pilot plant levels, using a novel type of enzymatic reactor to deal with a compressible immobilized preparation of Penicillin Amidase. Preparation of the immobilized enzyme (IME) from E.coli cells by a cross-linking method, as well as the preparation of Cephalosporin G are briefly described.

The characteristics of the laboratory novel enzymatic reactor used to determine the optimal conditions of the hydrolysis of the substract in respect to concentration, reaction pH, temperature and buffer molarity are described.

The optimal conditions for the isolation of 7-ADCA were determined and the results of the scale-up of the hydrolysis reaction are also presented.

The modeling of the hydrolysis reaction is attempted using a published model developed to characterize the enzymatic hydrolysis of penicillin G to 6-aminopenicillanic acid (6-APA) and pehnyl acetic acid (PAA) [1].

From the work presented here it is concluded that the industrialization of such a process is possible and is a valid alternative to the chemical hydrolysis used in our plant.

INTRODUCTION

7-Aminodeacetoxycephalosporanic acid (7-ADCA) is a key intermediate for the production of semisynthetic deacetoxycephalosporins namely cephalexin and cephradine. It can be prepared from cephalosporin C by hydrogenation to deacetoxycephalosporin C and subsequent deacylation [2]. Deacetoxycephalosporin C can be converted into 7-ADCA, but this is not economical [3].

In the early 1960's it was demonstrated that the thiazolidine ring of penicillin could be expanded, although in law yield, to the dihydrothiazine ring of cephalosporin [4]. The production of 7-ADCA from penicillin became preferable to other methods, when an efficient procedure with high yields ($>$70%)was worked out [5]. It involved three steps: Oxidation of benzyl penicillin to its sulfoxide; Conversion of the sulfoxide into phenylacetamido-deacetoxycephalosporanic acid (cephalosporin G) with transient protection of the carboxylic group; Deacylation of Cephalosporin G to 7-ADCA.

This process has been improved substantially and is nowadays used with quantitative yields in several companies. It consists in the formation of penicillin sulfoxide using peracetic acid followed by ring enlargement of the 5 member thiazolidine ring of penicillin to give cephalosporin G. The latter compound can be chemically hydrolysed in solution to 7-ADCA or it can be isolated as a crystalline product which can further be chemically hydrolysed to 7-ADCA. The chemical hydrolysis of cephalosporin G to 7-ADCA irrespective of isolating or not the former compound is a reaction very similar to the chemical hydrolysis of penicillin G to 6-APA and PAA. As a consequence, cephalosporin G can be enzymatically hydrolysed to 7-ADCA, using the enzyme penicillin G amidase (EC 3.5.1.11) which is used already for the production of 6-APA starting with benzyl penicillin. In this article we will describe the work done on the enzymatic hydrolysis of cephalosporin G to 7-ADCA both at laboratory and pilot plant levels using an immobilized penicillin amidase produced at our premises and already used industrially in our 6-APA plant. A novel type of biocatalytic reactor used in the laboratory and in the pilot plant, both capable of handling the compressible enzyme is described. The modelling of the hydrolysis of cephalosporin G to 7-ADCA and PAA is effected by using a two parameter model developed to model the hydrolysis of penicillin G to 6-APA and PAA [1].

MATERIALS AND METHODS

Materials

All the materials used for the production of penicillin amidase in fermentation, isolation and immobilization were of a technical grade. Reagents for analysis were of laboratory grade. Cephalosporin G was produced in the Chemical Synthesis Department of Cipan's Plant. The biocatalytic reactor used at laboratory scale was developed by the R & D Cipan's Department.

Preparation of immobilized penicillin amidase

Immobilized penicillin amidase was prepared from E.coli (ATCC 9637) cell suspension as follows: Cells were collected by centrifugation and disrupted in a Manton Gaulin homogeniser (model 15 M 8TA). The debris was separated by centrifugation and the protein was precipitated with isopropyl alcohol and redissolved in acetate buffer. The immobilized enzyme (IME) was obtained in a particulate form by cross-linking the soluble enzyme with glutaraldehyde in appropriate conditions. The IME particles were further classified by using a 2mm wire mesh sieve.

Preparation of cephalosporin G from benzyl penicillin

The production of cephalosporin G involved two steps: Oxidation of penicillin G to its sulfoxide; Conversion of the sulfoxide to the corresponding cephalosporin G with transient protection of carboxyl group. The reactions involved are shown in Figure 1.

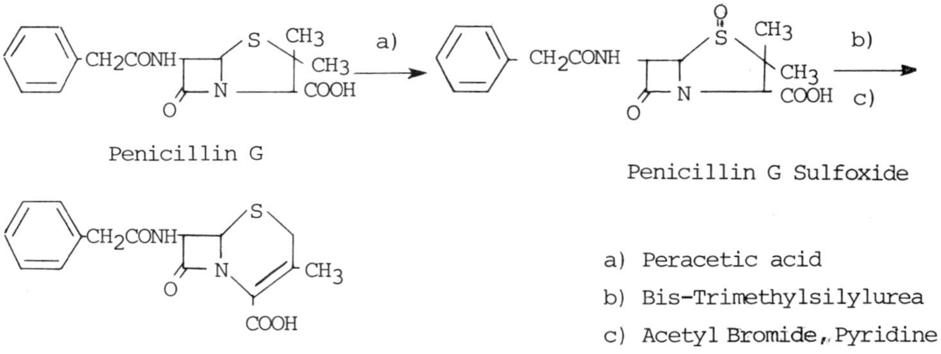

Penicillin G

Penicillin G Sulfoxide

a) Peracetic acid

b) Bis-Trimethylsilylurea

c) Acetyl Bromide, Pyridine

Cephalosporin G

Figure 1. Chemical Synthesis of Cephalosporin G by ring expansion from Penicillin G

Determination of enzyme activity

The activity of immobilized enzyme preparation was determined in a small batch stirred reactor with automatic pH correction, in a 4.0% cephalosporine G solution at pH 8.0 in 0.02M phosphate buffer at 35ºC. A weight/volume ratio of 10g IME per 100ml of cephalosporin G solution was used. The activity was calculated as initial velocity according to the expression:

$$\text{Activity in U.g}^{-1}\text{ IME} = \frac{\text{Volume of alkali (ml)} \times M \times 1{,}000}{(\text{g IME}) \times \text{time (min)}}$$

one unit of activity being the number of micromoles of 7-ADCA produced in one minute. M is the molarity of the alkali used.

Determination of IME stability

The stability of the IME was determined both in the laboratory and in pilot plant reactors during operation. At some time intervals, the enzyme was removed from the reactor, weighted and the activity determined in a representative 10 g sample.

Conditions of the enzymatic hydrolysis of cephalosporin G

The optimal conditions of the enzymatic hydrolysis of cephalosporin G in Figure 2 in respect to concentration, reaction pH, temperature and buffer molarity were determined at constant reaction time. The conversions were determined by the volume of a 4M sodium hydroxide solution consumed to keep the pH constant at 8.0.

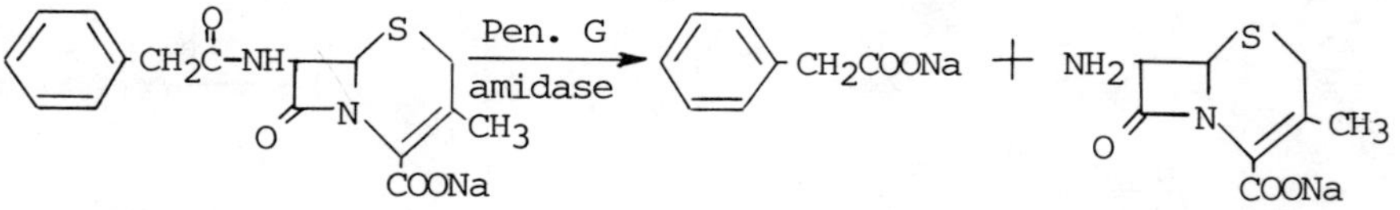

Cephalosporin G (Na salt) (PAA) (Na salt) 7-ADCA (Na salt)

Figure 2. Enzymatic hydrolysis of Cephalosporin G to 7-ADCA and PAA

Conditions of 7-ADCA isolation

The optimal conditions of 7-ADCA isolation were determined in respect to pH and temperature of precipitation.

HPLC assay of 7-ADCA, residual cephalosporin G and PAA

The concentration of 7-ADCA in converted solution was also determined by HPLC method using a Perkin Elmer series 10 liquid chromatograph. Residual cephalosporin G and PAA were also determined by HPLC, using the following analysis conditions:

7-ADCA		Residual cephalosporin G and PAA	
Column	RP C18	Column	RP C8
Temperature	40ºC	Temperature	35ºC
Mobile phase	15% CH_3CN/10mM $NH_4H_2PO_4$+5mM TBA, pH 5.7	Mobile phase	15% CH_3CN/10mM Na_2HPO_4+5mM NaCl
Flow rate	1.0 ml/min	Flow rate	1.5 ml/min
Detection	U.V. (220 nm)	Detection	U.V. (220 nm)

Modelling of the hydrolysis reaction

We have used a two parameter model to describe successfully the hydrolysis of benzyl penicillin potassium salt by immobilised penicillin G amidase (EC 3.5.1.11) [1].

Such a model represented by the equation $Et/V_oX=\alpha+\beta|(\ln(1-X))/X|$ was here assessed to describe the hydrolysis of cephalosporin G by the same immobilised enzyme since the reaction and inhibition effects are of the same type. For the sake of simplicity the reader is advised to consult reference [1] where the details of the data processing to use the above equation are outlined.

Principle of the novel type biocatalytic reactor

Since the IME particles displayed a gel like structure neither a normal packed column nor an agitated reactor could be used. A fluidised bed reactor also could not be used due to the small density difference between the IME and the substrate solution. We therefore had to develop a reactor capable of handling a very compressible and soft IME preparation and which could be conveniently scaled up. The laboratory biocatalytic reactor developed consists of a shallow bed cylindrical column with a jacket, made of stainless steel with capacity for about 300 ml of bulk enzyme (180 g of IME). The enzyme was confined in the column by means of two wire meshes with apertures of about 1mm, one at the top and the other at the bottom of the column. The bottom and top wire meshes were supported by a coarse wire mesh with about 5mm aperture so to provide, the bottom, channels for the drain of the substrate solution, the top, a better distribution of the substrate solution. The substrate solution was fed to and collected from the enzymatic reactor through a four tube distributor. Figure 3 is a schematic representation of the laboratorial biocatalytic reactor which has a height of 50 mm.

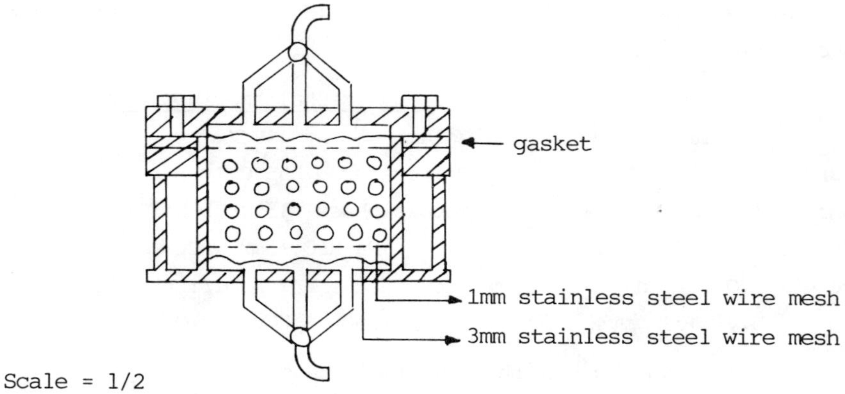

Figure 3. Laboratory enzymatic reactor

To scale-up the laboratory reactor, we used a Niagara type pressure filter with about 80 liters capacity (60kg of IME). This filter used as biocatalytic reactor is ilustrated in figure 4.

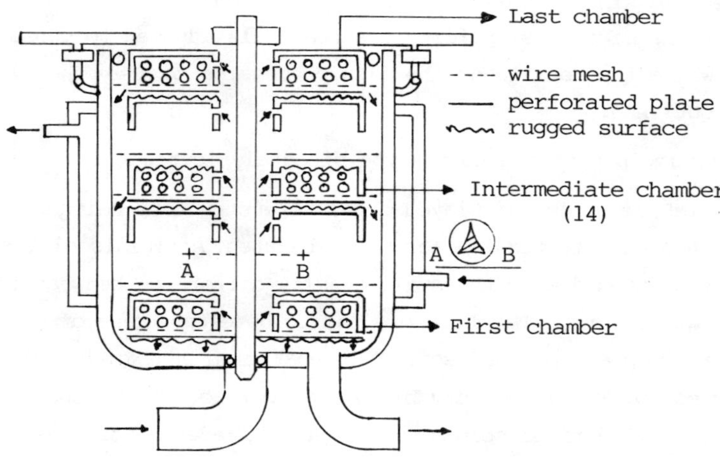

Figure 4. Pilot plant biocatalytic reactor

Description of the laboratory enzymatic plant

The enzymatic hydrolysis plant consists of: the enzymatic reactor itself; an agitated cylindrical jacketed stainless steel reactor of 2 litres capacity; a peristaltic pump capable of circulating about 50 l h^{-1}; a Metrohm pH--stat with a 645 Multi-Dosimat unit; a temperature control system; glassware; buckner type filters and vacuum dryer oven for the isolation of 7-ADCA from the converted solution and drying.

Description of the industrial enzymatic pilot plant

The enzymatic hydrolysis plant consists of the following parts:

- one Niagara type pressure filter, the biocatalytic reactor, with 80 litres solids capacity (about 60kg of wet enzyme). This filter has 16 plates each with a diameter of 50cm and height 5 cm. The total area of the filter is about 2.9 squared meters. To retain the IME in the filter each plate was covered with a stainless steel wire mesh about 1mm aperture;
- two stainless steel stirred jacketed vessels, with 500 litres working capacity, one for the make up of the cephalosporin G solution and the other for the circulating of the substrate through the enzymatic reactor;

a centrifugal pump capable of circulating about $30m^3h^{-1}$ at a pressure of 2kg cm^{-2}; a Watson Marlow peristaltic pump model MHRE; an automatic pH control system, from Instrumentation Laboratories with an electrode; a rotameter; a temperature control system; two 1,000 litres stainless steel jacketed vessels for the crystallization of 7-ADCA; one basket centrifuge for 7-ADCA crystals separation; one vacuum drier oven for the drying of 7-ADCA.

Figure 5 shows schematically a diagram of the laboratory and pilot plant installations for the preparation of 7-ADCA.

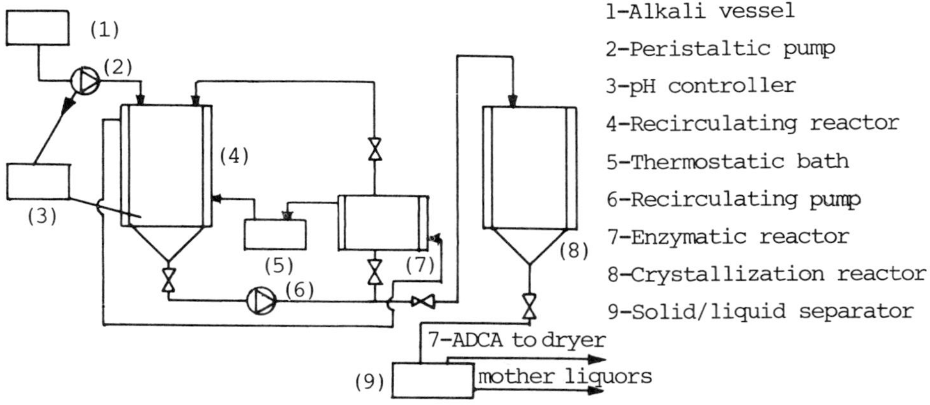

Figure 5. Flow sheet of enzymatic 7-ADCA laboratory/pilot plant installation

RESULTS

Results obtained in the laboratory pilot reactor

Table 1 presents the influence of reaction pH, temperature and molarity of buffer on the activity of the IME measured by the conversion attained at the end of 60 minutes. The constant conditions in all experiments were: Volume of substrate solution: 1,000 ml; Amount of IME (units): 2,780; Time of reaction: 60 min.; Concentration of substrate: 40 g/l.

TABLE 1

Influence of reaction parameters on the degree of conversion

pH Temp.^{re} 35ºC	X(%)	Temp.^{re} (ºC) pH=8.0	X(%)	Molarity(M) pH =8.0 Temp.^{re}=35ºC	X(%)
7.65	67.76	32	78.70	0.02	84.86
8.00	84.86	35	84.86	0.04	86.60
8.35	84.83	38	86.06	0.08	92.86

Although from Table 1 optimal temperature and buffer molarity could not be determined the figures underlined were those selected for all further laboratory reactions and scaling up purposes. This is so because stability of the IME decreases exponentially with temperature and because for buffer molarities greater than 0.04M crystallization yield decreases. Table 2 presents the influence of substrate concentration (in g/l) on the degree of conversion at the end of 60 minutes. The constant conditions used were: Temperature 35ºC pH of reaction 8.0; Volume of substrate 1,000 ml.

TABLE 2

Conversion as a function of substrate concentration in g/l at constant volume

Substrate concent. (g/l)	Conversion attained (%)	Amount of substrate converted (g)
20	90.10	18.02
40	84.86	33.94
60	73.30	43.98

As it can be observed from Table 2 the amount of cephalosporin G converted increases with substrate concentration. However, from concentrations greater than 40 g/l the final conversion attained when equilibrium is reached is not greater than about 95%. On the other hand, low substrate concentrations will imply low yiels of crystallization. We therefore selected as working concentration between 30 and 40 g/l. Table 3 presents the results achieved to evaluate the operational stability of the IME. The data of this Table was obtained by performing about 120 hydrolysis reactions using the conditions determined above that is, pH 8.0; temperature 35ºC; buffer molarity 0.04M;

substrate concentration 30 g/l and reaction volume 1,000 ml. Applying to the data of Table 3 the inverted linear model (I.L.M.) [6] and the exponential model (E.M.) [7] of enzyme decay we obtain respectively:

I.L.M. t 1/2 = 1,077.5 h, r = 0.961 E.M. t 1/2 = 874.3 h, r = 0.955

Although the correlation coefficient r is not very high, the ILM describes better the decay of IME. This had already been observed with the same IME working with Penicillin G [1] .

TABLE 3

Experimental data to calculate the operational stability of the IME

Time of operation (h)	Weight of IME (g)*	Specific activity ($U\ g^{-1}$)	Total activity (U)	Specific act. referred to the initial weigh ($U\ g^{-1}$)
0	162	17.16	2,780	17.16
190.75	162	13.03	2,170	13.02
340.39	145	14.46	2,100	12.96
403.97	145	14.02	1,991	12.29

* The decrease in weight is due to loss of water from the preparation.

Table 4 presents the results of the experiments performed for the determination of the best isolation conditions of 7-ADCA in respect to crystallization pH and temperature using a 4 hours period of crystallization. The hydrolysis reactions were performed according to the conditions indicated in page 8.

TABLE 4

Influence of pH and temperature of crystallization on the yield of 7-ADCA isolation

Crystallization pH (at 20ºC)	Yield (% of theoretical)	Crystallization Temp.^{re} (ºC) (at pH 3.5)	Yield (% of theoretical)
3.0	87.5	5	90.4
3.5	90.0	10	90.0
4.5	86.9	20	90.4

It can be seen that the optimal pH is about 3.5 and that temperature does not influence substantially the yield of 7-ADCA isolation. For economic reasons we will use 20ºC as the crystallization temperature.

Results obtained in the pilot plant reactor
(scale up factor of 300-400 times)

Using the laboratory optimal conditions for the hydrolysis of cephalosporin G (pH 8.0; temperature 35ºC and substrate concentration 33g/l) and for the isolation of 7-ADCA (pH 3.5; temperature 20ºC and crystallization time 4 hours) we performed several runs in the pilot plant reactor. Table 5 presents the results achieved in the two tests performed.

TABLE 5
Average results obtained in the pilot plant installation

Parameters	test 1	test 2
Amount of Cephalosporin G converted (kg)	342.5	461.0
Total time enzyme operation (hours)	171.16	199.0
Number of reactions	38	46
Mean conversion time (hours)	4.50	4.33
Mean conversion attained (%)	97.0	97.5
Amount of 7-ADCA produced (kg)	194.7	264.6
Yield of 7-ADCA (% of theoretical)	88.2	89.0
Initial activity (U)	981,240	1010,957
Final activity (U)	848,059	853,350
Half life (I.L.M.) (hours)	1089.9	1077.5

Fitting the experimental results of laboratory and pilot plant
by the α, β model

We were very interested in assessing the suitability of the α, β model to describe the hydrolysis of cephalosporin G by the immobilized penicillin acylase to 7-ADCA and PAA not only in the laboratory reactions but also in the pilot plant reactions. To this end we selected randomly some reactions at the laboratory and pilot plant levels to which we applied the α, β model. Tables 6 and 7 present for the runs selected respectively for the laboratory and pilot plant levels the α and β values and the correlation coefficients obtained.

TABLE 6

Fitting of the laboratory reactor data to the α, β model

Lab. run No.	Mean activity (U 10^{-3})	$\alpha \times 10^{-3}$ (min. U l^{-1})	$\beta \times 10^{-3}$ (min. U l^{-1})	r
AL	2.54	115.75	110.85	0.9916
BL	2.53	114.80	110.39	0.9921
CL	2.52	128.22	106.62	0.9874
DL	2.51	105.73	116.54	0.9946
	Average values	116.13	111.10	

TABLE 7

Fitting of the pilot plant reactor data to the α, β model

Pilot run No.	Mean activity (U 10^{-3})	$\alpha \times 10^{-3}$ (min. U l^{-1})	$\beta \times 10^{-3}$ (min. U l^{-1})	r
AP	888.864	143.97	154.57	0.9973
BP	863.630	148.57	161.04	0.9963
CP	857.380	158.41	150.24	0.9991
DP	854.690	169.26	170.85	0.9957
	Average values	155.05	159.18	

DISCUSSION AND CONCLUSIONS

It can be concluded that the scale up of the preparation of 7-ADCA from cephalosporin G yield results in accordance with those obtained at the labo ratory level.

In fact although the global yield at the laboratory level was about 90% and that at the pilot plant level 88.2 in the first test and 89.0 in the second test these small differences can be attributed to the difference in quality of the substrate.

In the pilot plant level a higher activity to substrate ratio was used to obtain the same conversion. At the same time the conversion times were also higher. This comparison is shown below in terms of initial activity used since the decay of the IME was equivalent at both scales.

Laboratory level	Pilot plant level (test 1)
2780 U/30 g substrate = 92.7 Ug^{-1}	981240U/9000g substrate=109.0 Ug^{-1}
Average conversion time 3.4 h	Average conversion time 4.5 h

Although the higher substrate concentration used at pilot plant level (33 gl^{-1} against 30 lg^{-1}) would imply a higher conversion time, these results show a lower efficient of the pilot plant reactor. This may be due to the fact that the linear velocity across each plate of the reactor was about 15 cm/min whereas the optimal value should be around 20 cm/min. The α, β model adequately described the enzymatic hydrolysis of cephalosporin G to 7-ADCA and PAA both at laboratory and pilot plant levels. The higher values of α and β also reflect the lower efficient of the pilot plant reactor. Despite this lower efficiency of the pilot plant reactor an economic evaluation shows that the process of preparation of 7-ADCA with the enzymatic hydrolysis step is a valid alternative to the preparation of this product by the complete chemical process.

ACKNOWLEDGEMENT

This work was developed in connection with a project supported by the Junta Nacional de Investigação Científica e Tecnológica.

REFERENCES

1. Cardoso, J.P. and Costa, M.B.C., Modellling of the hydrolysis of Benzylpenicillin to 6-Amino Penicillanic Acid and Phenyl Acetic Acid by an Immobilized Penicillin Amidase in a small Pilot Plant Batch Recirculated Reactor. British Polymer Journal, 1986, 18 (5), 323-32.

2. Stedman, R.J., Swered, K.H., and Hoover, J.R.E., 7-Aminodeacetoxycephalosporanic Acid and its derivatives. J. Med. Chem., 1964, 7, 117-119.

3. Liersch, M., Nuesch, J., and Treichler, M.J., Final steps in the biosynthesis of Cephalosporin C. In 2nd International Symposium on the Genetics of Industrial Microorganisms, 1974, ed. K.D. Mac Donald, Academic, London 1976, p.p. 179-95.

4. Morin, R.B., Jackson, B.G., Mueller, R.A., Lavagnino, E.R., Scnlon, W.B., and Andrews, S.L. Chemistry of Cephalosporin antibiotics III. Chemical correlations of penicillin and cephalosporin antibiotics. J. Am. Chem. Soc., 1963, 85, 1896-7.

5. Koning, J.J. de, Kooreman, H.J.; Tan, H.S., and Verweig, J. One step, high yield conversion of penicillin sulfoxides to deacetoxycephalosporins. J. Org. Chem., 1975, 40, 1346-7.

6. Cardoso, J.P. and Emery, A.N., A New Model to Describe Enzyme Inactivation. Biotechnol. Bioeng., 1978, 20, 1471.

7. Ho, L.Y. and Humphrey, A.E., Optimal Control of an Enzyme Reaction Subject to Enzyme Deactivation. I. Batch Process. Biotechnol. Bioeng., 1970 12, 291-311.

INTERNATIONAL CONFERENCE ON BIOREACTORS AND BIOTRANSFORMATIONS GLENEAGLES, SCOTLAND, UK: 9-12 NOVEMBER 1987

Paper I1

THE DESIGN OF A TUBULAR LOOP REACTOR FOR SCALE-UP AND SCALE-DOWN OF FERMENTATION PROCESSES

B. Kristiansen, Center for Industrial Research, P.O. 124 Blindern, 0314 Oslo 3, Norway

B. McNeil, Applied Microbiology, University of Strathclyde Glasgow G1 1XW, Scotland

ABSTRACT

The stirred tank fermenter is used on all stages from laboratory R&D work to full production scale. Inside the reactor circulation loops are set up from the impeller out to the wall and back again. The degree of mixing of reactor content is related to the time taken for the broth to travel through the loop. In small fermenters, circulation times of less than five seconds ensure that the reactor is well mixed. In large reactors, however, circulation times can be 100 secsonds or more, resulting in mixing being far from ideal.

Micro-organisms travelling through circulation loops in large fermenters are very likely to experience variations in their environment such as DO and pH gradients. The same changes are not experienced in small reactors, and it is suggested that herein lies one of the major reasons for the problems encountered when translating fermentation data from one scale to another.

One approach to study this problem is to look at the circulation loop itself. This paper concerns an attempt to simulate the circulation inside stirred tank reactors, using a loop reactor specially constructed for the purpose. The reactor carries a number of ports and process probes along its length for the determination of concentration gradients within. The broth is circulated around the loop by the use of peristaltic pumps and the circulation time is used as a measure of simulated reactor size.

The reactor system has been evaluated using the production of pullulan by Aureobasidium pullulans as a test process. Results will be presented which indicate that the loop reactor gives very good simulation of stirred tank reactor behaviour for the system chosen.

INTRODUCTION

Scale-up is considered to be a major bottleneck in fermentation technology. This is predominantly due to the near impossibility of reproducing the ideal conditions obtained in the small fermenters used for research in the much larger production vessels. Operating costs for the latter are also quite considerable, prohibiting the scope for full scale development work. To overcome these problems, scale-up was normally attempted by selecting one parameter which strongly influences the fermentation process and keeping the absolute value of this parameter constant on all scales.

Typical scale-up parameters were power input per unit volume, impeller tip speed, mixing time, oxygen transfer coefficient and the concentration of some specified substrate such as dissolved oxygen(1). There are examples of sucessful scaling up of fermentation processes based on physical parameters in literature. Increasingly, however, information on a number of process parameters and variables are included to ensure scale-up with no loss of yield as demanded by process economics (2,3). The total environmental approach was highlighted in a recent review, expressing the need to include biological constraints (4).

It is difficult, however, to measure micro-organisms reponse to local conditions in the turbulent and constantly changing environment inside bioreactors. The aim of our present investigation is to develop a method by which the physiology of micro-organisms can be taken into account when translating fermentation processes from one scale to another, be it increasing or decreasing in volume.

When considering the fermentation process itself, scale of operation seems to have an effect on the various rates involved. Table I gives typical values for a number of time constants which can be used to provide information on a process, in this case an actively growing batch culture of A.niger.

Table I Typical time constants for A.niger fermentation.

Time constant	Small scale	Large scale
Cell growth	5 h	5 h
Suger consumption	5 h	5 h
Oxygen transfer	10 s	18 s
Mixing time	10 s	100 s

The data in Table I refers to stirred tank reactors and it appears that the mixing is sufficiently fast in the small reactor to ensure an even supply of oxygen. In the large reactor, however, mixing is too slow to prevent dissolved oxygen gradients throughout the vessel. Inside STRs circulation loops are set up, two for each impeller, and cells travelling through such a loop in the large vessel will cycle through a gradient which may affect their physiology as the gradient may not necessarily be restricted to dissolved oxygen.

There have been attempts made to simulate this by enforcing cyclic dissolved oxygen variations on a growing culture (5), or use a two-compartment model in which cells cycle between an oxygen rich and an oxygen depleted region (6). The main problem with the former method is that the cells experience dissolved oxygen variations in an otherwise well mixed environment, in the latter method the cells encounter a step change in dissolved oxygen when travelling from one compartment to the other.

In STRs the departure from ideal mixing becomes more pronounced with scale. The ensuing gradients will partly be imposed on the micro-organisms as a result of the inadequate mixing and partly be a result of the micro-organism's own metabolic activity. Thus, the cells will experience a changing macro- as well as micro-environment. This may give rise to physiological variations which will be accentuated as the circulation loop increases with fermenter volume.

The two model systems have some merit, however, the reactor reported here resulted from a different approach to scale-up, particularly of STRs. In an effort to study the response of micro-organisms to the changing condition inside fermentors brought about by increasing (or decreasing) volume, our attention has been focused on the circulation loop itself. This report concerns an attempt to simulate these loops by constructing a tubular loop reactor.

The reactor body is made into a tubular loop of constant volume carrying a number of process probes along its length for the determination of gradients within. The broth is circulated around the loop using a pump and the circulation time should correspond to circulation in an STR of a given size.

Reports on tubular loop reactors have appeared in the literature, with small loops found to behave much like stirred tanks while larger loop reactors did not (7,8).

There are obvious differences between a stirred tank and a loop fermenter. In both reactors, however, cells will experience similar changes in their macro- and micro- environment, producing a response which should be indepedent of reactor type.

This report concerns the design and construction of a loop reactor intended for studies on scale-up of fermentation processes in which the broth is non-Newtonian. The reactor has been tested using batch cultivation of <u>Aureobasidium pullulans</u>, as a test system. This is a di-morphic fungus, which can be induced to produce considerable amounts of a bio-polymer (pullulan). The morphology and product formation capacity of the fungus is very sensitive to changes in the cultivating conditions, particularly with respect to pH and DO, hence its selection as the test system (9,10).

The reactor and fermentation process is to be considered as a tool for scale-up/down studies and for this reason details on the growth of the organism and subsequent production of pullulan is not considered relevant to this report. The aim is to obtain a tool which can predict bioreactor performance and scale-up reliably, cheaply and quickly. Therefore, the results concentrate on relationships between circulation time and pullulan production, using the latter as a measure of reactor performance.

MATERIALS AND METHODS

Loop fermenter

The fermenter body is made up of borosilicate glass sections with a nominal diameter of 50 mm. In the vertical riser and downcomer, sections with nominal diameter of 25 mm and 75 mm were also used. The sections are held together with backing flanges. Ports are placed along the length of the loop for probes, sampling, acid/ alkali addition etc., as shown in Figure 1. In the standard loop configuration, the riser, transverse section and downcomer all carry one pH and one DO probes, situated near the section midpoints. Temperature and foam sensors are placed in both the vertical legs of the reactor. This allows for easy determination of axial pH, DO and temperature profiles in the reactor. Total operating volume varies from 3.5 litres, using the vertical 25 mm sections, to 9 litres using the two 75 mm sections.

Broth circulation is promoted by two variable speed peristaltic pumps connected to the riser and downcomer sections by 15.9 mm dia silicone rubber tubing (alternatively 12.9 and 19 mm). To make the simulation as realistic as possible no attempts are made to degas the broth by introducing cyclones in the loop. Gas bubles are able to escape, however, at the top of both the riser and downenner.

Temperature is monitored both in the riser and the downcomer, through heat resistance probes via a controller/ indicator. One of the probes is also used for control purposes. When required, heating is supplied via two cartridge heaters in the riser. A cooling coil (PVC) is wrapped around the transverse section to provide cooling water. At the operating temperature chosen for the work reported here, 28^{0}C, this was rarely required.

pH is monitored independantly in the riser, downcomer and transverse sections. For the fermentation process reported here, the probe in the riser was also being used for pH control purposes with acid (I M HCl) and alkali (I M NaOH) added at a point halfway up the riser. As shown in Figure 2, this was sufficient to keep pH within the control limits of the fermentation reported in this study.

Antifoam probes are situated to act on each of the liquid surfaces at the top of the riser and downcomer respectively. Both probes are connected to the same controller for delivery of antifoam agent (Polypropylene glycol 2025) via a small persistable pump.

Three probes are used to monitor DO in the riser, transverse section, and downcomer respectively. The probes are connected to a multichannel recorder providing a direct readout of the DO profile in the loop. The

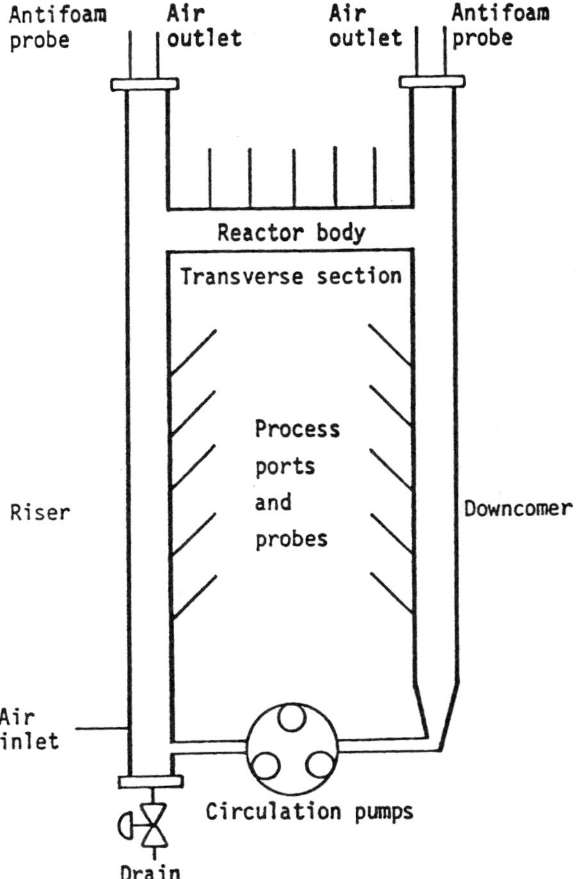

Figure 1. The tubular loop reactor

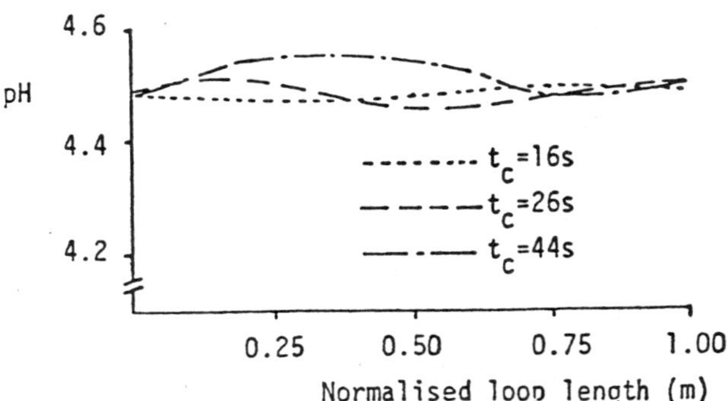

Figure 2. pH profiles along loop at different circulation times with apparent viscosity around 40 cP (measured at $63s^{-1}$).

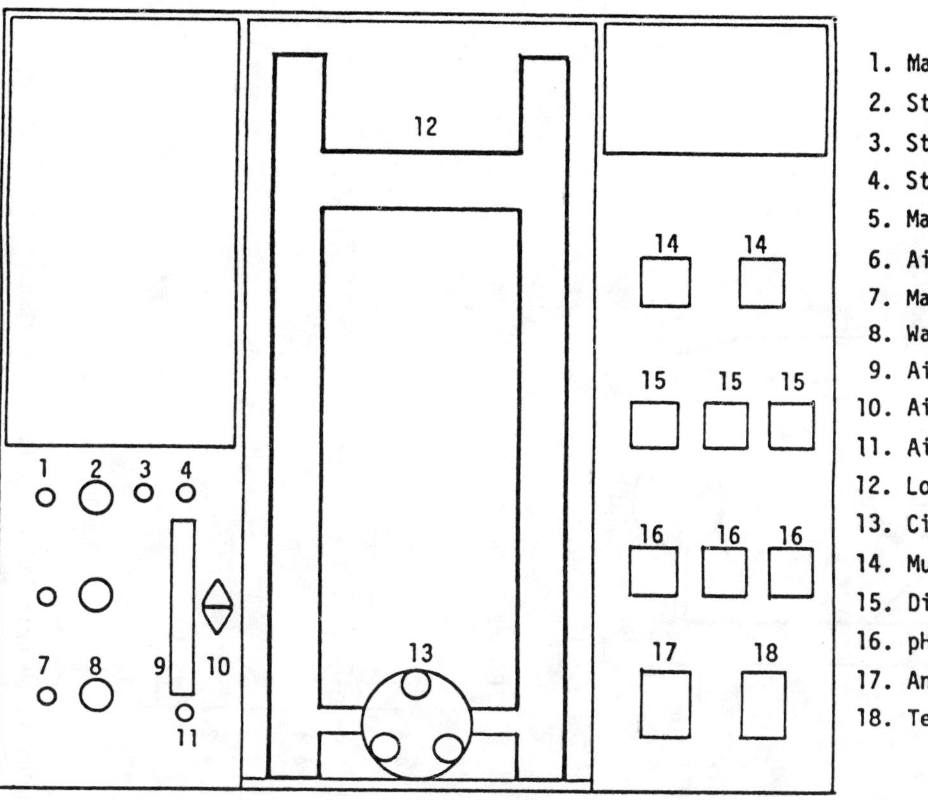

1. Mains steam valve
2. Steam pressure gauge
3. Steam valve
4. Steam/air valve
5. Mains air valve
6. Air pressure gauge
7. Mains water valve
8. Water pressure gauge
9. Air flow meter
10. Air filters
11. Air flow control valve
12. Loop reactor body
13. Circulation pumps
14. Multichannel recorders
15. Dissolved oxygen
16. pH indicators/controllers
17. Antifoam controller
18. Temperature indicator/controller

Figure 3. General arrangement of reactor services and controllers.

general arrangement of the services and controls is shown in Figure 3.

Services

The fermenter is connected to main steam, air and water, as shown in Figure 3.

Fermentation

Micro-organism: Aureobasidium pullulans

The medium composition and operating conditions were as follows:

Sucrose	30 kg/m^3
$(NH_4)_2SO_4$	0.6 "
KH_2PO_4	5 "
$MgSO_4$ x $7H_2O$	0.2 "
NaCl	1 "
yeast extract	0.4 "
Temperature	28^0C
pH	4.5
Air flow rate	1 vvm
Operating volume	3.5-9 litres

Broth circulation time was determined using the known reactor volume and calibrated pump flow rate. Mixing time was estimated by the conventional method of injecting alkali and noting the subsequent signal from a pH probe. In this case, the alkali was injected immediately upstream from the probe. Mixing time was also determined visually by injecting a die. This was not always satisfactory, however, as the pigment production towards the end of a pullulan fermentation was often very high, making visual observations difficult.

The mixing time quoted in the experiments reported here were determined prior to inoculating the reactor. During a run, the fermentation broth becomes very viscous and the effect on circulation and mixing time is shown in Table II.

Table II Effect of broth apparent viscosity on measured circulation and mixing times in the loop reactor.

(Apparent) visecosity (Cp)	Pump setting	Circulation time (s)	Mixing time (s)
1	99	15	72
23	99	15	67
41	99	15	66
70	99	15	54
70	75	19	68
70	50	30	142
70	25	59	157

The measurements given in the table above were carried out on a broth with an original apparent viscosity of 70 cP. Reduction in viscosity were obtained by diluting with water. This may have affected the buffering capacity of the broth producing the apparent reduction in mixing time in a broth of 70 cP. Ignoring this discrepancy, it is clear that broth circulation time and subsequent mixing time is not affected by rheological properties of the broth.

Analytical

Biomass was quantified by dry weight determination, separating the cells from the fermentation broth using GF/C (Whatman) filter papers. The addition of two volumes of alcohol to the filtrate resulted in the precipitation of the pullulan which was subsequently dried and weighed for dry weight estimations.

RESULTS AND DISCUSSION

The results from a pullulan fermentation in a 10 l stirred tank, operating at a stirrer speed of 300 rpm, and the loop reactor at a circulation time of 11 s, which is the minimum for this configuration of the reactor, are shown in Figure 4. It appears that the tubular loop reactor behaves much like a stirred tank if the circulation time is sufficiently small. The biomass curves are comparable and the pullulan curves show similar trends although the slightly higher rate of production in the stirred tank led to more product in this system.

In a recent report it was established that stirrer speed has a marked influence a pullulan production (11). It is feasible, therefore, to select a stirrer speed which will give a pullulan production comparable to the loop results in Figure 4. At the present stage of the investigation, however, it suffices to ascertain if the loop reactor approaches stirred tank performance in the small scale fermentations which serve as reference points for the study. According to Figure 4, this seems to be the case.

The influence of the circulation time on the pullulan production is shown in Figure 5, where each curve represent a batch fermentation carried out

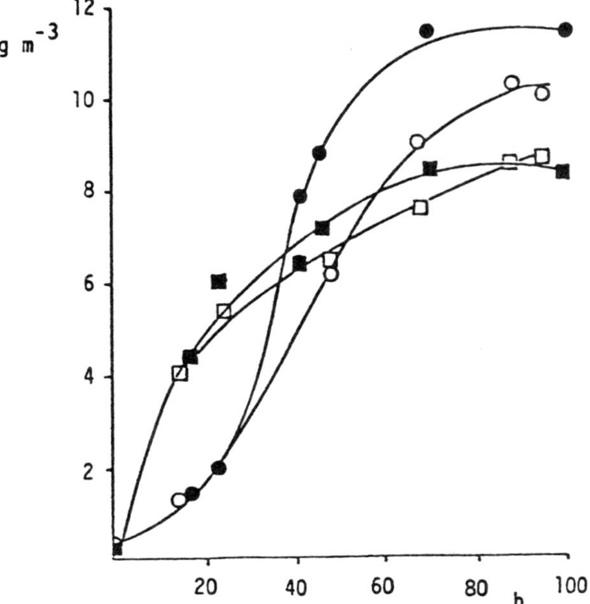

Figure 4. The pullulan fermentation in a 10 litres stirred tank (closed symbols) and the loop reactor at circulation time 11s (open symbols).
□ = biomass; ○ = pullulan

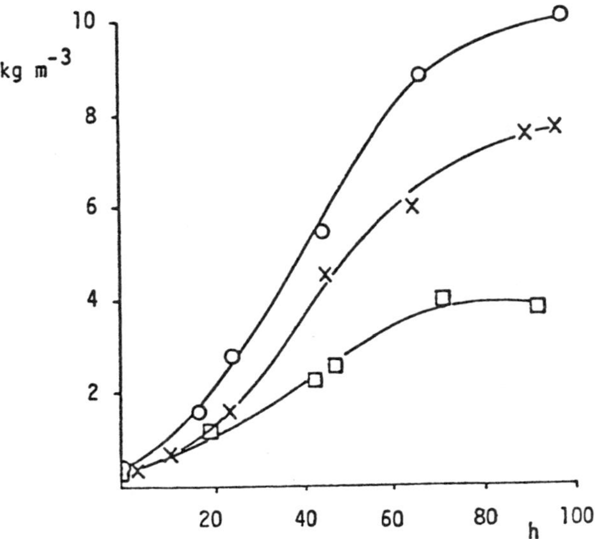

Figure 5. Production of pullulan in the loop reactor at circulation times of; 15s (○); 30s (X) and 62s (□).

at a predetermined circulation time. Clearly, the circulation time has a marked effect on pullulan production. As the circulation time is taken to be a measure of reactor size, it appears that the pullulan fermentation is a good choice as a test system, being sensitive to scale.

The variations in circulation time in Figure 5 was achieved by altering the pump speed. As the cells travel through the pump head they will be subject to shear. There are few indications that A. pullulans is a particularly shear sensitive organism, but it is important to determine if the shear variation have had some influence on the fermentations.

To investigate the effect of shear, a series of experiments were carried out in which the pump speed was kept constant and the circulation time was altered by using glass sections of different diameter in the reactor. The results are shown in Figure 6. Again the trend of reduced pullulan production with increasing circulation time, as seen in Figure 5, can be observed.

The result of keeping the circulation time constant by the use of both different diameters glass sections and pump speed, are given in Figure 7. It appears that one curve is sufficient to describe pullulan production in the various fermentations. It must be concluded, therefore, that the different shear regimes experienced had little effect on the organism and the differences in the pullulan curves shown in Figure 5 is a response to variations in circulation times only.

If the loop reactor is intended to simulate the circulation loop experienced in stirred tank reactors, the circulation time will be a measure of reactor size. Other parameters such as number of impellors and power input will also influence the relationship between loop circulation time and reactor size, but that does not detract from the general argument. From the results presented in Figures 5-7, it appears that the test system is scale sensitive. The relationship between products concentration and circulation time is given in Figure 8. This includes a number of experiments where both tube diameter and pump speed is used to vary the circulation time. The results quoted are maximum amounts of biomass and pullulan produced during batch cultivation, with the circulation time having a pronounced effect. Also included are results from STRs in which the circulation time was determined as described.

The existence of gradients, in this case DO gradients in the loop, is shown in Figure 9, where fermentation time is used as a parameter. The loop length is measured from the air inlet and the figure shows that parts of the loop suffer oxygen exhaustion as early as 7 hours into the run. It also indicates the DO limitations which can be experienced in large fermentors despite apparently high DO readings from certain regions in the fermentor.

It is suggested that the relationship between pullulan concentration and circulation time is linear, in which case the pullulan fermentation has turned out to be a remarkable good choice as a test system. From the comparative experiments in 10 and 200 litres STRs carried out so far, it appears that the loop fermentor does behave as the corresponding stirred tank, representing a valuable tool in both scaling up and scaling down of fermentation processes.

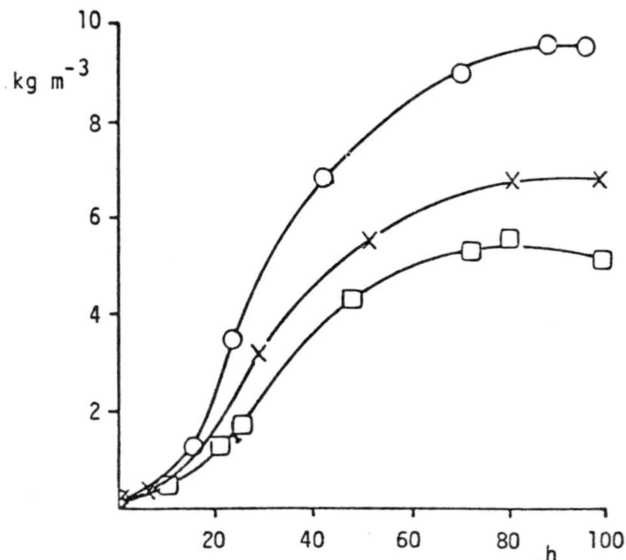

Figure 6. Production of pullulan in the loop reactor at constant pump speed and circulation time of 21s (O); 39s (X) and 50s (□).

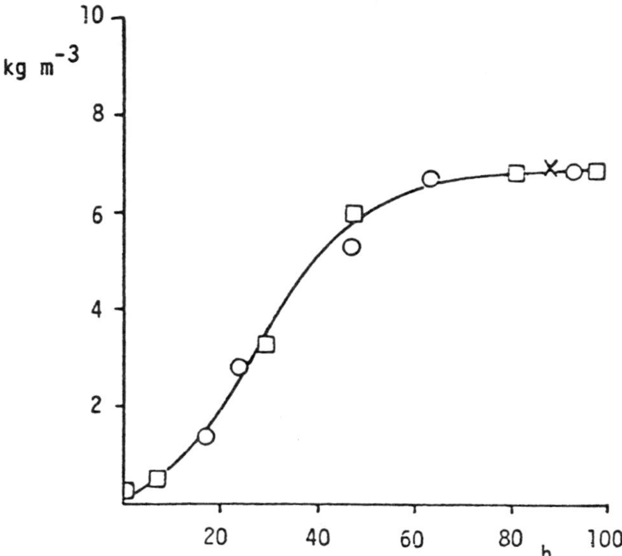

Figure 7. Production of pullulan in the loop reactor at circulation time of 40s. (X) =tube dia 25mm; (O)= tube dia 50mm and (□)=tube dia 75mm.

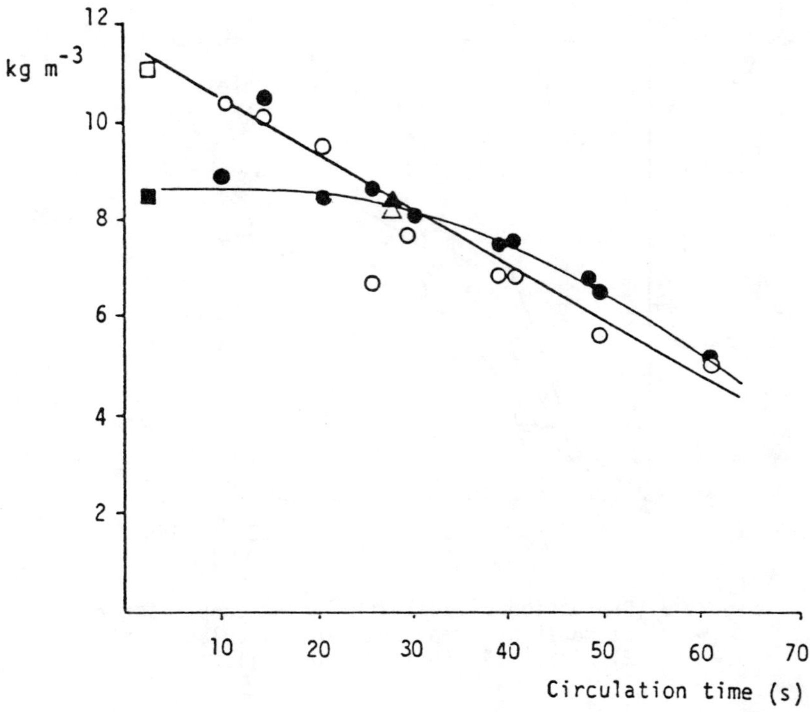

Figure 8. Influence of circulation time on the production of biomass (●) and pullulan (○) in the loop reactor.

■=biomass and □=pullulan in a 10 litres STR.
▲=biomass and △=pullulan in a 200 litres STR.

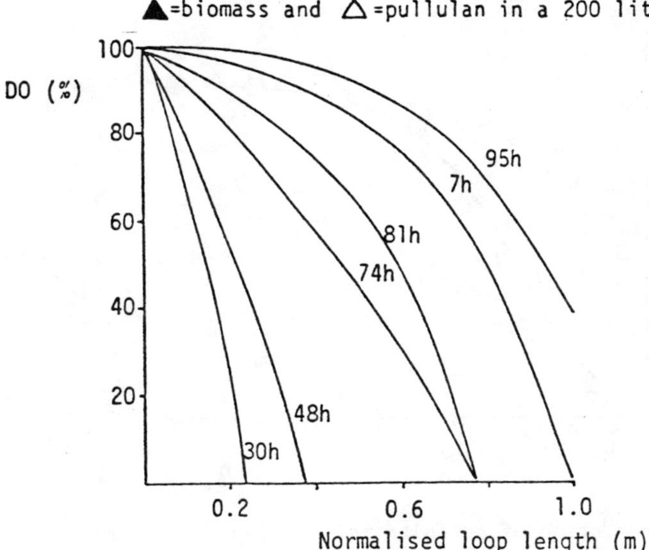

Figure 9. Dissolved oxygen profiles during a run at circulation time of 39s.

High μa ?

CONCLUSION

A tubular loop fermentor has been designed with view to simulate circulation loops found in stirred tank reactors. The loop reactor has been tested using the production of pullulan by Aureobasidium pullulans as a test system. It appears that the reactor can be used to simulate stirred tanks, thus providing a cheap method for studies on scale-up. The gradients in stirred tanks set up as a result of unadequate mixing are well demonstrated in the loop reactor. Thus, the latter may be used to determine maximum scale of operation for avoiding detrimental gradients and substrate exhaustion.

ACKNOWLEDGMENT

Support for this work was provided through the NEL Bioreactor project.

REFERENCES

1. Wang, D.C., Cooney, C.L., Demain, A., Dunhill, P., Humphrey, A.H. and Lilly, M.D., "Fermentation and Enzyme Technology", Wiley, New York, 1975.

2. Buckland, B.C., The translation of scale in fermentation processes: The impact of computer process control. Bio/Technology, Oct. 1984, pp 875-883.

3. Einsele, A., Scaling up bioreactors. Proc. Biochem.,July 1978, pp 13-14.

4. Young, T.B. III, Fermentation scaleup: Industrial experience with a total enviromental approach, Annals New York Acad. Sciences, 1979, 326, 165-180.

5. Vardar, F. and Lilly, M.D., The effect of cycling dissolved oxygen concentration on product formation in penicillin fermentation. Eur. J. Microbiol Biotech. 1982, 14, 203-211.

6. Oosterhuis, N.M.G., Groesbeek, N.M., Kossen, N.W.F., Olivier, A.P.C. and Shenk, E.S., Scale up/down of bioreactors. Proc. 3rd European Congress on Biotechnology Vol II, 1984, pp 277-282.

7. Ziegler, H., Meister, D., Dunn, I.J., Blanch, H.W. and Russel, T.W.F., The tubular loop fermentor: Oxygen transfer, growth kinetics and design. Biotech. Bioeng., 1977, 19, 507-525.

8. Ziegler, H.,Dunn, I.J., and Bourne, J.R., Oxygen transfer and mycelial growth in a tubular loop fermentor. Biotech. Bioeng., 1980, 22, 1613-1635.

9. Heald, P.J. and Kristiansen, B., Synthesis of polysaccharide by yeast-like forms of Aureobasidium pullulans. Biotech. Bioeng., 1985, 27, 1516-1519.

10. Kristiansen, B. and McNeil, B., Effect of regulating dissolved oxygen levels in the pullulan fermentation by Aureobasidium pullulans. National Engineering Laboratory Bioreactor Project Research Symposium 1, 1985, pp 65-70.

11. McNeil, B. and Kristiansen, B., Influence of impeller speed upon the pullulan fermentation, Biotech. Letts., 1987, 9, 101-104.

INTERNATIONAL CONFERENCE ON BIOREACTORS AND BIOTRANSFORMATIONS
GLENEAGLES, SCOTLAND, UK: 9-12 NOVEMBER 1987

Paper I2

PNEUMATICALLY AGITATED BIOREACTOR DEVICES: EFFECTS OF FLUID HEIGHT AND FLOW AREA SHAPE ON HYDRODYNAMIC AND MASS TRANSFER PERFORMANCE

M.Y. Chisti and M. Moo-Young
Department of Chemical Engineering
University of Waterloo
Waterloo, Ontario N2L 3G1
Canada

ABSTRACT

The effects of static liquid height or reactor height on gas holdup and gas-liquid mass transfer performance of a bubble column (d_c = 0.243 m, h_L = 1.5 to 5.875 m) and three split cylinder (d_c = 0.243 m, Ad/Ar = 0.411, L_b = 1.6 to 4.8 m) airlift reactors were examined in water and cellulose fibre suspensions. Additionally, the effect of flow area shape, whether rectangular or circular, was investigated using two bubble columns with similar hydraulic diameters (~0.238 m).

No influence of h_L was found in the bubble column for any fluid. In airlifts, gas holdup in water declined by up to 27% with increasing reactor height. In viscous non-Newtonian suspensions, the holdup first increased with the split-cylinder height and then declined with further height increase.

The gas holdup in water in bubble columns was independent of the flow area shape. In suspensions, however, the rectangular geometry produced lower gas holdup relative to the circular vessel for otherwise identical conditions.

ABBREVIATION LIST

A_d	Downcomer cross sectional area	m^2
A_r	Riser cross sectional area	m^2
a_L	Gas-liquid interfacial area per unit liquid volume	m^{-1}
BC	Bubble column	
C*	Saturation concentration of dissolved oxygen	kgm^{-3}
C_o	Initial dissolved oxygen concentration	kgm^{-3}
C_L	Dissolved oxygen concentration in the liquid	kgm^{-3}
DO	Dissolved oxygen	
d_c	Column diameter or equivalent hydraulic diameter	m

d_d	Equivalent hydraulic downcomer diameter	m
d_h	Sparger hole diameter	m
d_r	Equivalent hydraulic riser diameter	m
g	Gravitational acceleration	$m^2 s^{-1}$
H	Overall reactor height	m
H'	Henry law constant	$m^2 s^{-1}$
h_L	Height of liquid	m
h_p	Axial location of dissolved oxygen probe above reactor base	m
k_L	Mass transfer coefficient	ms^{-1}
L	Reactor length	m
L_b	Height of baffle	m
L_c	Baffle clearance from reactor base	m
n_h	Number of sparger holes	(-)
P_G	Power input	W
P_h	Head space pressure	$N\ m^{-2}$
P_T	Total pressure	$N\ m^{-2}$
SC	Spilt cylinder airlift	
SF	Solka-Floc	
t	Time	s
Usgr	Superficial gas velocity based on riser	ms^{-1}
V_L	Liquid volume	m^3
W	Reactor width	m
W_B	Baffle width	m
Y_A	Mole fraction of oxygen in air	(-)

Greek Symbols

ε	Overall gas holdup	(-)
ρ_L	Liquid density	kgm^{-3}

INTRODUCTION

A bioreactor is the core of any bioprocess. Reactor design, performance and limitations affect both the nature and the extent of pretreatment and also the downstream processing requirements. While the stirred tank fermenter remains the work horse of the biotechnology industry, it is increasingly being displaced by newer reactor types for well known reasons [1], particularly in animal and plant tissue culture applications. Pneumatic reactors in which compressed air (or oxygen), in addition to satisfying the oxygen needs, provides the requisite agitation, are being developed. Some of these bioreactors -- bubble columns and airlift devices -- have been studied, for example by Chisti et al [1,2] among others; however, the knowledge available for the design of this type of vessels remains limited.

In the following sections we report on two aspects of pneumatic reactor design which have not been examined previously: (1) The comparative effects of unaerated liquid height in a circular bubble column and and in split cylinder internal loop airlift devices; and (2) the influence of bubble column cross sectional shape -- rectangular and circular -- on performance. Hydrodynamic and oxygen transfer performance of those reactors is investigated in gas-liquid and gas-liquid-solid systems, the latter carefully formulated [1,2] to simulate certain mycelial fermentations.

THEORY

The fundamental theoretical aspects of gas-liquid mass transfer, hydrodynamic parameters such as gas holdup and gas residence time in pneumatically agitated reactors have been discussed recently [1-4] and need not be repeated here. The following will be noted with little comment:

The overall volumetric mass transfer coefficient, $k_L a_L$, may be calculated using

$$\ln \frac{(C^* - C_o)}{(C^* - C_L)} = k_L a_L t \qquad (1)$$

based on the transient gassing-in method [1,2] for a backmixed reactor.

The steady state dissolved oxygen concentration, C*, is a function of pressure which varies with the depth of liquid. Theoretically, C* is given by

$$C^* = Y_A \; P_T \, / \, H' \qquad (2)$$

where

$$P_T = P_h + \rho_L g(h_L - h_p) \qquad (3)$$

The specific power input is obtained as explained elsewhere [3].

TABLE 1

Reactor dimensions and sparger details

Reactor	H (m)	d_c or equivalent hydraulic diameter (m)	d_r (m)	d_d (m)	L (m)	W (m)	h_L (m)	W_B (m)	$\frac{A_d}{A_r}$ (-)	L_b (m)	L_c (m)	SPARGER Type	n_h (-)	d_h (m) $\times 10^3$	Pitch (m)	Free area %
BC 1	2.146	0.232	-	-	0.46	0.155	1.372	-	0	-	-	Perforated plate	50	1.0	0.025 x 0.045 rectangular	0.055
BC 2	7.825	0.243	-	-	-	-	1.50 to 5.875	-	0	-	-	Perforated plate	106	1.0	0.02 triangular	0.18
SC 1	7.825	0.243	0.19	0.102	-	-	2.227	0.229	0.411	1.6	0.102	Perforated pipe 0.008 m i.d.	38	1.5	-	0.204
SC 2	7.825	0.243	0.19	0.102	-	-	~3.5	0.229	0.411	3.2	0.102	As for SC 1				
SC 3	7.825	0.243	0.19	0.102	-	-	4.976 to 6.0	0.229	0.411	4.8	0.102	As for SC 1				

EXPERIMENTAL

Experiments were conducted with batchwise liquid (or slurries) and continuous air flow. Air from 20MPa mains routed through a filter, pressure reducer/regulator, flow control valve and rotameter arrangement was supplied to the reactors at ambient temperature. The fluids used included hard tap water, aqueous sodium chloride (0.15 kmol m^{-3} NaCl), 1,2 and 3 dry wt./vol.% suspensions of Solka-Floc cellulose fibre grade KS-1016 (James River Corporation) in aqueous salt (0.15 kmol m^{-3} NaCl) solution. The properties of these media were reported elsewhere [1,2] where they were also compared with mycelial broths. All experiments were carried out at 20 ± 2°C with the reactors open to atmosphere.

The reactors (Fig. 1) used were made of prespex to allow visual observations. A rectangular bubble column (BC1), a tall (H = 7.825 m) circular bubble column (BC2) and three split cylinder type airlifts (SC1, SC2, and SC3) were examined. Perforated plate or perforated pipe type gas spargers were used throughout. Complete details of the reactors and spargers are given in Table 1.

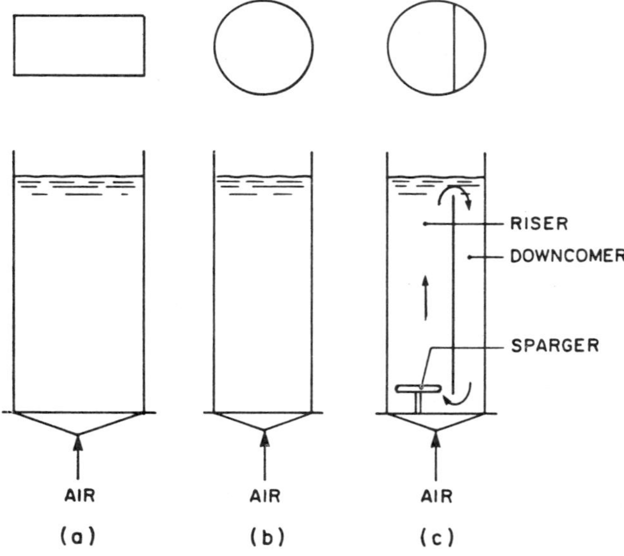

Figure 1. Reactors (schematic): (a) rectangular bubble column (BC 1), (b) circular bubble column (BC 2) and (c) split-cylinder air-lifts (SC 1-3).

The design and operation of the reactors used ensured that the correct vertical alignment of the vessels was maintained to avoid the possible deliterious effects of small departures from vertical that were reported [5,6] by other investigators.

The overall volumetric mass transfer coefficient, k_La_L, was determined by the dynamic gassing-in technique [1,2]. Equation (1) was used to calculate k_La_L. Rapid response (>95% in under 10s) dissolved oxygen electrodes (Yellow Springs, YSI 5739 with standard membrane) connected to YSI model 57 dissolved oxygen meters were used. Fully backmixed liquid (slurry) phase was assumed and, as discussed elsewhere [1], electrode-time delay was ignored in k_La_L calculations.

The overall gas holdup (ε) was determined by measuring the difference between the unaerated and aerated dispersion heights.

RESULTS AND DISCUSSION

The performance parameters (e.g. ε, k_La_L) were typically expressed in terms of the specific power input. The kinetic energy input being always less than 1% of the total power.

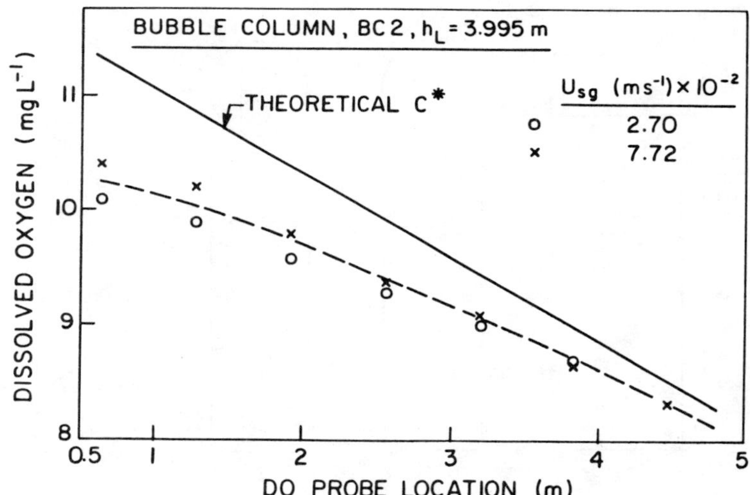

Figure 2. Variation of C* with DO electrode location (h_p) above the base of BC 2 in water at 22.8°C and 99.5 kPa absolute atmospheric pressure.

Height Effects

The axial steady state dissolved oxygen (C*) concentration profile in bubble column (BC 2) is shown in Fig. 2. C* profile is not flat, and in a batch liquid bubble column which is radially well mixed a flat profile would be theoretically impossible. As expected, C* increased with fluid depth (Fig. 2); however, the experimental C* values were consistantly lower than the theoretical (Eq. 2) plot. The experimental data was for hard tap water whereas the theoretical curve was for distilled water. Presence of dissolved salts is known to reduce the solubility of oxygen in water [7]. The increasing deviation of the experimental curve from the Henry's law prediction (Fig. 2) which was observed as the oxygen electrode moved deeper into the fluid was associated with the well known [8] breakdown of Henry's law at higher pressures. The change in fluid mixing or agitation intensity achieved by increasing the superficial gas velocity from 0.0270 to 0.0772 ms^{-1} had almost no affect on C* profile (Fig. 2). The marginally higher C* recorded at the higher gas flow rate could be explained as being due to the inherent flow sensitivity [9] of the polarographic dissolved oxygen electrodes used in measurements. Unlike in the bubble column, quite flat axial C* profiles were found in the risers of the split cylinders even in highly viscous SF suspensions over the entire range of gas flows used. This is shown in Fig. 3 for the tallest split cylinder SC 3 in 1 dry wt/vol% SF suspension. In keeping with our observation, some earlier theoretical model investigations of airlift devices [10, 11] predicted fairly flat axial oxygen concentration profiles for those reactors. The differences in the C* profiles in BC 2 and SC 3 indicated very different mixing characteristics of the two types of reactors despite their similar height-to-diameter, h_L/d_C, ratio (BC 2: h_L/d_C = 16.4; SC 3; h_L/d_C = 20.6).

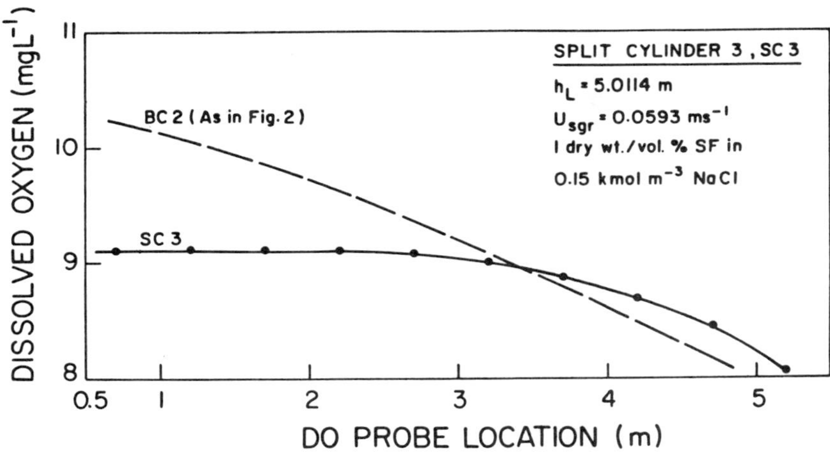

Figure 3. Variation of C* with h_p in split-cylinder (SC 3) at 20.8°C and 99.8 kPa absolute atmospheric pressure.

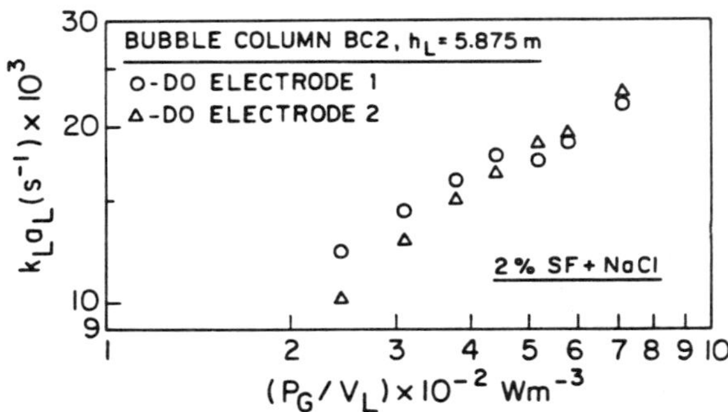

Figure 4. Effect of DO electrode location on k_La_L.

Although C* values varied axially, particularly in tall bubble columns such as BC 2, the k_La_L obtained at different axial locations were quite comparable. Fig. 4 shows k_La_L calculated by taking axial C* changes into account and with the assumption of uniform instantaneous bulk dissolved oxygen concentration. The data shown was obtained in a 2 wt/vol% SF suspension in BC 2 with an unaerated slurry height of 5.875 m. Measurements conducted with DO electrodes 1 and 2 located at the column axis at 5.752 and 0.635 m, respectively, above the reactor base agreed closely as shown in Fig. 4. Although in viscous suspensions the fluid mixing was visibly poor, for mass transfer purposes the "well mixed" assumption did not break down because the mass transfer rate (i.e. k_La_L) was correspondingly slow.

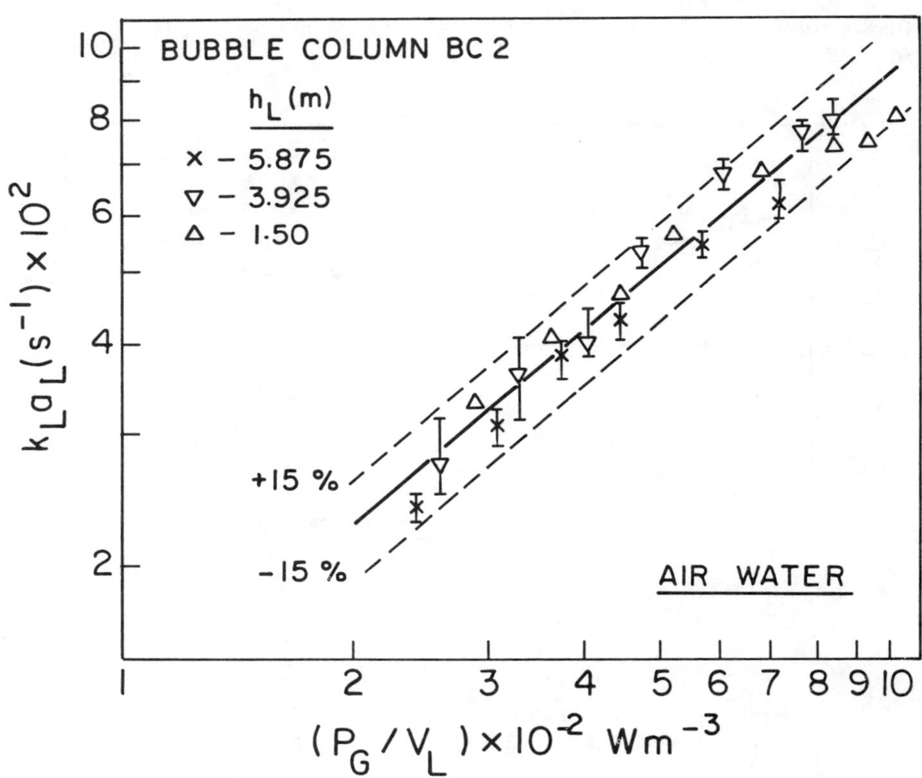

Figure 5. Effect of unaerated liquid level (h_L) on k_La_L in BC 2.

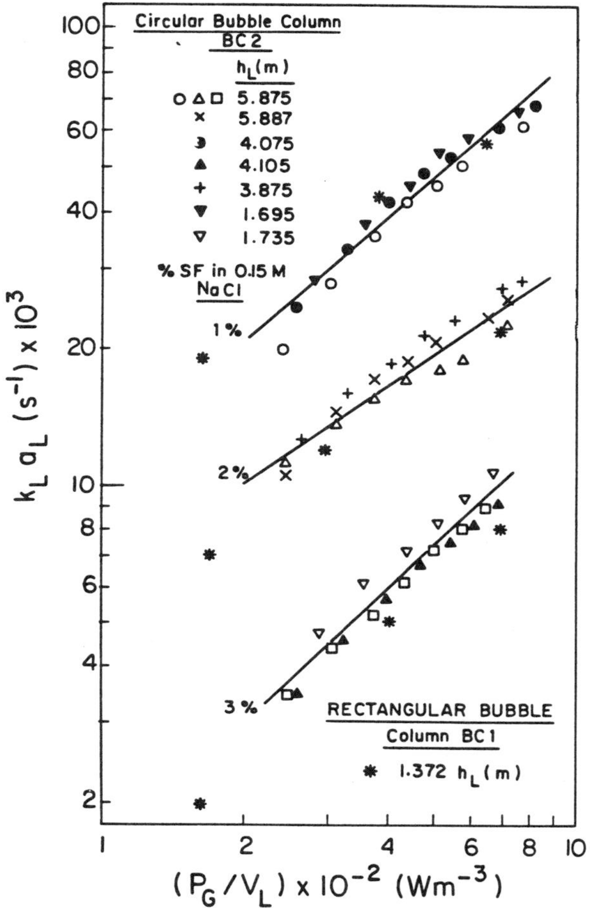

Figure 6. Effect of liquid height on k_La_L in BC 2; effect of column shape on k_La_L.

No effect of unaerated liquid height (1.50 to 5.875 m) or h_L/d_C (6.2 to 24.2) on k_La_L was found either in suspensions or in water in the bubble column (BC 2) as shown in Figs. 5 and 6. The air-water line shown in Fig. 5 followed the equation:

$$k_La_L = 2.393 \times 10^{-4} \ (P_G/V_L)^{0.859} \tag{4}$$

within ± 15%. Equation (4) compares very well with

$$k_La_L = 2.49 \times 10^{-4} \ (P_G/V_L)^{0.82} \tag{5}$$

which was proposed, in a different form, by other investigators [12] for air-water in smaller bubble columns. The good agreement between equations (4) and (5) lent further credence to the fluid mixing assumption we used and our experimental measurements and confirmed the absence of any h_L/d_C effects on k_La_L.

The decline of k_La_L and gas holdup with Solka-Floc solid contents of the slurry has been discussed by us for other pneumatic reactor types [1,2] elsewhere and will not be repeated here.

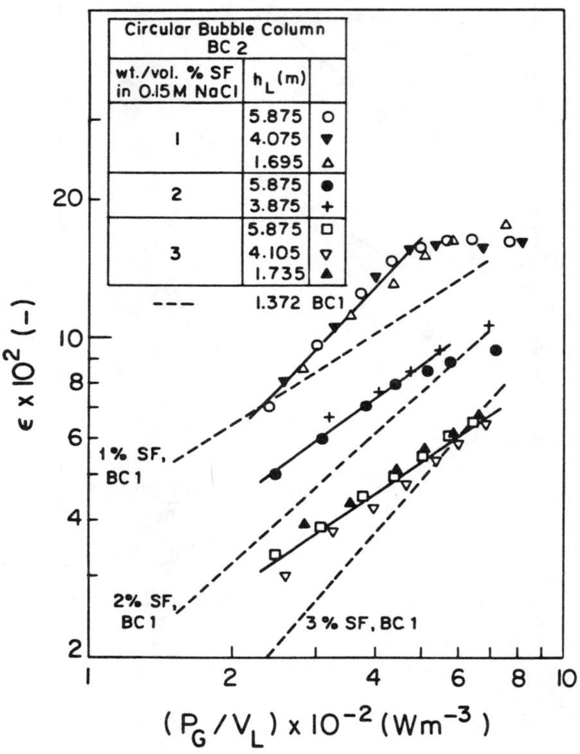

Figure 7. Effect of liquid height on gas holdup in BC 2; effect of column shape on gas holdup.

In bubble column, BC 2, the gas holdup in aqueous solutions as well as in suspensions was independent of unaerated fluid height, as shown in Fig. 7 for SF slurries. Other investigators [13,14] have reported similar observations in Newtonian homogeneous media. An exception being the theoretical correlation of Azbel and Zeldin [15] which predicted a small decline in gas holdup with increasing liquid height. A possible explanation for this discrepancy has been discussed [3]. In viscous non-Newtonian liquids in tall bubble columns, the bubble size has been claimed [16] to increase significantly with liquid height. This leads to the expectation that in these fluids the gas holdup (and k_La_L) should be height dependent. This was not the case, however, for SF suspensions.

The three split-cylinder internal loop airlifts (SC 1, SC 2 and SC 3) had identical column diameter, sparger and A_d/A_r ratio (Table 1); the only difference being in the splitting baffle height, L_b. The baffle height ranged from 1.6 to 4.8 and the corresponding L_b/d_c from 6.6 to 19.8, making them some of the largest vessels of their type reported anywhere.

For solid-free liquids gas holdup declined slightly with split cylinder reactor height in contrast to the observations in bubble columns. As shown in Fig. 8, for identical power input, the gas holdup in SC 3 was up to 27% lower than in SC 1 (h_L = 2.227 m). The data for SC 2 (~3.5 m tall), not shown in Fig. 8 for the sake of clarity, lay inbetween the lines shown for SC 1 and SC 3. The reducing gas holdup with reactor height was a result of the increase in liquid circulation velocity with height which has been well documented [17] for water-like fluids. The liquid velocity superimposed an additional component on the bubble rise velocity which lead to gas holdup reduction.

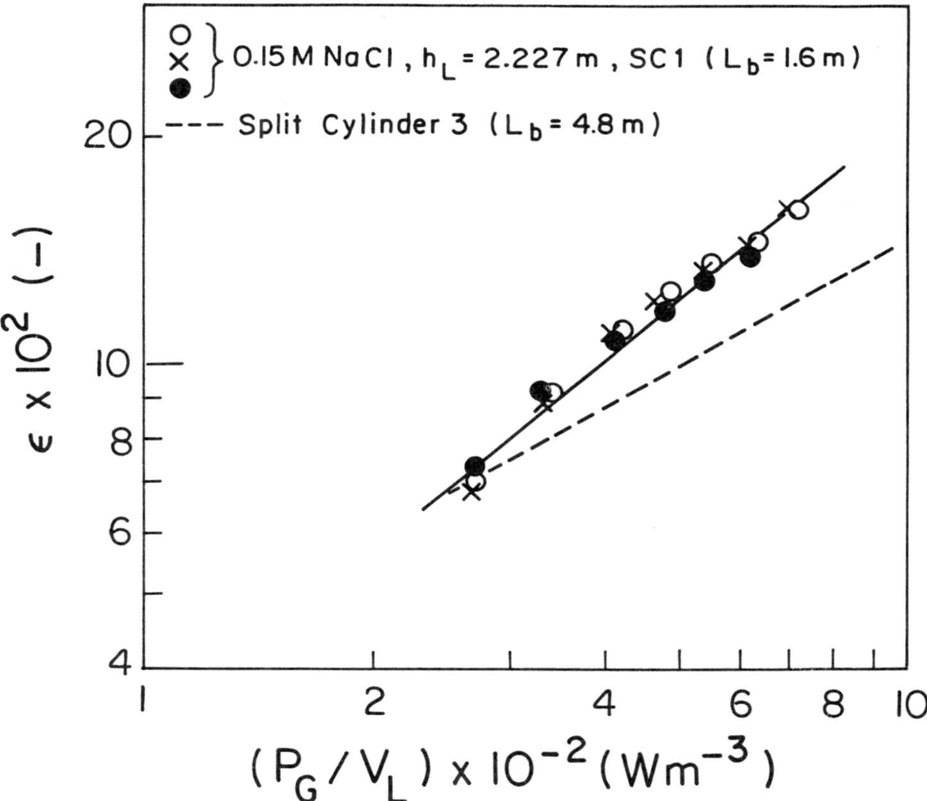

Figure 8. Gas holdup in split cylinders SC 1 and SC 3: effect of baffled reactor height, L_b. The dashed line is based on data in tap water, 0.15 $kmolm^{-3}$ NaCl and 1% SF suspension.

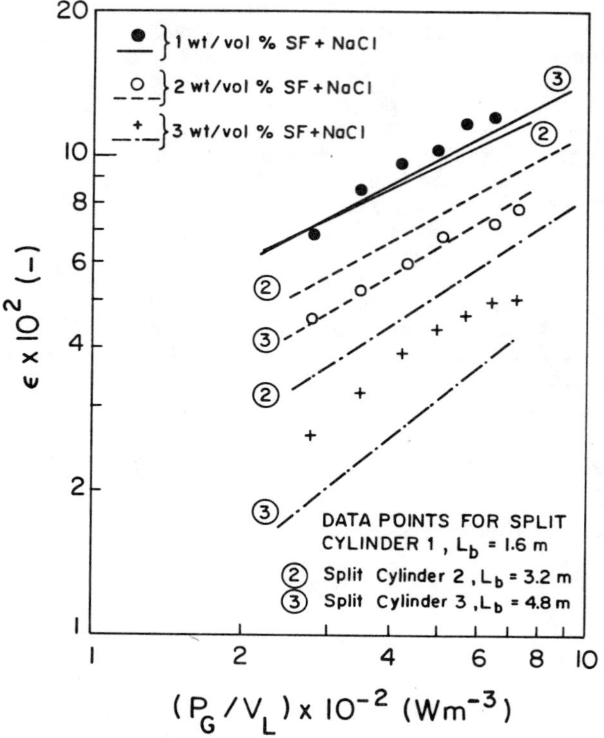

Figure 9. Effect of split cylinder height (L_b) on gas holdup in SF suspensions.

In Solka-Floc suspensions the effect of split cylinder height on gas holdup was more complex: As shown in Fig. 9 for 1 dry wt./vol.% SF suspension, gas holdup in the three airlifts (SC 1-3) was almost identical at identical specific power input. For this fluid (almost water-like) variations in reactor height from 2.227 to 6.0 m did not affect gas holdup. On the other hand, for 2 and 3 dry wt./vol.% SF suspensions the holdup in the shortest airlift (SC 1, 2.227 m tall) was inbetween that observed in the intermediate (SC 2, 3.453 m tall) and the tallest (SC 3, 4.976 to 6.0 m tall) vessels; the vessel of intermediate height performing best (Fig. 9). Thus in more concentrated suspensions gas holdup first increased with reactor height and then declined with height at otherwise identical power input (Fig. 9). Furthermore, this effect of height became more pronounced with increase in suspended solids concentration from 2 to 3 dry wt./vol.% (Fig. 9). This phenomenon probably was the result of two conflicting mechanisms: as has been discussed [17] for low viscosity Newtonian liquids, so also for suspension, the velocity of liquid circulation in the airlift vessel should increase with square root of reactor height when skin friction in the flow channel may be neglected. The addition of solids, on the other hand, may have increased the skin friction between the reactor walls and the fluid due to the abrasive action of solids. The resulting frictional pressure drop per unit length-to-diameter ratio of the flow channel was apparently a strong non-linear function of the solid content of the slurry and it must have increased with solid contents and suspension viscosity. The sum of those two conflicting mechanisms lead to liquid velocity maxima in the split cylinder of intermediate height (SC 2) for 2 and 3 dry wt./vol.% SF suspensions relative to SC1 and SC3. Consequent changes in the degree of turbulence lead to corresponding gas holdup maxima. The gas holdup in 3 dry wt./vol.% SF suspension plotted against h_L/d_C at conditions of equal specific power input in split cylinder vessels is shown in Fig. 10 where the very prominent overall gas holdup maxima are clearly seen.

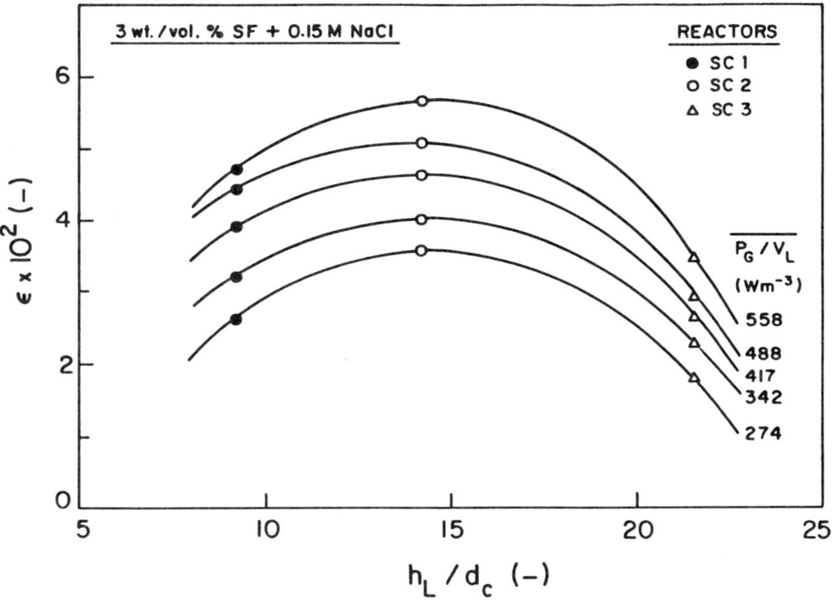

Figure 10. Effect of h_L/d_C on gas holdup in split cylinder airlifts in 3% SF suspension.

Bubble Column Cross-sectional Shape Effects

The rectangular (BC 1) and circular (BC 2) bubble columns had nearly the same equivalent hydraulic diameters at 0.232 and 0.243 m, respectively. These values being greater than the 0.14 m diameter up to which wall or diameter effects have been reported [14]. In water or aqueous salt solution no influence of cross-section shape was found [4] on gas holdup. A report [14] of similar gas holdup in bubble columns with square and circular cross-sections also exists. As shown in Fig. 6, k_La_L values in BC 1 and BC 2 were also quite similar for otherwise identical conditions. No stagnant zones were observed in the rectangular column (BC 1) even in the 3 wt./vol.% SF suspension which was very viscous. The advantage of the rectangular shape (BC 1) being that it provided almost 1.7-fold more potential heat transfer surface than a circular bubble column with an equivalent hydraulic diameter identical to that of BC 1.

Unlike the observation in water, the column cross-sectional shape influenced gas holdup in SF suspensions (Fig. 7). For otherwise similar conditions the holdup in the rectangular column (BC 1) was slightly lower (Fig. 7) than in the circular (BC 2) device. This probably meant less intense shear fields in the rectangular shape relative to the circular one.

CONCLUSION

In radially mixed bubble columns with zero net liquid flow, the saturation oxygen concentration is a function of axial position as can be theoretically predicted. Significantly, the axial C* profiles are independent of the state of axial mixedness. In airlift devices, on the other hand, the axial C* profiles over most of the riser height are much flatter. Irrespective of the nature of C* profiles, fully backmixed liquid assumption is satisfactory for k_La_L calculation in tall airlifts and batch bubble columns where radial liquid mixing is complete. The assumption is valid even in relatively viscous fluids where the mixing may be visibly poor.

In bubble columns gas holdup and k_La_L are independent of static fluid height (or h_L/d_c) in water-like, as well as in non-Newtonian media. In contrast, the gas holdup in airlifts declines with reactor height in water-like fluids; in viscous non-Newtonian systems, on the other hand, the holdup may initially increase with reactor height and then decline with further increase in height.

In bubble columns the shape of column cross-section, whether rectangular or circular, does not influence gas holdup in water-like fluids. In non-Newtonian suspensions, however, a rectangular vessel with the same hydraulic diameter as a circular device may produce slightly lower gas holdup. The rectangular bubble column geometry has heat transfer advantages and it is possibly more suitable for low shear fermentations.

ACKNOWLEDGEMENTS

This work was supported by a grant from the Natural Sciences and Engineering Research Council of Canada.

REFERENCES

1. Chisti, M.Y., Fujimoto, K. and Moo-Young, M., Hydrodynamic and oxygen mass transfer studies in bubble columns and airlift bioreactors. Paper 117a presented at the AIChE Annual Meeting, Miami Beach, November 2-7, 1986.

2. Chisti, M.Y. and Moo-Young, M., Hydrodynamic and oxygen transfer in pneumatic bioreactor devices. Biotechnol. Bioeng., 1987. Submitted for publication.

3. Chisti, M.Y. and Moo-Young, M., Airlift reactors: Characteristics, application and design considerations. Chem. Eng. Commun., 1987. Submitted for publication.

4. Chisti, M.Y. and Moo-Young, M., Gas holdup in pneumatic reactors. Chem. Eng. J., 1987, accepted.

5. Valdes - Krieg, E., King, C.J. and Sephton, H.H., Effects of vertical alignment on the performance of bubble and fractionation columns. AIChE J., 1975, **21**, 400.

6. Tinge, J.T. and Drinkenburg, A.A.H., The influence of slight departures from vertical alignment on liquid dispersion and gas hold-up in a bubble column. Chem. Eng. Sci., 1986, **41**, 165.

7. Bailey, J.E. and Ollis, D.F., Biochemical Engineering Fundamentals, McGraw-Hill, New York, 1977, pp. 418.

8. Alberty, R.A., Daniels, F., Physical Chemistry, 5th Edition, John Wiley and Sons, New York, 1979, pp. 111-112.

9. Lee, Y.H. and Tsao, G.T., Dissolved oxygen electrodes. Adv. Biochem. Eng., 1979, **13**, 35.

10. Ho, C.S., Erickson, L.E., Fan, L.T., Modelling and simulation of oxygen transfer in airlift fermentors. Biotech. Bioeng., 1977, **19**, 1503.

11. Merchuk, J.C. and Stein, Y., A distributed parameter model for an airlift fermentor. Effects of pressure. Biotech. Bioeng., 1981, **23**, 1309.

12. Shah, Y.T., Kelkar, B.G., Godbole, S.P. and Deckwer, W.-D., Design parameters estimations for bubble column reactors. AIChE J., 1982, **28**, 353.

13. Deckwer, W.-D., Burckhart, R. and Zoll, G., Mixing and mass transfer in tall bubble columns. Chem. Eng. Sci., 1974, **29**, 2177.

14. Akita, K. and Yoshida, F., Gas holdup and volumetric mass transfer coefficient in bubble columns. Ind. Eng. Chem. Process Des. Develop., 1973, **12**, 76.

15. Azbel, D.S. and Zeldin, A.N., Bubble layer corrected for dissipative forces. Theoreticheskie Osnovy Khimicheskoi Tekhnologii, 1971, **5** (6), 863.

16. Buchholz, H., Buchholz, R., Lücke, J., Schügerl, K., Bubble swarm behaviour and gas absorption in non-Newtonian fluids in gas sparged columns. Chem. Eng. Sci., 1978, **33**, 1061.

17. Chisti, M.Y., Halard, B. and Moo-Young, M., Liquid circulation in airlift reactors. Chem. Eng. Sci., 1987, Submitted for publication.

INTERNATIONAL CONFERENCE ON BIOREACTORS AND BIOTRANSFORMATIONS
GLENEAGLES, SCOTLAND, UK: 9-12 NOVEMBER 1987

Paper I3

MASS TRANSFER IN AIR-LIFT REACTORS: EFFECTS OF GAS RECIRCULATION

Marc H. Siegel and Jose C. Merchuk
Department of Chemical Engineering
Ben-Gurion University of the Negev
Beer-Sheva, Israel

ABSTRACT

The effects of gas recirculation on gas hold-up and mass transfer rates were studied in a split-vessel air-lift reactor of rectangular cross section. The gas recirculation was varied by changing the geometrical design of the gas-liquid separation section. The evaluation of the gas recirculation ratio, QD/QR, and the definition of a true superficial gas velocity in the riser made possible the correlation of both gas hold-up and mass transfer coefficients as a function of such velocity. The data of mass transfer were also correlated to the specific power input, which allows comparison with results published by other authors. The addition of an auxiliary sparger ("two sparger systems") at the entrance of the downcomer was also studied. It was found that introducing gas into the downcomer improved the overall mass transfer in reactor.

NOTATION

A	=	cross-sectional area (cm^2)
De	=	equivalent diameter (cm)
g	=	gravitational acceleration (cm/s^2)
HL	=	liquid height in reactor (m)
JGD	=	downcomer superficial gas velocity (cm/sec)
JGR	=	true riser superficial gas velocity (cm/s)
KLA	=	mass transfer coefficient (1/hour)
P	=	pneumatic power of gas input (Kwatt)
QG	=	volumetric gas flow rate (cm^3/s)
QD/QR	=	ratio of the gas flow rate in the downcomer to the gas flow rate in the riser (-)
R^2	=	R-square value describing closeness of curve fit
UL	=	liquid velocity (cm/s)
VD	=	dispersion volume (m^3)
VL	=	liquid volume (m^3)

Indices

D	=	downcomer

L	=	liquid
R	=	riser
s	=	sparger

Greek Letters

ρ	=	density (kg/m^3)
ϕ	=	gas hold-up (-)

INTRODUCTION

The air-lift reactor has been attracting increasing attention in the field of Biotechnological processes over the past years. This family of reactors is looked upon as an alternative reactor design to the more conventional bubble columns and mechanically stirred tank reactors (STR) currently predominating in both chemical and biological processes. The appeal of air-lift reactors in many processes is that they are capable of producing efficient mixing and gas-liquid mass transfer in the reacting system without any moving parts. Further, air-lift reactors exhibit relative mild and uniform turbulence in the system, compared to conventional bubble columns and STR's. This trait of air-lift reactors is of particular importance in cell fermentations where flow conditions in the reactor may influence cell morphology and metabolism.

Air-lift reactors are gas-liquid reactors which exhibit a clear and defined cyclic flow pattern through interconnected channels, riser, downcomer, and gas-liquid separator. The riser of the air-lift is the section where gas and liquid flow upward. The gas is usually introduced at the lower end of the riser. The gas-liquid separator is the section where separation of the gas and liquid phases takes place. The extent of separation dictates the amount of gas recycled to the downcomer. The downcomer is the section where the liquid or liquid-gas mixture flows downward from the gas-liquid separator to the bottom of the reactor, where the stream is returned to the riser. The overall behavior of the air-lift is determined by the sum of these three parts.

The influence of the gas-liquid separator on the bioreactor performance has received little previous attention. The liquid flow rate in the air-lift is due to the mean or pseudo density difference (ie. the hydrostatic pressure difference) between the riser and downcomer. The extent of the gas disengagement in the separator will determine the pseudo density of the gas-liquid dispersion in the downcomer, thus the pressure differences. This in turn will determine the liquid velocity for any given geometric configuration and gas input flow rate. Therefore, the extent of gas recirculation will have a strong influence on the entire process since it essentially determines the gas hold-up the downcomer, unless a second gas sparger is used to introduce gas to the downcomer.

This paper will present a discussion of the work we have conducted examining the influence on gas-liquid separator design and, subsequently, downcomer operating conditions on the parameters of gas hold-up, liquid velocity, and gas-liquid mass transfer. The data published in the literature for gas hold-up and mass transfer in air-lift reactors has varied greatly [1,2]. This wide variation can be attributed to differences in column size and geometry, gas recirculation rate, liquid phase physiochemical properties and gas sparger types. In the case of mass transfer, this variation can be further explained by differences in experimental technique (sodium sulfide versus degassing methods). Further, there appears to be a controversy as to the contribution of the downcomer to the overall mass transfer in the reactor. Some authors claiming negligible mass transfer contribution in the downcomer [1,3] and others that the downcomer plays a significant role in the overall mass transfer process [4,5].

The research reported here involved studying the effects of changing various operating parameters in the air-lift, in particular the gas-liquid separator configuration, and the resulting effects on the gas hold-up and mass transfer coefficient (KLA). Further, the addition of an auxiliary gas sparger at the entrance of the downcomer was also examined. This work emphasizes the importance in considering downcomer and gas-liquid separator design and operation when analyzing the overall behaviour of air-lift reactors.

MATERIALS AND METHODS

The split vessel air-lift reactor used in these experiments was of rectangular cross-section with a total volume of approximately 300 liters. The gas sparging system was designed in such a way that gas could be introduced in either and/or both sides, and at any level, of the divided chamber. Thus, either side of the split vessel could be used as riser or downcomer. This was done in the gas hold-up experiments. In the gas absorption experiments only the wider side was used as the riser.

The overall dimensions of the air-lift were:

Riser/Downcomer 1	0.09 m X 0.25 m X 4.0 m
Riser/Downcomer 2	0.07 m X 0.25 m X 4.0 m
Gas-Liquid Separator	1.0 m X 0.25 m X 0.6 m

The gas-liquid separator was designed with a baffle system which enabled the manipulation of the fluid residence time in this section. The width could be changed on either riser or downcomer sides. Either a straight baffle or a T-baffle was used as an extension of the wall separating the riser and downcomer. Also, the ungassed liquid level above the baffle separating riser and downcomer was varied.

The local gas hold-up was measured in both the riser and downcomer by a system of 9 inverted differential manometers located in each chamber. Liquid velocity was measured by a Signet "Paddlewheel Flosenser", which was calibrated by salt tracer experiments.

The overall volumetric mass transfer coefficient was determined by a physical dynamic method using a polarographic probe (Yellow Springs Instruments combination probe #5739 with standard membrane). The oxygen concentration in the water was measured as a function of time after a step change in the influent gas from nitrogen to air. In order to maintain the hydrodynamic conditions in the air-lift during the switch from nitrogen to air, the gas flow rate was uninterrupted. A small aliquot of bubble-free liquid was constantly withdrawn from the bottom of the air-lift, its oxygen concentration measured, and returned to the reactor.

An implicit expression for the local gas hold-up was derived based on Wallis's pseudo-homogeneous model for pressure drop in a two phase flow regime. The mean or average gas hold-up of the riser or downcomer was calculated by integrating the local gas hold-up over the length of the chamber.

The overall mass transfer coefficient was calculated using the model developed by Weiland and Onken [7]. This model assumes perfect mixing of both the liquid and gas phases, as well as considers the lag in the electrode response due to oxygen electrode dynamics.

Further details of the materials and methods used in these experiments can be found elsewhere [2,5].

RESULTS

Gas-Liquid Separator and Downcomer Flow Patterns

Visual and photographic observations of the gas flow in the reactor showed that geometric configurations and operating conditions (such as decreased liquid level in the gas-liquid separator) which produced shorter fluid residence times in the separator, lead to greater gas recirculation in the downcomer. This was further confirmed by an increase in the measured gas hold-up in the downcomer. Thus, operating the air-lift with the straight baffle extension between the riser and downcomer and the gas-liquid separator closed by baffles flush with the riser and downcomer gave high gas recirculation and high downcomer gas hold-up. On the other hand, operating the reactor with the T-baffle between the riser and downcomer, and the separator fully open, there was near complete separation of the gas and liquid phases in the gas-liquid separator.

Different gas flow configurations were observed in the downcomer dependent on the downcomer liquid velocity. Straight bubble flow was observed in the downcomer when the downcomer relative gas-liquid velocity exceeded the terminal free rise velocity of like size bubbles in stagnant liquid. Thus, a relative gas-liquid velocity in excess of approximately 30 cm/sec was required for there to be straight bubble flow the downcomer.

When operating the reactor so that the liquid velocity was of the order of the terminal free rise of the gas bubbles, an oscillating bubble flow pattern was observed. The oscillating flow pattern exhibited essentially three different stages:

1. "Bubble Front": A front of bubbles stratified and stagnated in the downcomer without reaching the bottom of the reactor and recirculating to the riser. The liquid velocity was sufficient to entrain small gas bubbles into the downcomer which subsequently coalesce into larger bubbles and stratify towards the top of the downcomer.
2. Swirling Flow: Downcomer liquid velocity is sufficiently high for smaller gas bubbles with lower terminal free rise velocities to recycle through the downcomer to the riser. However, coalesce of gas bubbles in the downcomer persists causing larger bubbles to oscillate and swirl in the downcomer as they rise countercurrently.

3. Wavy Flow: Prior to a switch in flow pattern from oscillating to straight flow the bubbles in the downcomer will exhibit a "wavy" pattern. This is due to short, sporadic flow oscillations.

4. Straight Bubbly Flow: Gas bubble circulation is characterized by a clear and defined downward flow pattern, free of flow oscillations. Gas bubble size is rather uniform, with bubble diameter size approximately 4 to 6 mm.

Gas Recirculation Rate

A method was developed to measure the gas recirculation rate in the reactor. This was accomplished by using the T-baffle configuration, with a 15 cm water level above the baffle, in the gas-liquid separator which enabled near complete separation of the two phases. An auxiliary gas sparger was then placed 10 cm from the entrance of the downcomer. Gas was first introduced to the riser, after the liquid circulation was fully established, gas was also injected via the downcomer sparger. It was thus possible to generate a correlation relating the downcomer superficial gas velocity as a function of the measured downcomer gas hold-up.

The downcomer gas flow rate, gas recirculation rate, and the true riser superficial gas velocity (influent gas flow rate plus recycled gas flow rate) could then be determined in one-sparger and/or two-sparger systems with gas recirculation by measuring the gas hold-up and calculating the downcomer superficial gas velocity using the correlation found using the above method, where:

downcomer gas flow rate = QG_D = $JGD \times A_D$ (1)

true riser gas flow rate = QG_R = $QG_D + QG_S$ (2)

true riser superficial gas velocity = $JGR = QG_R/A_R$ (3)

gas recirculation rate = QD/QR = QG_D/QG_R (4)

The downcomer gas flow rate will be comprised of both recycled and freshly entrained air; the latter is relatively much smaller than the former.

Different rates were obtained using different gas-liquid separator configurations and liquid levels [2]. It was shown that above a certain value of JGR, recirculation rate remains fairly constant for changing gas flow rates in the riser. This indicates that the recirculation rate is largely determined by the geometric configuration of the gas-liquid separator and the liquid level in the separator.

Gas Hold-Up, Liquid Velocity, and Mass Transfer Correlations

Gas hold-up, liquid velocity, and mass transfer correlations were developed which described the gas hold-up, liquid velocity and mass transfer in terms of the true riser superficial gas velocity, JGR, for the various operating conditions and geometric configurations. A very close fit was obtained using exponential multiple regression on the data points from the gas hold-up experiments.

The results of the mass transfer experiments indicate that the reactor conditions could be divided into two major groupings, depending on gas-liquid separator configuration and gas injection point. Mode B includes operation conditions in the reactor with a straight plane baffle separating the riser and downcomer with gas being introduced via two sets of spargers, at the entrances of both the riser and downcomer. Mode A includes all other experimental conditions, either single sparger systems or two sparger systems using a T-baffle in the gas-liquid separator between the riser and downcomer.

The correlations are summarized in Table 1:

TABLE 1
Gas hold-up, liquid velocity, and overall mass transfer correlations as a function of the true superficial gas velocity, JGR

Gas Hold-up Correlations			
Riser, Straight baffle			
$\phi = 0.0176(JGR)^{0.71}(De_R/De_D)^{0.32}$		$R^2 = 0.94$	(5)
Riser, T-Baffle			
$\phi = 0.0136(JGR)^{0.69}$		$R^2 = 0.99$	(6)
Downcomer (two-sparger system, stable state)			
$\phi = 0.0311(JGD)^{0.68}$		$R^2 = 0.94$	(7)
Riser Liquid Velocity Correlation			
$UL = 33.87(JGR)^{0.40}(De_R/De_D)^{-0.41}$		$R^2 = 0.97$	(8)
Overall Mass Transfer Coefficient Correlations			
Mode A	$KLA = 16.0(JGR)^{0.90}$	$R^2 = 0.92$	(9)
Mode B	$KLA = 37.6(JGR)^{0.82}$	$R^2 = 0.92$	(10)
Modes A & B	$KLA = 15.8(JGR)^{0.94}$	$R^2 = 0.79$	(11)

Equivalent diameter: $De = 2[(a)(b)]/[a + b]$
where; a and b are the sides of the cross-section of the chamber.

The definition of the true superficial gas velocity allowed the formulation of the above correlations. Attempts to correlate gas hold-up or mass transfer coefficients as a function of the superficial gas velocity conventionally defined as the sparged gas flow rate, Q_S, divided by the riser cross-sectional area, A_R, have been largely unsuccessful. On the other hand, Weiland [8] obtained satisfactory correlations using the superficial gas velocity defined as $QG_S/(A_R+A_D)$. This is unexpected, since such hypothetical superficial gas velocity is not related to the actual fluid dynamics phenomena in the reactor. The origin of such approach seems to be the lumping of the air-lift reactor into a "black box bubble column". It is well recognized that superficial gas velocity is a very effective variable in correlating mass transfer and gas hold-up in bubble columns. The reason for the success of this approach can be understood by assuming that isothermal expansion the pneumatic power of the gas input can be given by:

$$P = g\rho_L QG_S HL \qquad (12)$$

Since the total liquid volume in the reactor, neglecting the gas-liquid separator, is $VL = HL(A_R+A_D)$, then;

$$P/VL = g\rho_L QG_S/(A_R+A_D) \qquad (13)$$

Therefore, the superficial gas velocity based on the total area is proportional to the power per unit volume, which is obviously related to the hydrodynamics in the reactor from the point of view of energy balance. However, it seems more conceptually correct to directly use the specific power input to correlate the results, as done by many researchers, rather than a hypothetical superficial gas velocity.

In order to compare our results with others previously published correlations were also developed which described the mass transfer coefficient as a function of the pneumatic power of gas input per total dispersion volume, P/VD [5]. A close fit was obtained using exponential multiple regression on the data points from the experiments. These correlations are given in Table 2:

Table 2
Overall Mass Transfer Coefficient Correlations
as a Function of the P/VD

Mode	Correlation	R^2	Eq.No.
A	$KLA = 389(P/VD)^{0.97}$	0.93	(14)
A	$KLA = 465(P/VD)^{0.93}(QD/QR)^{0.10}$	0.95	(15)
A	$KLA = 913(P/VD)^{1.04}(UL)^{-0.18}$	0.94	(16)
B	$KLA = 630(P/VD)^{0.90}$	0.91	(17)
B	$KLA = 849(P/VD)^{0.96}(QD/QR)^{0.14}$	0.93	(18)
B	$KLA = 1893(P/VD)^{1.05}(UL)^{-0.23}$	0.93	(19)
A & B	$KLA = 453(P/VD)^{1.04}$	0.84	(20)
A & B	$KLA = 589(P/VD)^{0.99}(QD/QR)^{0.15}$	0.87	(21)
A & B	$KLA = 1562(P/VD)^{1.15}(UL)^{-0.27}$	0.85	(22)

DISCUSSION AND CONCLUSIONS

Until recently, the behaviour of the gas-liquid separator in air-lift reactors has largely gone unstudied. The work presented here demonstrates that the gas-liquid separator will influence the overall reactor performance and deserves further investigation. The gas recirculation rate is largely determined by the gas-liquid separator's geometric configuration and the liquid level in the separator. Configurations of the gas-liquid separator which increase fluid residence time, allow for greater gas-liquid phase separation and thus lower gas recirculation.

Traditionally, the superficial gas velocity has been defined as the influent gas flow rate divided by the riser cross sectional area, excluding the gas flow in the reactor due to gas recirculation. This lead to a wide variation of the data presented in the literature for gas hold-up as a function of superficial gas velocity. Reactors with higher gas recirculation would give greater values of gas hold-up for similar values of superficial gas velocity. By using the method described in these experiments to determine the "true" superficial gas velocity, it was possible to correlate the data for varying reactor geometries and rates of gas recirculation. It appears that for the rectangular air-lift used in these experiments, the dependence of the gas hold-up on the superficial gas velocity can be approximated by the proportionality, $\phi = b(JG)^{0.7}$, regardless of the direction of flow, where b is a constant dependent on the physical properties and geometry of the system.

As can be seen from the gas hold-up and liquid velocities correlations the ratio of downcomer and riser cross sectional areas (AD/AR) will have a significant influence. This is in agreement with other researchers [1,8] and is due to the combined effect of AD/AR on the liquid velocity in both the riser and the downcomer and on bubble disengagement in the gas-liquid separator. It has also been shown that physically lowering the liquid velocity by placing a constriction in the downcomer (ie. a partially closed valve) will have a similar effect of lowering the riser liquid velocity and increasing the gas hold-up [9,10,11].

The above correlation for liquid velocity is in general agreement with the literature for the rate of change of the riser liquid velocity as a function of the riser superficial gas velocity [12]. It appears that the dependence of the riser liquid velocity on the riser superficial gas veloctiy can be approximated by the proportionality, $UL = b(JG)^{0.4}$. The constant b is dependent on the physiochemical properties of the system and the geometric design of the air-lift (eg. AD/AR, obstructions in the loop, connection zone design, etc.).

As can be seen from the above correlations, KLA is essentially directly proportional to P/VD, with QD/QR having a positive effect and the liquid velocity (UL) in the riser having a negative effect. The liquid velocity can be directly measured, while QD/QR can be calculated from an appropriate correlation, if available. The constants in the correlations depend on the physical properties of the system and geometry of the reactor.

It would appear that for the rectangular air-lift reactor the mass transfer coefficient is proportional to approximately, $KLA = b(P/VD)^{1.0}$, where b is a constant dependent on the physical properties of the system, orifice design, and the geometric design of the reactor. The direct relationship between the KLA and P/VD indicates a flexibility in air-lift reactor design for a wide range of process operations, depending on the microorganism oxygen demand.

The correlations in Table 2 show the influence of the ratio of the gas flow rate in the downcomer to the gas flow rate in the riser (QD/QR). Conditions which increase the gas flow rate in the downcomer (either high gas recirculation or two sparger systems) will increase QD/QR and subsequently the overall KLA at the same power input. Reducing the gas-liquid separator liquid volume, and thus fluid residence time will increase gas recirculation and KLA.

The most interesting feature of the air-lift bioreactor with respect to energy, is that the aeration efficiency (oxygen transfer to the degassed liquid volume per pneumatic power of gas input) remains constant for a wide range of specific power input. This can be seen in Figure 1 which shows typical curves for the air-lift operated with the straight baffle and one sparger (Mode A) or two spargers (Mode B). This in contrast to conventional bubble columns and mechanically stirred tank reactors which exhibit a sharp decrease in aeration efficiency with increasing oxygen transfer rate [13,14].

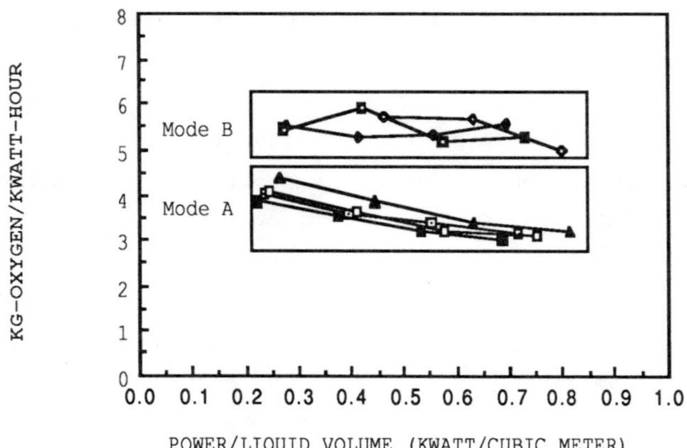

Figure 1. Aeration efficiency

These counterbalancing effects indicate a flexibility in air-lift reactor design for different operating objectives. The liquid volume can be increased while maintaining the required KLA to P/VD relationship. For example, systems with high oxygen consumption can be designed to maximize gas recirculation and/or introduce gas via two spargers. The fluid residence time in the gas-liquid separator can also be increased to remove oxygen depleted gas in this zone, and fresh gas introduced at the downcomer entrance. Systems with low oxygen requirements, can be designed by maximizing the dispersion volume while maintaining the required KLA for proper system performance. In the latter situation, power requirements can be reduced further by introducing part of the gas at the downcomer entrance.

In addition to the strategies mentioned above, the effect of liquid velocity on the KLA can also be exploited in air-lift design by changing the downcomer to riser cross-sectional area ratio. In systems requiring a high KLA, such as single-cell protein production and fungal growth, the liquid velocity can be physically decreased by reducing this ratio. This would increase the downcomer liquid velocity, thus decrease microorganism residence time in the downcomer, and increase gas entrainment. In systems requiring relatively lower mass transfer rates at maximum energy efficiency and liquid volume, such as wastewater treatment and algal systems, a ratio closer to unity would be more efficient. Further work is needed to optimize this relationship for actual reactor applications.

This work indicates that the downcomer does play an influential role in the overall mass transfer process. This would contradict the assumption of Bello et al.[1] that only the volume of the riser contributed to the mass transfer. Their explanation of this phenomenon is that most of the bubbles carried down by the liquid in the downcomer have a negligibly small relative velocity. However, in the experiments reported here, systems which increase the gas hold-up in the downcomer, either by increasing the gas recirculation rate or introducing gas via a second sparger in the downcomer, may improve the mass transfer.

This work shows that the gas recirculation has a major influence on the entire reactor behaviour and must be considered in the design and scale-up of air-lift reactors, mainly by an adequate design of the gas-liquid separator. This work also demonstrates the potential and flexibility of air-lift reactors over a very wide range of applications due to the relatively high and constant aeration efficiency over a wide range of gas flow rates. Thus, the air-lift reactor design can be either optimized for maximum oxygen transfer rate or maximum liquid volume, depending on the desired application. Further study is needed of both the behaviour of the gas-liquid separator and it's influence on potential applications of air-lift reactors.

ACKNOWLEDGMENTS

We wish to acknowledge the support of this work by the Bundesministerium fur Forschung und Technologie (BMFT) of West Germany and the Ministry of Energy and Infrastructure National Council for Research and Development (NCRD) of Israel within the framework of the Joint German-Israeli Research Projects. We would also like to thank Moshe Golden for his help with computer programming.

REFERENCES

1. Bello, R.A., Robinson, C.W. and Moo-Young, M., Gas holdup and overall volumetric oxygen transfer coefficient in airlift contactors. Biotechnol. Bioeng., 1985, **27**, 369-381.

2. Siegel, M.H., Merchuk, J.C. and Schugerl, K., Air-lift reactor analysis: Interrelationships between riser, downcomer, and gas-liquid separator behavior, including gas recirculation effects. AIChE J., 1986, **32**, 1585-1596.

3. McManamey, W.J. and Wase, D.A.J., Relationship between the volumetric mass transfer coefficient and gas holdup in airlift fermentors. Biotechnol. Bioeng., 1986, **28**, 1446-1448.

4. Koide, K., Horibe, K., Kitaguchi, H. and Suzuki, N., Contributions of annulus and draught tube to gas-liquid mass transfer in bubble columns with draught tube. J. Chem. Eng. Japan, 1984, **17**, 547-549.

5. Siegel, M.H. and Merchuk, J.C., Mass transfer in a rectangular air-lift reactor: Effects of geometry and gas recirculation. Biotechnol. Bioeng., 1987, Submitted for publication.

6. Wallis, G.B., One-dimensional two-phase flow. McGraw-Hill, New York, 1969.

7. Weiland, P. and Onken U., Fluid dynamics and mass transfer in an airlift fermenter with external loop. Ger. Chem. Eng., 1981, **4**, 42-50.

8. Weiland, P., Influence of draft tube diameter on operation behaviour of airlift loop reactors. Ger. Chem. Eng., 1984, **7**, 374-385.

9. Merchuk, J.C. and Stein, Y., Local hold-up and liquid velocity in air-lift reactors. AIChE J., 1981, **27**, 377-388.

10. Merchuk, J.C., Gas hold-up and liquid velocity in a two- dimensional air lift reactor. Chem. Eng. Sci., 1986, **41**, 11-16.

11. Weiland, P. and Onken, U., Differences in the behaviour of bubble columns and airlift loop reactors. Ger. Chem. Eng., 1981, **4**, 174-181.

12. Siegel, M.H., Hallaile, M., and Merchuk, J.C., Air-lift reactors: Design, operation, and applications. In Advances in Biotechnological Processes, ed. A. Mizrahi, Alan R. Liss, New York, 1987, in press.

13. Wang, D.I.C., Hatch, R.T. and Cuevas, C., Engineering aspects of single-cell protein production from hydrocarbon substrates: the airlift fermentor. Proceedings: 8th World Petroleum Congress, Moscow, 1971, 149-156.

14. Orazem, M.E. and Erickson, L.E., Oxygen transfer rates and efficiencies in one- and two-stage airlift towers. Biotechnol. Bioeng., 1979, **21**, 69-88.

INTERNATIONAL CONFERENCE ON BIOREACTORS AND BIOTRANSFORMATIONS
GLENEAGLES, SCOTLAND, UK: 9-12 NOVEMBER 1987

Paper I4

HYDRODYNAMICS, AXIAL DISPERSION AND GAS-LIQUID OXYGEN TRANSFER IN AN AIRLIFT-LOOP BIOREACTOR WITH THREE-PHASE FLOW

P. Verlaan* and J. Tramper
Agricultural University,
Department of Food Science,
Food and Bioengineering Group,
De Dreijen 12, 6703 BC Wageningen,
the Netherlands.

(* To whom correspondence should be addressed)

ABSTRACT

Hydrodynamics, axial dispersion and oxygen transfer in a pilot plant airlift-loop bioreactor (0.165 m^3) with a three-phase flow have been studied in order to investigate the influence on the physical properties of an airlift-loop reactor (ALR). The third phase consisted of polystyrene or calcium alginate beads both with a density of ρ= 1050 kg/m^3 and diameters ranging from 2.4 to 2.7 mm, being good representatives for immobilized biocatalysts. It was found that the overall reactor performance is strongly influenced by the presence of the solid phase. The maximum bead loading at which the ALR could be operated was 40 volume-procent. At this loading the liquid velocity declined to 60% of the initial two-phase value independent of the gas injection rate while the gas hold-up decreased from 80% to 20% of the two-phase value depending on the gas injection rate. The essential mixing parameter, the Bodenstein number, tended to a 40% higher value at this loading indicating a better established plug flow. The influence of the solid phase on the volumetric oxygen transfer coefficient k_la was investigated to a maximum bead loading of 20 volume procent. In this case, the k_la-value decreased with 40% compared to the two-phase value.

INTRODUCTION

An airlift-loop reactor is a so called second generation type of bioreactor in which efficient oxygen transfer and mixing is combined with a controlled liquid flow while the shear rate can be very low. These properties make the ALR a suitable reactor for shear sensitive organisms requiring a controlled dissolved oxygen concentration. An example of such an application is the production of secondary metabolites by plant cells [1]. In many cases immobilized biocatalysts or micro-organisms growing in aggregates are used in biotechnological production processes. This means that the biophase in the reactor is concentrated in or on beads with diameters up to several millimeters. Also in this case an ALR seems a suitable reactor having excellent suspension characteristics due to the high liquid velocity.

Little research has been reported yet on the influence of relatively large (2-3 mm) particles with a neutral buoyancy, like gel-entrapped biocatalysts, on bioreactor performance. Recently, Frijlink [2] published results on the influence of calcium alginate beads (ρ= 1050 kg/m^3, d= 2.2 mm) on oxygen transfer in a stirred-tank reactor. The author found that the volumetric oxygen transfer coefficient decreased proportional with the bead loading. For a bead loading of 37 vol-procent the decrease amounted to 55%-59%, depending on the gas-flow rate. Metz [3] reported results on the influence of yeast pellets on oxygen transfer in a bubble column. A pellet loading of 20% diminished the $k_l a$-value with 20-30 %.

For ALRs no such data is available. Therefore the aim of this article is to give a concise overview of the physical ALR properties and the interaction with relative large solid particles in order to provide essential information for three phase ALR design. Results are reported on the physical influence of neutral buoyant polystyrene or calcium alginate beads with diameters ranging from 2.4 to 2.7 mm, on ALR performance at pilot plant scale.

MATERIALS AND METHODS

The experiments have been carried out in a pilot plant ALR with external loop as shown in figure 1. The ALR has a reactor volume of 0.165 m^3 and an aerated height of 3.23 m. The upflow and downflow sections, also called riser and downcomer, were constructed of borosilicate glass pipe sections with diameters of 0.2 m and 0.1 m, respectively. The gas sparger, situated at the bottom of the riser, produces bubbles with the same diameter as the equilibrium diameter of air bubbles in tap water. The topsection of the ALR was designed such that complete deaeration occurs during operation and no gas entrains into the downcomer. The ALR was filled with Wageningen tap water and its temperature was maintained on a constant value of 30° C. More details about the ALR and measuring methods of the hydrodynamic parameters are given elsewhere [4].

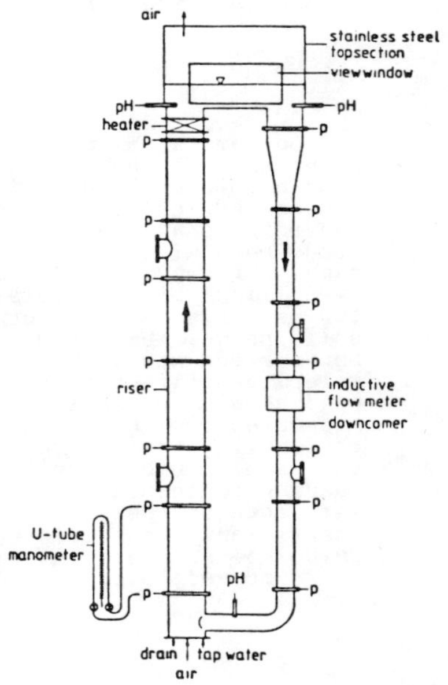

Fig.1 The airlift-loop reactor

The mixing performance of the ALR was characterised by estimating the axial dispersion number on the basis of pulse respons measurements using acid and base as tracers. Detection of these tracers by pH-electrodes was not disturbed by the presence of air bubbles or solid beads. A mathematical description and detailed information about the experimental method have been published earlier [5,6].

The typical ALR mixing characteristics allowed us to treat the modelling of oxygen transfer in two different ways. On the one hand

the ALR behaves like a loop reactor with relative high circulation rates and a short mixing time. From this point of view the reactor can be modelled as an ideally stirred tank reactor (STR) [7]. On the other hand the ALR is a tube reactor in which the liquid phase as well as the gas phase behaves like plug flow which has been experimentally verified earlier [5]. In this work, $k_l a$-experiments in the three phase flow have been carried out by the STR-method for reasons of simplicity.
The solid phase used consisted of calcium alginate or polystyrene spheres with a particle density and diameter of ρ= 1050 kg/m³ and d= 2.35- 2.7 mm, respectively. The polystyrene spheres have been used in the hydrodynamic and axial dispersion measurements. Both the polystyrene and the calcium alginate spheres have been used in the oxygen transfer experiments. The calcium alginate spheres were produced by a new method described by Hulst et al. [8] which makes it possible to produce large quantities of beads in a relative short time.

RESULTS AND DISCUSSION

Hydrodynamics

Figure 2 shows the results of the hydrodynamic experiments together with model evaluations for both the liquid velocity in the downcomer and the gas hold-up in the riser. The model calculations were derived from an iterative procedure which has been described by Verlaan et al. [4].

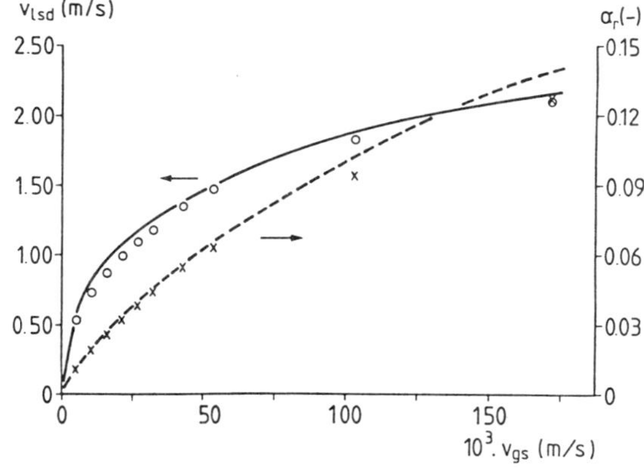

Figure 2. The downcomer liquid velocity and the riser gas hold-up as a function of the superficial gas velocity in the riser: (o,x) experimental; (——,---) simulation

For low gas input rates the liquid velocity and gas hold-up are very sensitive to changes in the gas input rate. For high input rates on the other hand only a minor increment of the liquid velocity or the gas hold-up is observed when the gas velocity is increased. The model gives an adequate prediction of the flow behaviour in the ALR with an accuracy of at least 5-10%.

When the polystyrene particles were added to the ALR up to a loading of 40% the liquid velocity decreased gradually to 40% of the initial two phase value as is shown in figure 3. This was also the maximum loading at which the ALR could be operated. When the reactor was stopped it was not possible to restart the liquid circulation at this loading mainly due to the fact that the packed bed volume of the particles approximated the aerated riser volume. The decrease in velocity is caused by a decrease in gas hold-up and an increased friction. The decrease in gas hold-up is clearly shown in figure 4 and, in contrast to the liquid velocity, strongly affected by the gas injection rate. Obviously, the presence of the particles increases the collision frequency due to the diminished flowed area for the air-water mixture. As a result the coalescence process will be stimulated which on its turn reduces the gas hold-up. For high gas velocities and gas hold-ups, when bubbles already interact, this effect will be of less importance than for low gas velocities. Hence, for low gas velocities a reduction of 60% is achieved at a bead loading of 20% while for high gas input rates the gas hold-up is reduced about 20% at a bead loading of 40%.

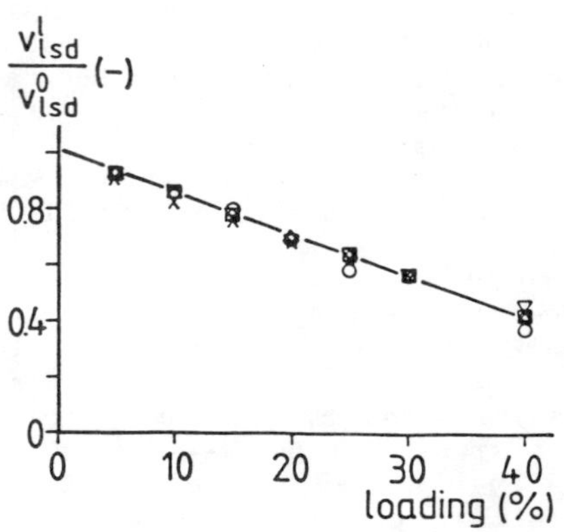

Figure 3. The relative liquid velocity as a function of the particle loading and the superficial gas velocity as a parameter.

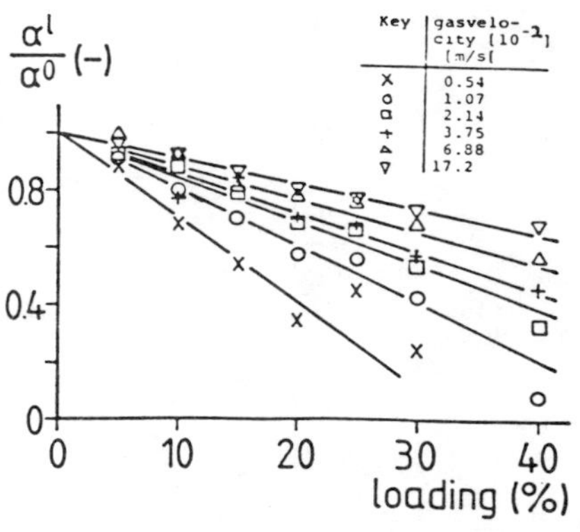

Figure 4. The relative gas hold-up as a function of the particle loading and the superficial gas velocity as a parameter

Of course, the coalescence process also depends on the local solids concentration and the particle size. Epstein [9] reviewed the mechanisms reported in literature which could be responsible for bubble characteristics and therefore on gas hold-up in a three-phase system. Agreement exists on the assumption that small particles increase the bubble coalescence rate due to the enhanced viscosity of the pseudohomogeneous three-phase medium. For large particles on the contrary several theories are introduced to account for bubble disintegration. As in our case the particles are neutral buoyant and easily follow the liquid motion, the effect of turbulence induced by the particles on bubbles will be of minor importance. We believe that in our system bubbles will break up if the solid particles have sufficient inertia to penetrate the surface of a bubble, when the Weber number $We = \rho v^2 d/\sigma$, the numerical criterium for break-up, exceeds about 3 [9,10]. As in our system the Weber number is about three, it is assumed that neither the bubble coalescence nor the bubble disruption according to

the above theories, contribute significantly. These findings are in accordance with the results of Brück and Hammer [11] who concluded that solid beads with densities less than 1050 kg/m³ and diameters ranging from d=0.06 to d=4.35 mm, cause a decrease in gas hold-up. The authors explained this by the increased solid hold-up and the increased suspension viscosity while they also support the criterium for bubble break-up mentioned above.
As the liquid velocity and the gas hold-up are unambiguously related to each other according to Verlaan et al. [4], the results in figure 3 and figure 4 might at first view seem discrepant in relation to the context mentioned above. The relationship between the liquid velocity and the gas hold-up can be mathematically formulated by:

$$\rho g \alpha L = \tfrac{1}{2} K_f \rho v^2 \qquad (1)$$

where ρ is the liquid density, g the gravitational constant, α the gas hold-up in the riser, L the aerated length, v the superficial liquid velocity and K_f the overall friction coefficient. From equation 1 it should be expected that the dependency of the gas hold-up, shown in figure 4 also should occur in the results shown in figure 3. However, figure 5 demonstrates that in contrast to gas-liquid flow [4], friction in a three phase flow is severly influenced by the gas injection rate. This happens in such a way that for high gas velocities the increased friction counterbalances the decrease in the relative influence of the gas injection rate on the liquid velocity.
As is also shown in figure 5 friction increases with an increasing particle loading. Both phenomena can be explained by the fact that for an increased gas injection rate or an increased bead loaloading, bubbles tend to concentrate in the middle of the column which has been verified by visual observation. As a result the solid phase concentration at the wall of the column will increase, thus enlarging friction. This phenomenom has been experimentally demonstrated by Linnenweber and Blaβ [12] in a bubble column. They found that the solid hold-up at the tube wall of a bubble column with gas hold-ups ranging from α=0.05 to α=.1, is twice as high as the solid hold-up in the centre of the tube. The authors also report that this effect becomes less significant at high gas injection rates which in our case corresponds to the results in figure 5.

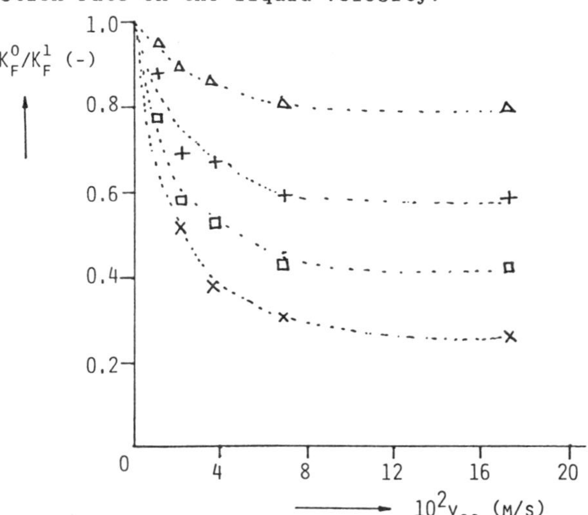

Figure 5. The friction coefficient of the three phase flow relative to the friction coefficient of the two phase flow as a function of the superficial gasvelocity and the bead loading as a parameter (× 40%, □ 30%, + 20%, △ 5%)

Mixing

An important parameter to quantify axial dispersion characteristics in a tubular reactor is the dimensionless Bodenstein number (Bo) which represents the ratio of convective mass transport and mass transport by axial dispersion. Results for gas-liquid flow in the pertinent ALR are shown in figure 6 which are obtained from the results of Verlaan et al. [6]. As is shown in figure 6, the Bodenstein values lie in between 50<Bo<60 depending on the gas injection rate and indicating a plug flow character. Addition of a solid phase consisting of polystyrene spheres significantly enhances the Bodenstein number up to 50% for a bead loading of 40% (figure 7). Obviously, the presence of polystyrene spheres in the gas-liquid flow damps the small edies which are, apart from other mechanisms, responsible for the axial dispersion. In the literature, there is no agreement on this subject. In his literature overview, Frijlink [2] concludes that sometimes particles are said to dampen the turbulence in the continuous liquid phase while in other cases they are supposed to increase turbulence intensities, depending on particle size and particle density. Epstein [9] and Kelkar [14] conclude that small particles in a three phase flow do not significantly influence axial dispersion as solid dispersion is mainly determined by the liquid dispersion. When particle sizes become larger solid and liquid phase dispersion start to differ. Kato et al. [13] gives an empirical correlation by which solid dispersion can be calculated from liquid dispersion. Epstein [9] stated that it is not unreasonable to assume that when particle size or density increase up to the point were the liquid and solid dispersion start to differ, the flow regime in effect is moving from a regime of slurry-column

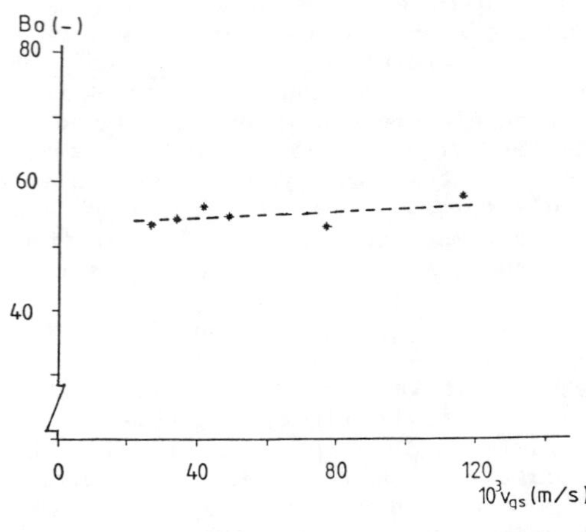

Figure 6. The Bodenstein number as a function of the superficial gas velocity

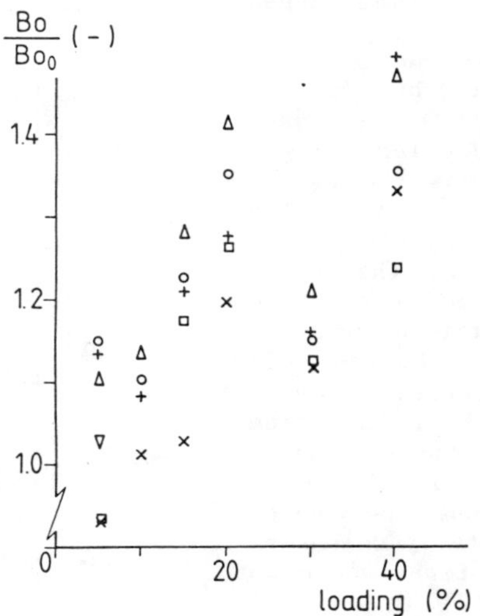

Figure 7. The relative Bodenstein number as a function of the relative bead loading

operation to three-phase fluidisation.
In the present case obviously three-phase fluidisation is involved and we propose that two mechanisms are responsible for the three-phase mixing behaviour of the ALR. On the one hand as bubble size grows, as explained in the previous section, the bubble rise velocity increases and the amount of liquid which can be transported in the form of liquid wakes decreases. This phenomenom results in a decrease in the axial dispersion coefficient [14]. On the other hand the ratio of particle diameter to scale of turbulence is considered as a measure for assessing fluid-particle interaction. The strongest mutual influence is to be expected if the size of the phase elements are of the same order. As the particle diameter lies in between 2-3 mm, turbulence on this scale and even on a smaller scale will be damped, which makes the explanation given above a plausible one.
Another conclusion which can be drawn from figure 4 and figure 7 is the fact that axial dispersion decreases more than proportional to the Bodenstein number (Bo= v.L/D) at an increasing gas injection rate as the liquid velocity simultaneously decreases (figure 4) thus effecting the ratio of mass transport by convection and mass transport by dispersion.

Oxygen transfer

The results for the volumetric oxygen transfer coefficient, k_la, estimated by three different methods are shown in figure 8. Two methods are based on the plug flow characteristics of the ALR for both the liquid and the gas phase, the first method being a non isobaric, steady-state, plug-flow model [7], the second method being a dynamic, non-isobaric, plug-flow model on the basis of which also the dissolved oxygen concentration control was performed [15]. The third method consists of an isobaric model predicting the dissolved oxygen concentration in the liquid phase of an ideally-stirred-tank reactor [7]. For the present ALR, the k_la-values obtained by all three methods harmonize rather well notwithstanding both different ways of approximating the (gas-) liquid flow in the ALR. However, as already stated in the previous section, it is allowed within certain restrictions, to model the ALR as being a STR due to its high circulation rate. The following results were obtained by the STR method as this method appeared to be a reliable and fast response estimation method requiring little computing time.

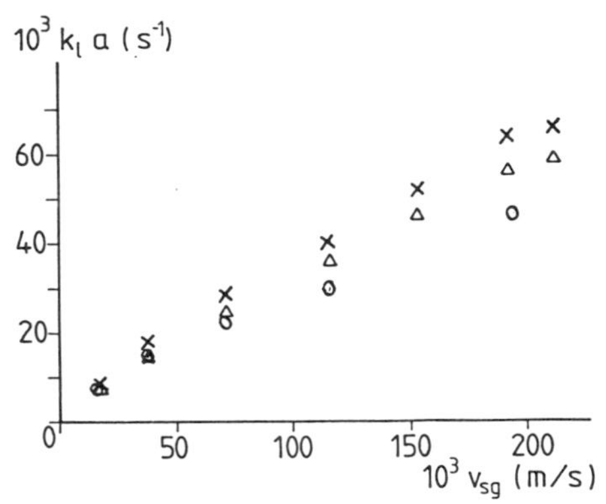

Figure 8. The volumetric oxygen transfer coefficient as a function of the superficial gas velocity (x plug flow model 1, o plug flow model 2, STR model)

The presence of the solid phase negatively influences aeration for both the polystyrene and calcium alginate particles as shown in figure 9 and figure 10. In literature many results are reported on the influence of small particles on aeration [10,11,16-18] and agreement

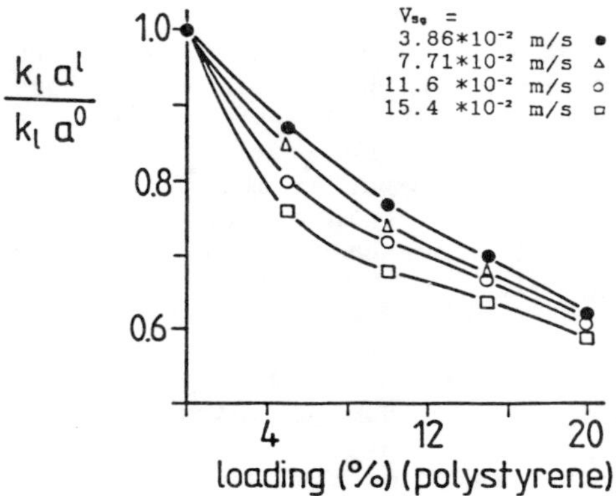

Figure 10. The relative volumetric oxygen transfer coefficient as a function of the bead loading

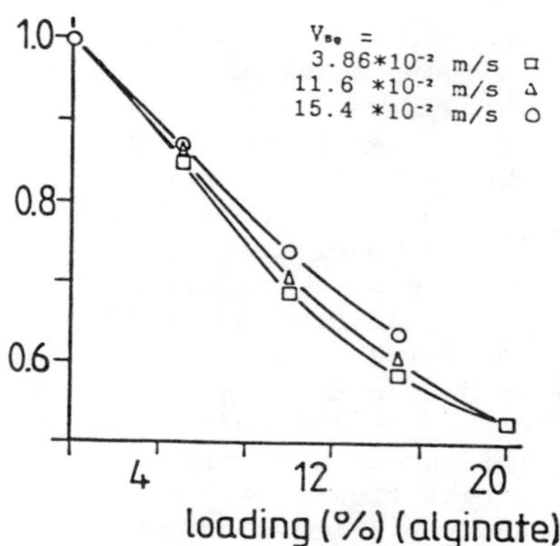

Figure 9. The relative volumetric oxygen transfer coefficient as a function of the bead loading

exists on the mechanism responsible for the change in k_la. It is reported that for low particle loadings a slight enhancement for k_la occurs and that no dramatic change in k_la can be expected until a bead loading of 20 vol-%. It is proposed that for these low concentrations the small particles do not change the viscosity of the water but enhance the surface renewal and mobility thereby increasing the value of k_la. Higher concentrations increase the viscosity of the slurry thereby increasing coalescence as a result of which k_la decreases. This has been experimentally verified in the literature mentioned and a sharp decrease in k_la for particle loadings greater than 20 vol-% is reported. In our case, for large particles, the sharp reduction in k_la is probably due to a larger extend to a reduction in the specific area a, as a result of the coalescence process which has been discussed in the first section of this paragraph. The effect on the mass transport coefficient k_l will be of minor importance as for these large particles the increase in apparant viscosity for high loadings only has its effect on macro (reactor) scale but not on micro-scale where mass transfer takes place. Therefore the apparent viscosity has no effect on oxygen transfer. On the contrary it is reasonable to suppose that the k_l value is slightly increased by surface renewal due to coalescence of the bubbles. On the other hand as mentioned in the previous section, turbulence is damped by the particle, having a negative influence on k_l thus counterbalancing the effect of surface renewal on k_l.

Our results agree with the results of Frijlink [2] who measured k_la in a STR with a three-phase flow, the third phase being calcium-alginate beads and comparable to the beads in our research. The author found a slow linear decrease of k_la in relation to the bead loading and compared his results with the present results in the ALR. Frijlink suggested that the decrease in k_la as a function of the bead loading in the ALR could be a result of reduction of turbulence intensity resulting in larger stable bubbles, the effect being much stronger in a system with low energy input such as the ALR than in a system with high energy input such as the STR. However this explanation is in contrast to our discussion in the hydrodynamics and mixing section. In the latter section it was suggested that small eddies were damped by the particles while the larger eddies determine particle motion due to

the negligible difference in density with water, as suggested in the first section. As the bubbles are larger in diameter than the particles are, the larger eddies which are not damped by the particles are responsible for bubble break up. As these eddies are hardly influenced by the particles no effect on bubble break up will occur. In fact the mechanism responsible for the reduction in the interfacial area, a, in an ALR is, apart from the mechanism in a STR as mentioned by Frijlink, also responsible for the decrease in the interfacial area in a STR. As the slenderness of the ALR is much larger than that of a STR the flowed area for the air-water mixture will be less in a STR than in an ALR, resulting in less coalescence and therefore less decrease of the interfacial area.
The reduction for alginate beads, shown in figure 9, being perfectly wetted is slightly more significant than for polystyrene beads, being poorly wetted. These findings are in accordance with the results of Kelkar and Shah [14] who reported that solids wettability was found to enhance the coalescence tendencies in the liquid phase thereby, in our case, reducing oxygen transfer.

CONCLUSION

The liquid velocity and the gas hold-up of a gas-liquid flow in an ALR can be easily modelled with a sufficient accuracy. Neutral buoyant particles with a diameter of 2-3 mm reduce the liquid velocity and the gas hold-up in an ALR significantly. The decrease in liquid velocity is caused by the decrease in gas hold-up and an increased friction. The gas hold-up is reduced mainly because the presence of the particles increases the collision frequency thereby increasing coalescence due to the diminished flowed area for the air-water mixture. In comparison to a gas-liquid flow axial dispersion is reduced in the three phase flow as the presence of the particles damps the small edies which are, apart from other mechanisms, responsible for the axial dispersion. Moreover, the increased coalescence also contributes to a decrease in axial dispersion. The presence of the particles negatively influences aeration due to a reduction in the gas-liquid interfacial area as a result of the coalescence process. The effect of the increase in apparent viscosity in the ALR was not supposed to contribute to the decrease in the aeration process.

REFERENCES

[1] A.C. Hulst, P. Verlaan, H. Breteler and D.H. Ketel. Thiophene production by Tagetes patula in a pilot plant airlift-loop reactor. Proc. 4th Europ. Congress on Biotechnology, 1987, vol 2, pp 401-404, (O.M. Neijssel, R.R. van der Meer, K.Ch.A.M. Luyben eds.) Elsevier Science Publishers B.V., Amsterdam.
[2] J.J. Frijlink. Physical aspects of gassed suspension reactors. Thesis, University of Technology, Delft, 1987.
[3] B. Metz. From pulp to pellet, Thesis, University of Technology, Delft. 1976.
[4] P. Verlaan, J. Tramper, K. van ´t Riet and K.Ch.A.M. Luyben. A hydrodynamic model for an airlift-loop bioreactor with external loop. Chem. Eng. J. 33 (1986) B43-B53.
[5] P. Verlaan, J. Tramper, K. van ´t Riet and K.Ch.A.M. Luyben. Estimation of axial dispersion in individual sections of an airlift-loop reactor. Submitted.

[6] P. Verlaan, J. Tramper, K. van ´ t Riet and K.Ch.A.M. Luyben. Hydrodynamics and axial dispersion in an airlift-loop bioreactor with two and three phase flow. Proc. Int. Conf. on Bioreactor Fluid Dynamics (BHRA) Cambridge, England, 15-17 april, 1986.

[7] P. Verlaan and J. Tramper. Influence of nearly floating particles on the behaviour of a pilot plant airlift-loop bioreactor. Proc. 4th Europ. Congress on Biotechnology, 1987, vol. 1, pp 101-104 (O.N. Neijssel, R.R. van der Meer, K.Ch.A.M. Luyben eds.) Elsevier Science Publishers B.V., Amsterdam.

[8] A.C. Hulst, J. Tramper, K. van´ t Riet and J.M.M. Westerbeek. A new technique for the production of immobilized biocatalysts in large quantities. Biotechnol. Bioeng. 27, pp 870-876 (1985).

[9] N. Epstein. Three phase fluidisation: some knowledge gaps. Canadian J. of Chem. Eng. 59, 1981, pp 649-657.

[10] W.-D. Deckwer and A. Schumpe. Transporterscheinungen in Dreiphasen-Reaktoren mit fluidisiertem Feststoff. Chem.-Ing.-Techn. 55, 1983, pp 591-600.

[11] F.J. Brück and H. Hammer. Intensivierung des Stoffaustausches in Blasensäulen-Reaktoren durch suspendierten Feststoffen. Chem.-Ing.-Techn. 58, 1986, pp 60-61.

[12] K.-W. Linnenweber and E. Blaβ. Messung örtlicher Gas- und Feststoffgehalte in Blasensäulen mit in der Flüssigkeit suspendiertem Feststoff. Chem.-Ing.- Techn. 54, (1982) pp 682-683.

[13] Y. Kato, A. Nishiwaki, T. Fukuda and S. Tanaka. Chem. Eng. Japan, 5, 1972, p 112.

[14] B.G. Kelkar, Y.T. Shah and N.L. Carr. Hydrodynamics and axial mixing in a three-phase bubble column. Effects on slurry properties. Ind. Eng. Chem. Process. Des. Dev. 29, 1984. pp. 308-313.

[15] A.K.M. Krolikowski, M.H. Zwietering, D.P. van den Akker and P. Verlaan. Control of the dissolved oxygen concentration (DOC) in an airlift-loop bioreactor. Proc. 4th Europ. Congress on Biotechnology, 1987, vol. 3 pp. 149-152 (O.M. Neijssel, R.R. van der Meer, K.Ch.A.M. Luyben eds.) Elsevier Science Publishers, Amsterdam.

[16] M. Miyachi, A.Iguchi, S. Uchida and K. Koide. Effect of solid particles in liquid-phase on liquid-side mass transfer coefficient. Canadian J. Chem. Eng. 59, 1981, pp 640-641.

[17] R.S. Albal, Y.T. Shah, A. Schumpe and N.L. Carr. Mass transfer in multiphase agitated contactors. Chem. Eng. J. 27, 1983, pp 61-80.

[18] G.E.H. Joosten, J.G.M. Schilder and J.J. Jansen. The influence of suspended material on the gas-liquid mass transfer in stirred gas-liquid contactors. Chem. Eng. Sci. 32, 1977, pp 563-566.

SYMBOLS

a	interfacial area	[m²]
α	gas hold-up	[-]
D	dispersion coefficient	[m²/s]
d	diameter	[m]
ρ	density	[kg/m³]
g	gravitational constant	[m/s²]
k	mass transfer coefficient	[m/s]
$k_l a$	volumetric mass transfer coefficient	[s^{-1}]
K_f	friction coefficient	[-]

L	length	[-]
v	velocity	[m/s]

Subscripts

l	liquid
s	superficial
d	downcomer
g	gas

Superscripts

o	concerning the two-phase system
1	concerning the three-phase system with a bead loading 1

INTERNATIONAL CONFERENCE ON BIOREACTORS AND BIOTRANSFORMATIONS GLENEAGLES, SCOTLAND, UK: 9-12 NOVEMBER 1987

Paper J1

IMMOBILIZATION OF RECOMBINANT ESCHERICHIA COLI IN SILICONE BEADS

Patrick Oriel and Jayashree Nayini
Michigan State University
Department of Microbiology and Public Health
East Lansing, Michigan, 48824

ABSTRACT

An immobilization system has been developed based on polymerization of cell-bearing silicone prepolymer emulsion droplets. Viable microbial colonies maintained in cavities throughout the beads released free cells and plasmid-encoded amylase. Immobilization was found to inhibit overgrowth of variants lacking plasmid when organisms were grown in the absence of antibiotic.

INTRODUCTION

Bioprocesses based on immobilization of bacterial cells have attracted attention for some time arising from expectations of simplified reactor design, higher conversion efficiency, and ease of product recovery (see, for example 1). More recently, stabilization of a multicopy plasmid following immobilization of *E. coli* was reported (2), providing an additional potential advantage. The most developed immobilized cell systems are those which utilize a single biochemical step and for which sustained cell viability is not required (see, for example, 3). As pointed out by Mattiasson, one of the principal challenges for increased utilization of viable immobilized cells in aerobic processes is the provision of adequate gaseous transport, requirements for which are established by the biomass, particle size, and gaseous diffusion rates within the support (4). In the frequently used polysaccharide entrapment methods, viable cells are limited to a thin layer on the outer bead periphery, presumably due to limited oxygen availability (5).

We have initiated studies to relate the behavior of immobilized cells to the physical characteristics of synthetic polymer supports in the hope of developing immobilization systems with improved properties. Our objective is the development of support materials with the following properties:

1) Physical properties allowing use with both mesophilic and thermophilic bioprocesses.

2) High gaseous diffusion rates for oxygen and carbon dioxide.
3) Low toxicity to living cells by the immobilization procedure.
4) Viable biomass retention over prolonged periods.

Our first approach utilized styrene/divinylbenzene macroporous ion exchange beads because of their high surface area and thermal stability. The need for adequate cell immobilization required beads of high porosity which were also fragile, confining their productive use to airlift reactors (6). More recently, we have focused on silicone polymer systems because of their high gaseous diffusion rates, which has led to their use in artificial lungs and bioreactor oxygenation devices. This paper summarizes progress on the development and optimization of a silicone polymer system for bacterial immobilization, which is still in progress.

MATERIALS AND METHODS

Silicone polymer beads containing viable bacterial cells were formed by dispersing 0.5 g of an aqueous paste of centrifuged log phase E. coli into two ml fresh L broth (10 g bactotryptone, 5 g yeast extract, 5 g NaCl per 1). This supension was emulsified by stirring into a mixture containing 20 g Dow Corning® Sylgard 184 prepolymer, 2 g Sylgard 184® catalyst, and 2 g Dow Corning 200® silicone fluid. Droplets of the emulsion were formed by pouring into 100 ml fresh L broth prewarmed to 37°C and stirring with a magnetic stir bar, the speed of which controlled the droplet size. Polymerization of the droplets into beads took place with continued stirring overnight.

Measurement of viable biomass utilized a modification of the INT (2-(p-iodophenyl)-3-(p-nitrophenyl)-5-phenyl-tetrazolium chloride) assay of Acuri (7). This assay measures the conversion of INT to formazan dye following uptake and reduction by viable bacteria. In our procedure 300 mg of beads were chopped finely with a razor blade and added to 1 ml of 0.25 percent INT in M9 salt solution (g/l, Na_2HPO_4, 6; KH_2PO_4, 3; NaCl, 0.5; $MgSO_4$, 0.24; $CaCl_2$, 0.011). After shaking for 30 minutes at 37°C, the INT solution was removed and the particles incubated with 1 ml acetone for 30 minutes to extract the formazan. The acetone extract was clarified in a microcentrifuge and dye absorbance measured at 490 nm.

Amylase and cell release experiments in semicontinuous culture were carried out at 37°C in L broth containing 100 micrograms/ml ampicillin. At each time interval, a 10 ml free cell culture was initiated in a 50 ml shake flask using 0.1 ml innoculum of the preceeding culture. Bead culture was maintained by washing 3 g beads twice with L broth and transferring to a 50 ml shake flask containing 10 ml fresh L broth with 100 micrograms/ml ampicillin. Amylase was measured with a starch-iodine assay (8).

Plasmid stability experiments in semicontinuous culture were carried out similarly, except that M9 salts containing 0.4 percent glycerol and 0.1 percent thiamine HCl with no antibiotic were used with 10 g of beads. Because of slower growth in this medium, new cultures were initiated at 24 hour intervals.

Continuous culture experiments utilized a cylindrical glass reactor 54 mm in diameter and 85 mm high, in which an 8 mm ID overflow tube was placed 20 mm from the base. Stirring was carried out with a stir bar. L broth containing 100 micrograms/ml ampicillin was introduced with a Pharmacia peristaltic pump from a 1 liter reservoir through a silicone sponge closure which formed the top of the reactor. Experiments were carried out in a 37°C constant temperature room.

Loss of amylase expression was examined by plating on L agar containing 0.5 percent starch. Following cell growth at 37°C, starch degradation around colonies was visualized by flooding with a solution containing 0.3 percent KI and 0.3 percent I_2. For examination of cells within the beads, beads were washed with M9 salts, chopped finely with a razor blade, and 3 g were extracted into 10 ml M9 salts with shaking. Replica plating onto L starch plates with and without 100 micrograms/ml ampicillin was used in examinations of plasmid loss.

RESULTS

Bead Formation. Although silicone polymers were selected because of high oxygen and carbon dioxide diffusion and solubility, their inert and hydrophobic nature presented a significant challenge for attachment or incorporation of microbes. Silicone polymers have been utilized as matrices for slow release of proteins and other materials (9). Release from the system has been attributed to microchannels in the polymer matrix formed by leaching of the the incorporated material in a process called transient diffusion (9). In order to similarly weaken the silicone polymer structure to allow nutrient access to the immobilized cells, low molecular weight silicone fluid was incorporated into the polymer mixture. Methods were then developed to form cell-bearing emulsion droplets and polymerize them into beads at the temperature of optimum cell growth. The first attempts using a single component system were not successful due to release of acetic acid during the polymerization. Utilization of prepolymers which release methanol during polymerization by an organometallic catalyst (Figure 1), worked successfully, yielding rubbery beads of 1-5 mm diameter which were very resistant to breakage by compression and could be stirred at high speed mith a magnetic stirring bar for prolonged periods with no evidence of degradation.

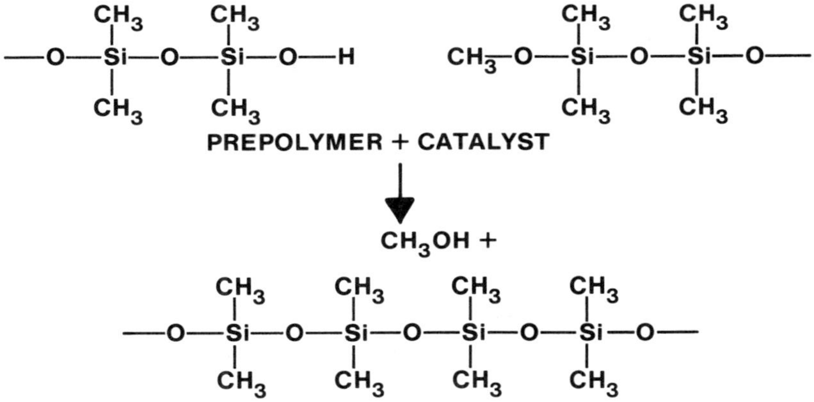

$$-O-\underset{CH_3}{\overset{CH_3}{Si}}-O-\underset{CH_3}{\overset{CH_3}{Si}}-O-H \qquad CH_3-O-\underset{CH_3}{\overset{CH_3}{Si}}-O-\underset{CH_3}{\overset{CH_3}{Si}}-O-$$

PREPOLYMER + CATALYST

$$\downarrow$$

$$CH_3OH +$$

$$-O-\underset{CH_3}{\overset{CH_3}{Si}}-O-\underset{CH_3}{\overset{CH_3}{Si}}-O-\underset{CH_3}{\overset{CH_3}{Si}}-O-\underset{CH_3}{\overset{CH_3}{Si}}-O-$$

Figure 1. Crosslinking of silicone prepolymer

Microbial Test System. As a test system for immobilization we chose an E. coli containing a multicopy plasmid pOS101 (Figure 2) containing the genetic information for B. stearothermophilus amylase. In this DH1 strain, this amylase has the unusual property of extracellular release, allowing the organism to grow on starch as a sole carbon source (8).

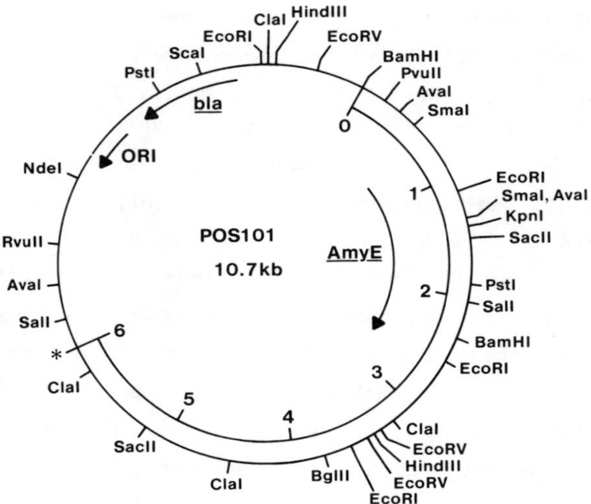

Figure 2. Plasmid pOS101. Double line indicates Bacillus MboI insert into BamHI site of plasmid pBR322. The amylase structural gene is designated.

Bead Colonization. Following formation of beads containing the E. coli recombinant, it was noted that INT penetrated the beads slowly, requiring periods up to 8 hours for complete staining. If beads were cut before INT addition, microscopic observation indicated rapid staining of bacterial colonies located in cavities throughout the beads (data not shown). Examination of these colonies by scanning electron microscopy revealed that they are open in nature (Figure 3). Fissures were found between cavities and connecting cavities to the bead exterior, but because of their shallow nature and since bacterial cells were not observed in them, it is possible that they are an artifact produced by the fixation procedure.

When beads were incubated in semicontinuous culture, the viable immobilized cells measured by INT conversion of cut beads decreased over the first 24 hours, and subsequently reached a constant value which corresponds to viable count of 2.5×10^6 CFU/g (Figure 4). It should be noted that this value is an underestimate, since only a small fraction of the cavities were exposed by cutting.

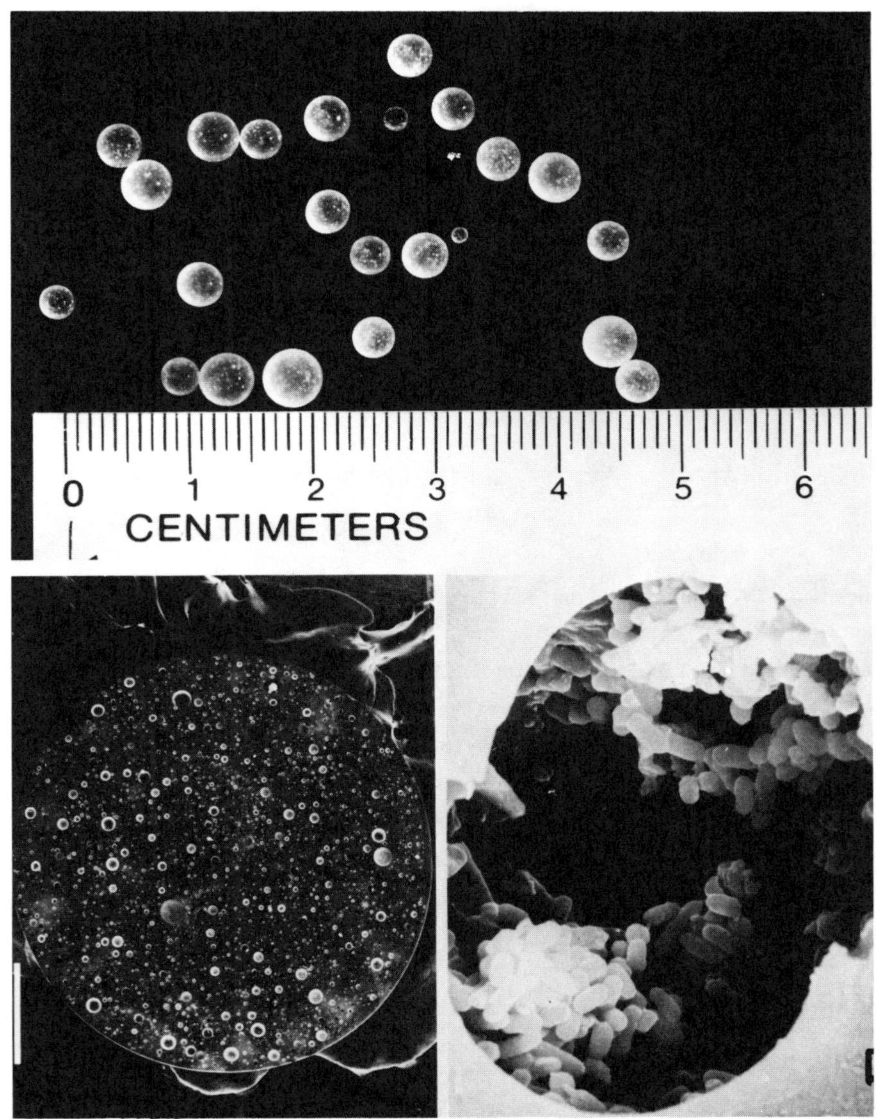

Figure 3. Upper, photograph of silicone beads. Lower left, scanning electron micrograph of cut silicone bead. Bar indicates 1000 micrometers. Lower right, bacterial colony within bead cavity. Bar indicates 1 micrometer.

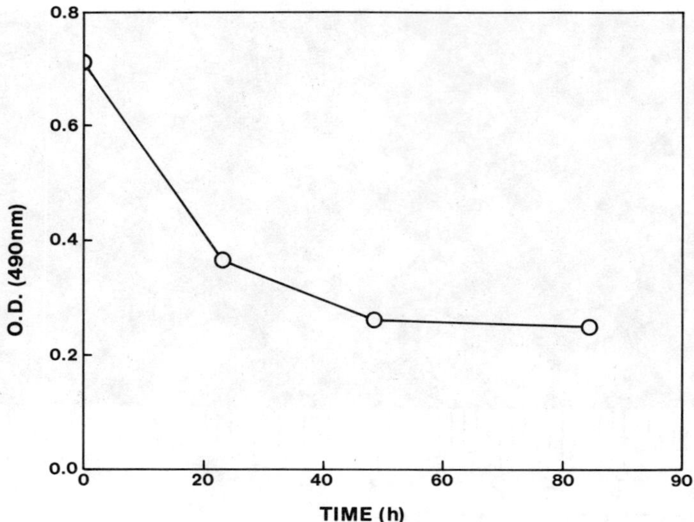

Figure 4. INT conversion by silicone beads in semicontinuous culture.

Free Cell and Amylase Production in Semicontinuous Culture. In addition to establishment of a stable level of viable biomass within the beads, bacterial cell outgrowth and amylase release from the beads into the medium were observed in semicontinuous cultures. Free cell densities from bead outgrowth were similar to those reached with semicontinuous culture of free cells alone (Figure 5). Greater amounts of amylase were observed in the supernatants of the immobilized cell culture, however. In both cases decrease in amylase production was noted over the course of measurement. Although the cause of this decrease was not investigated, loss of amylase expression in recombinant *E. coli* has been observed by other investigators and traced to plasmid modification by deletion (10).

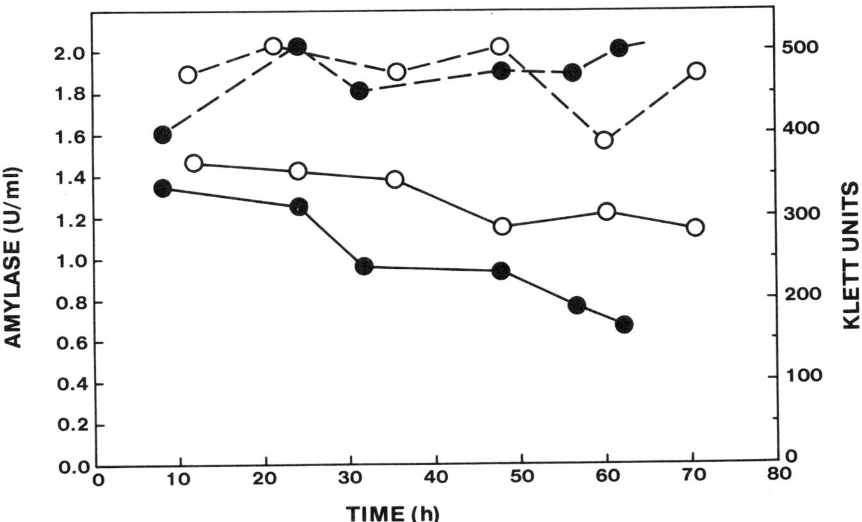

Figure 5. Cell growth and amylase production in semicontinuous culture. Amylase production is indicated by solid lines and cell growth by broken lines. Filled circles indicate free cell culture and open circles indicate immobilized cell culture.

Table 1

Production of Amylase by Free and Immobilized Continuous Culture

	Dilution Rate (h^{-1})	Amylase (U/mlxh)
Free cells	0.41	0.47
	0.66	0.63
	0.83	0.83
	1.15	1.16
Immobilized cells		
Experiment 1	5.4	2.1
Experiment 2	5.3	2.1

Amylase is expressed per ml reactor volume for free cells and per ml bead volume for immobilized cells. 8 ml and 13 ml of beads were used in experiments 1 and 2, respectively.

Amylase Production in Continuous Culture. Amylase production by free and immobilized cells was also measured in continuous culture. For continuous cultures containing silicone beads, dilution rates above maximum growth rate of 1.3 hr^{-1} were utilized to minimize the growth and amylase production from progeny of free cells released from the beads. As seen in Table 1, production of amylase in free cell culture increased with dilution rate, but did not reach that of the immobilized cells measured at high dilution rate.

Plasmid Stability. Many multicopy plasmids such as pBR322 do not maintain genetic control of partition, but instead distribute statistically between daughter cells. The probability of plasmidless cell formation is therefore highly dependent on the plasmid copy number, which is governed by the cell, plasmid, and growth conditions (11). du Poet et al. (2) have reported that immobilization of a recombinant E. coli strain with carrageenan offers protection against loss of plasmid expression in the absence of antibiotic selection. It was of interest to examine the extent of this effect with a different plasmid and immobilization method.

Measurements of plasmid stability were made with silicone beads in M9/glycerol medium without antibiotic. In this medium, cells lacking amylase expression have a significant growth advantage (T. Schwacha and P. Oriel, unpublished). As seen in Figure 6, amylase producing capability was rapidly lost in free cell semicontinuous culture, whereas in this time period the immobilized cells showed little loss. Free cells released from the beads showed higher amylase retention than the free cell culture, but less than the immobilized cells, presumably reflecting some overgrowth by plasmidless cells after release. Amylase negative cells from free cell and immobilized cultures were examined for ampicillin resistance to determine whether loss of amylase expression resulted from loss or alteration of plasmid. In 40 colonies from free cell culture, 50 colonies from free cells released from beads, and 7 colonies from cells within the beads, all cells that did not produce amylase had also lost ampicillin resistance, indicating that plasmid loss rather than plasmid modification had occurred.

DISCUSSION

The polymerization of silicone prepolymer emulsions provides a gentle bacterial immobilization method which yields beads with physical properties superior to those of calcium alginate or carrageenan. Although viable cells were retained throughout the bead and significant amylase released, slow INT penetration into the bead indicates slow diffusion rates for larger molecules, providing an opportunity for improved performance with further bead modifications. Current efforts are focusing on this aspect together with gaining a better understanding of the cell and amylase release mechanism.

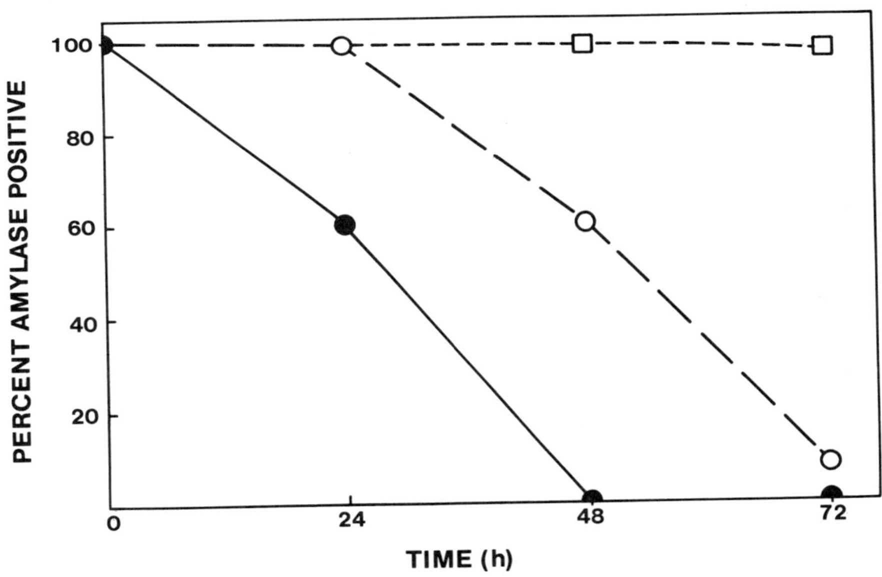

Figure 6. Loss of amylase expression in semicontinuous culture. Short broken line indicates cells extracted from the beads and longer broken lines indicates free cells released from the beads. Solid line represents free cell culture.

Although the effect of wall adsorption in preventing overgrowth of rapidly growing variants in continuous culture has been recognized for some time (12), very few studies have been carried out to examine the utility of immobilization in providing a pool of plasmid-bearing cells to limit overgrowth by plasmidless variants. In addition, the probability of plasmidless cell formation within the beads resulting from outgrowth may vary significantly from that of free cells due to the altered micro-environment and growth rates. The silicone immobilization system provides an opportunity for further studies of these effects.

ACKNOWLEDGEMENT

This project was supported by a State of Michigan Research Excellence and Economic Development Grant.

REFERENCES

1. Chabata, I. and Tosa, T., Immobilized cells: historical background. In Applied Biochemistry and Bioengineering, Vol. 4., ed. L.B. Wingard, Jr., E. Katchalski-Katzir, and L. Goldstein. Academic Press, New York, 1983, pp. 1-9.

2. du Poet, P.T., Dhulster, P., Barbotin, J.W., and Thomas, D., Plasmid inheritability and biomass production comparisons between free and immobilized cell cultures of E. coli (pTG201) without selective pressure. J. Bacteriol., 1986, 165, 871-877.

3. Linko, P. and Linko, Y.-Y., Applications of immobilized microbial cells. In Applied Biochemistry and Bioengineering Vol. 4., ed. L. B. Wingard, Jr., E. Katchalski-Katzir, and L. Goldstein. Academic Press, New York, 1983, pp. 54-136.

4. Mattiasson, B., Oxygenation of processes involving immobilized cells., In Immobilized cells and organelles, ed. B. Mattiasson. CRC Press, Boca Raton, 1984, pp. 41-60.

5. Gossman, B., and Rehm, H.J., Oxygen uptake of microorganisms entrapped in calcium alginate. Appl. Microbiol. Biotech., 1986, 23, 163-167.

6. Glassner, D.A., Grulke, E.A., and Oriel, P.J., Characterization of an immobilized biocatalyst system for production of an enzyme using a thermophilic bacterium. Presented at the Minneapolis AICHE Conference, August, 1986.

7. Acuri, E.J., Continuous ethanol production and cell growth in an immobilized cell bioreactor employing Zymomonas mobilis. Biotech. Bioeng., 1982, 24, 595-602.

8. Oriel, P., and Schwacha, T., Growth on starch and extracellular production of a thermostable amylase by Escherichia coli. (Submitted for publication).

9. Hsieh, D.S.T., Shiang, C.C., and Desai, D.S., Controlled release of macromolecules from silicone elastomer. Pharmaceutical technology June 1985, pp. 39-49.

10. Mitzutani, S., Fakuzono, S., Tsukagoshi, N., Udaka, S., and Kobayashi, T., Stability of a recombinant plasmid containing alpha-amylase gene in a chemostat. J. Chem. Eng. Japan, 1985, 18, 220-224.

11. Nordstrom, K., Control of plasmid replication, In Plasmids in Bacteria, ed D.R. Helinski, S.N. Cohen, D.B. Clewell, D.A. Jackson, and A. Hollaender. Plenum, New York, 1985, pp. 185-215.

12. Dykhuizen, D.E., and Hartl, D.L., Selection in chemostats. Microbiol. Rev., 1983, 47, 150-168.

INTERNATIONAL CONFERENCE ON BIOREACTORS AND BIOTRANSFORMATIONS GLENEAGLES, SCOTLAND, UK: 9-12 NOVEMBER 1987

Paper J2

SECRETIVE FERMENTATION OF γ-LINOLENIC ACID PRODUCTION USING CELLS IMMOBILISED IN BIOMASS SUPPORT PARTICLES

Hideki Fukuda and Hisashi Morikawa

Kanegafuchi Chemical Industry Co., Ltd.
1-8 Miyamae, Takasago, Hyogo 676,
JAPAN

ABSTRACT

γ-Linolenic acid (GLA) production by *Mucor ambiguus* IFO 6742, immobilised in Biomass Support Particles (BSPs), has been investigated in a fluidised-bed fermenter in the presence of nonionic surfactants. In this system, repeated batch cultivation was achieved at higher yield and productivity than by the conventional methods, since microbial lipids including GLA were significantly secreted into the culture broth and/or on the surface of the cell wall.

INTRODUCTION

Considerable interest has arisen in recent years regarding the production of γ-linolenic acid (GLA; 6,9,12-octadecatrienoic acid) from the pharmacological view-point. The most common source of this fatty acid is evening primrose (*Oenothera binnis*) seed oil. However, the productivity of GLA from the seed oil is extremely low, since both a long period and a huge area for harvesting seed are required. Consequently, much research effort has been directed towards the development of microbial production, and some suitable strains have been proposed as show in Table 1[1-6]. There are, however, problems regarding industrial application, since the lipids are normally accumulated within cells. This may limit the potential for greater yield coefficient and productivity of GLA, and also reduce the yield coefficient of GLA in the following purification steps.

We found that some part of the lipids within cells were effectively secreted into the culture broth and/or on the surface of the cell wall without damage to cell growth in the presence of nonionic surfactants.

TABLE 1
Various strains for GLA production.

Absidia spinosa	Hoffmann *et al.*(1978)
Cunninghamella echinulata	Hoffmann *et al.*(1978)
Basidiobolus meristosporus	Tyrell(1967)
Mortierella isabellina	Hoffmann *et al.*(1978), Suzuki *et al.*(1981)
Choanephora cucurbitarum	Shaw(1965), Deven *et al.*(1976)
Rhizopus stolonifer	Shaw(1965)
nigricans	Tyrell(1967)
Phycomyces blakeeanus	Shaw(1965)
Mucor javanicus	Shaw(1965)
globosus	Mumma *et al.*(1970)
Helicostylum pyriforme	Shaw(1965)
Pythium debaryanum	Shaw(1965), Tyrell(1967)
Saprolegnia litoralis	Shaw(1965)
Entomophthora coronata	Shaw(1965)

There has also been considerable interest recently in the use of immobilised cells as a means of achieving enhanced fermenter performance. The technique of cell immobilisation using porous Biomass Support Particles (BSPs)[7] is a "passive" one and involves minimal preparation and handling. This technique has been successfully applied to a wide variety of microbial systems[8].

The work presented in this paper is based on the enhancement of GLA production in a system using both nonionic surfactants and cells immobilised in reticulated foam BSPs.

MATERIALS AND METHODS

Microorganism and medium

All experiments were performed using *Mucor ambiguus* IFO 6742, and the base medium used contained in 1 litre of tap water: glucose, 30 g; KH_2PO_4, 3.0 g; $(NH_4)_2SO_4$, 2.0 g; urea, 1.0 g; NaCl, 0.1 g; yeast extract, 0.4 g; malt extract, 0.4 g; polypeptone, 0.2 g; $MgSO_47H_2O$, 0.5 g; $FeSO_47H_2O$, 10 mg; $CaCl_22H_2O$, 10 mg; $CuSO_45H_2O$, 0.5 mg; $ZnSO_47H_2O$, 1.0 mg; $MnCl_24H_2O$, 1.0 mg; nonionic surfactants, (Rheodol TW-0320: polyoxyethylene sorbitan fatty acid ester)(Kao Corp., Japan), 20 g.

The fluidised-bed fermenter system with BSPs

The 10 litre fluidised-bed fermenter arrangement was used for repeated batch culture. A schematic flow diagram of the fluidised bed fermenter system is shown in Figure 1. The fermenter vessel (1), made of glass, consisted of: a 10 cm-long, 2.5 cm-diameter nozzle section (set a perforated plate of 5 mm hole diameter); a conical base 10 cm in height; a 60 cm-long, 15 cm-diameter cylindrical section; and a 60 cm-long, 5 cm-diameter vapour cooling section. A side-arm for sampling

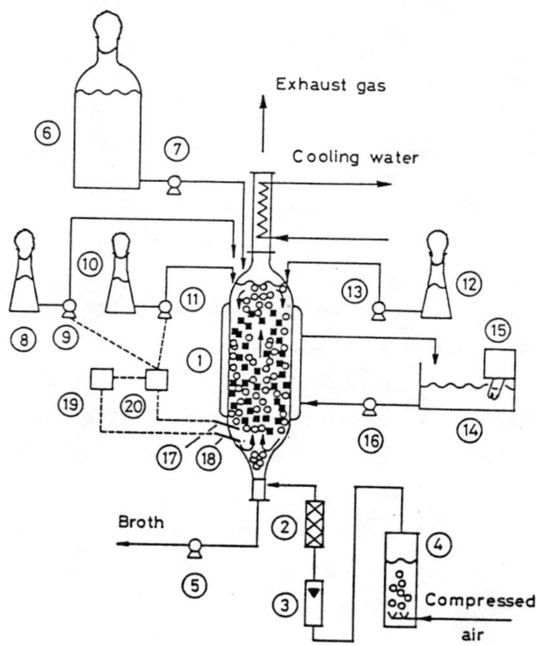

Figure 1. Schematic flow diagram of the fluidised-bed fermenter system. (1) fluidised-bed fermenter; (2) air filter; (3) rotameter; (4) water-saturated tower; (5) broth withdraw pump; (6) medium storage bottle; (7) medium feed pump; (8) NaOH storage bottle; (9) NaOH feed pump; (10) H_2SO_4 storage bottle; (11) H_2SO_4 feed pump; (12) antifoam storage bottle; (13) antifoam feed pump; (14) water bath; (15) temperature controller; (16) circulation pump; (17) pH probe; (18) thermometer; (19) recorder; (20) pH controller.

particles as described elsewhere[9] was provided on the cylindrical section. The overall height of the fermenter was 140 cm, and its working volume was maintained at 6-8 litre. Air was introduced at a rate of 3.0-5.0 vvm via a water-saturated tower (4), a rotameter (3) and an air filter (2) into the nozzle section. Temperature within the fermenter was controlled at 30°C by recycling the water from a water bath (14) through the jacket. The pH of the culture broth in the fermenter was controlled at 4.0 automatically by adding NaOH(8) or H_2SO_4(10) solutions. Foam control was effected manually with a polyglycol liquid by a peristaltic pump (13) when required.

The fermenter, containing the base medium and 15,000 empty BSPs, was inoculated aseptically with approximately 500 ml mycelial suspension from a shake flask. After the first batch cycle, the culture broth was almost completely withdrawn using a pump (5) and then fresh medium was fed intermittently. Surfactants were added to the base and fresh media. Seven such batch cycles were carried out.

6-mm cubic polyurethane foam BSPs (Bridgestone Co., Ltd., Tokyo, Japan) having a porosity of around 0.97 and a pore size of 50 pores per linear inch were used in this study. Sampling of particles was performed as described elsewhere[9].

Extraction procedure

Approximately 100 ml of culture broth was filtered onto 7.0 cm filter paper, and the wet mycelia were immediately transferred to a homogenizer (Nihonseiki Kaisha Ltd., Japan; Type: Ace Homogenizer AM-8) containing 50 ml glass beads (Nippon Rikagaku Kikai Co., Ltd., Japan; Type: GMB-40), where the mycelia were extracted at a rotation speed of 10,000 rpm with chloroform-methanol (2:1 v/v) twice for 30 min. Lipids in the supernatant were extracted with approximately 50 ml chloroform. These solvents were washed with NaCl solution, evaporated to dryness in tared flasks, weighed and redissolved in 2 ml chloroform. Mycelia within BSPs were treated with an aqueous solution of sodium hypochlorite (approximately 10% v/v) to remove them from BSPs, and after filtration and washing the wet mycelia were also extracted by the same procedure.

Separation of fatty acid methyl esters

Lipids in chloroform were saponified by refluxing with methanolic NaOH, and the methyl esters were prepared by transesterification at 60°C for approximately 5 min with 12.5% BF_3 in methanol. The fatty acid esters were extracted by adding approximately 4 ml of *n*-hexane, then 4 ml of saturated NaCl solution, shaking briefly, and centrifuging until both layers were clear. They were quantitatively analyzed by a gas chromatograph (Shimadzu Seisakusho Ltd., Japan, Model GC-7A) fitted with a flame ionization detector, and the operating conditions were as follows: column, 3.0 mm x 3.1 m glass columun packed with Unisole 3000 (Gasukuro Kogyo Inc., Japan); column temperature, 210°C; temperature of injector and detector, 270°C; carrier gas, nitrogen (flow rate: 25 ml/min). A digital integrator (Shimadzu Seisakusho Ltd., Japan, Type Chromatopac C-R1A) was run for quantitative analysis. Chloroform solutions of the authentic methyl esters (Sigma Chemical Co., U.S.A.) were injected for use as standards, which gave the following retention times relative to palmitic acid (16:0): lauric (12:0), 0.41; myristic (14:0), 0.63; palmitoleic (16:1), 1.13; stearic (18:0), 1.76; oleic (18:1), 1.90; linoleic (18:2), 2.17; γ-linolenic (18:3), 2.39; α-linolenic acid (18:3), 2.59. Arachidic acid (20:0) was completely separated from γ-linolenic acid under the above conditions.

γ-Linolenic acid extracted from both cells and culture broth was precisely identified by glass capillary gas chromatography and gas chromatography-mass spectrometry, and this will be reported elsewhere.

Biomass and glucose concentrations

The biomass concentration within BSPs was determined as described elsewhere[9], except for washing BSPs with acetone to remove surfactants. Freely suspended cell concentration was determined by filtering a sample of culture broth on filter paper, washing with water, and drying at 120°C for 24 hours before weighing. Glucose concentration was determined by a glucose analyzer (Beckmann Instruments Inc., U.S.A.; Type: Glucose Analyzer 2).

RESULTS AND DISCUSSION

Figure 2 shows the time courses in repeated batch culture of seven batch cycles with nonionic surfactants and BSPs. During the first batch cycle the freely suspended cell concentration, Xs, increased rapidly, while the cells were scarcely immobilised within BSPs. The GLA concentration extracted from cells, $(GLA)_C$, increased with cell growth but stopped sharply at almost the point of maximum cell concentration. The GLA concentration secreted into the culture broth, $(GLA)_B$, also increased with cell growth until the point of maximum cell concentration.

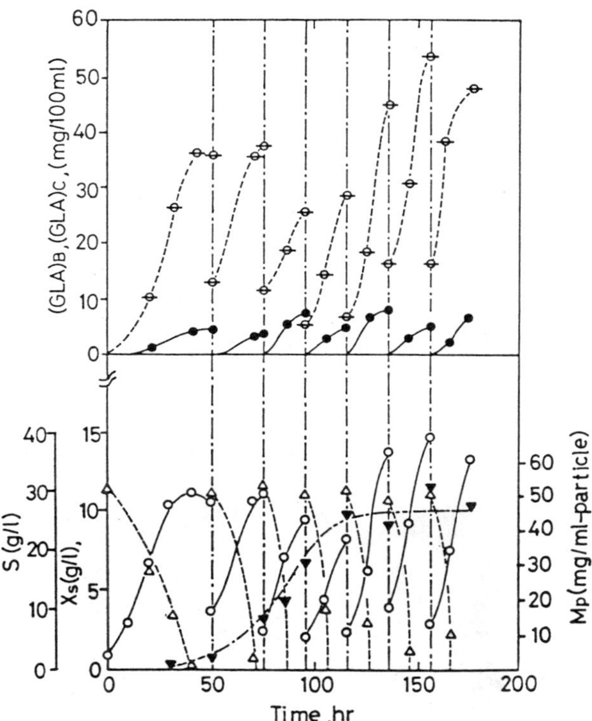

Figure 2. Time courses in repeated batch culture of seven batch cycles with surfactants and BSPs.
Xs(O),freely suspended cell concentration;
Mp(▼),cell concentration within BSPs;
S(Δ),glucose concentration;
$(GLA)_C$(-⊖-),GLA concentration extracted from cells;
$(GLA)_B$(●),GLA concentration secreted into the culture broth.

During the second and fourth batch cycles, Xs, $(GLA)_B$ and $(GLA)_C$ reached their maximum values in a much shorter time than during the first batch cycle, since the amount of cells in the fermenter at the commencement of each cultivation was much higher than that of the first batch cycle. The cell concentration within BSPs, Mp, gradually increased and reached a maximum value of around 46.3 mg/ml-particle at the end of the fourth batch cycle. This value lies between those obtained in the systems of spouted bed fermenter/stainless steel BSPs [9] and of circulating bed fermenter/ foam BSPs[10]. However, since Mp should be directly affected by fluid mechanical shear[10], this could be improved under optimal hydrodynamic conditions within the fermenter.

After the fifth batch cycle, Mp kept a constant value, and the ratio of free cells to immobilised cells was around 0.75. The values of $(GLA)_C$ varied during the seven cycles, i.e. between 3.5 and 7.5 mg/100ml. This may be caused by the difference of physical conditions in the fermenter such as surface tension of culture broth and shear force around the cell wall. Figure 3 shows a cross section of a BSP, where microorganisms were partially immobilised in a BSP, since oxygen and/or substrate were limited in the central part.

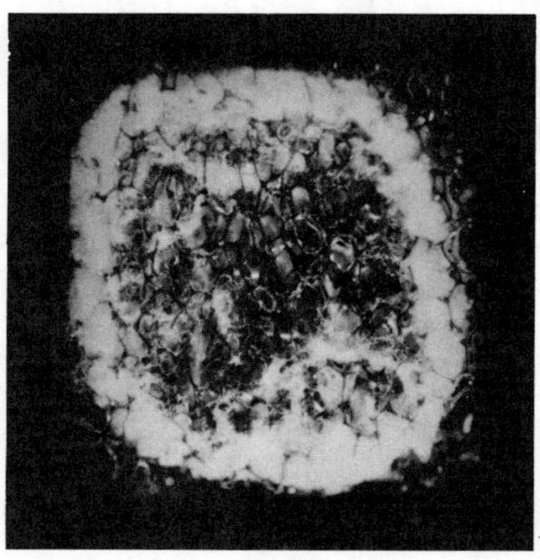

Figure 3. A cross section of a BSP.

The total amount of GLA within BSPs at the end of the seven batch cycles was 3.70g. Thus, the fermenter productivity and the yield coefficient of GLA on glucose medium through the seven batch cycles reached 17.1 mg/l.h and 0.0142 mgGLA/mg-glucose, respectively.

It was found from shake flask experiments that among various types of nonionic surfactants, some kinds of polyoxyethylene sorbitan fatty acid ester and polyethylene glycol fatty acid ester could be effectively lead to the secretion of GLA into the culture broth, and that the values of $(GLA)_C$ with these surfactants were greater by as much as 30-50% compared with the values without surfactants. This suggests that the excessive lipids including GLA adhered considarably to the surface of the cell wall, and the improvement in yield coefficient of GLA with surfactants may be related to the release from feedback regulation by the product in some degree.

Although other kinds of surfactants such as anionic, cationic and amphoteric surfactants have been investigated in a wide range of concentration, none of them gave satisfactory results.

A comparison of this system in terms of yield coefficient and fermenter productivity with the conventional processes (data not shown) is shown in Table 2. It should be noted that the yield coefficients of this system and a batch culture with nonionic surfactants were greater by as much as 45-50% compared with batch and repeated batch cultures without surfactants. The fermenter productivity of this system was 84% higher than in the batch culture (II), and, in addition, the productivity obtained in the repeated batch culture(I) was 75% greater than in the batch culture(I). Thus, the repeated batch operation mode with BSPs significantly contributed to the improvement of fermenter productivity.

TABLE 2
Comparison of performance of GLA production processes.

Fermentation process	Fermentation time (h)	Fermenter productivity (mgGLA/l.h)	Yield coefficient (mgGLA/mg-glucose)
Batch (I)*1	45*5	6.4	$9.6x10^{-3}$
Batch (II)*2	45*5	9.3	$14.0x10^{-3}$
Repeated batch (I)*3	160	11.2	$9.2x10^{-3}$
Repeated batch (II)*4	175	17.1	$14.2x10^{-3}$

*1 Without either surfactants or BSPs.
*2 With nonionic surfactants and without BSPs.
*3 Seven batch cycles without surfactants and with BSPs.
*4 Seven batch cycles with both surfactants and BSPs (Fig. 2).
*5 Time taken to reach maximum GLA production.

The performance of GLA production by *Mucor ambiguus* IFO 6742 may not be directly compared with that of other strains, since shake flask experiments have been applied in most cases. However, considering both the comparison in terms of yield coefficient and productivity on glucose medium, and the high maximum specific growth rate obtained in this study (0.13 h^{-1}), *Mucor ambiguus* IFO 6742 seems to be suitable for industrial production of GLA.

REFERENCES

1. Hoffmann, B. and Rehm, H.J., Degradation of long-chain-alkanes by Mucorales. Euro. J. Appl. Microbiol.,1978, 5, 189-195.

2. Tyrell, D., The fatty acid compositions of 17 entomophthora isolates. Can. J. Micro., 1967, 13, 755-760.

3. Suzuki, O., Yokochi, T. and Yamashita T., Studies on production of lipids in fungi,II. YUKAGAKU, 1981, 30, 863-868.

4. Shaw, R., The occurence of γ-lonolenic acid in fungui. Biochim. Biophys. Acta, 1965, 98, 230-237.

5. Deven, J.M. and Manocha, M.S., Effect of various cultural conditions on the fatty acid and lopid composition of *Choanephora cucurbitarum*. Can. J. Micro., 1976, 22, 443-449.

6. Mumma, R.O., Fergus, C.L. and Sekura, R.D., The lipids of thermophilic fungui: Lipid composition comparison between thermophilic and mesophilic fungui. Lipids, 1970, 5, 100-103.

7. Akinson, B., Black, G.M., Lewis, P.J.S. and Pinches, A., Biological particles of given size, shape, and density for use in biological reactors. Biotechnol Bioeng,1979, 21, 193-200.

8. Webb, C. and Black, G.M., An immobilisation technology based on biomass support particles. In Process engineering aspects of immobilized system., eds. C. Webb, G.M. Black, B. Atkinson, The Institution of Chemical Engineers, Rugby,1986, pp277-285.

9. Webb, C., Fukuda, H. and Atkinson, B., The production of cellulase in a spouted bed fermentor using cells immobilized in biomass support particles. Biotechnol Bioeng,1986, 28, 41-50.

10. Black, G.M., Webb, C., Mattews, T.M. and Atkinson, B., Practical reactor systems for yeast cell immobilization using biomass support particles. Biotechnol. Bioeng.,1984, 26, 134-141.

INTERNATIONAL CONFERENCE ON BIOREACTORS AND BIOTRANSFORMATIONS
GLENEAGLES, SCOTLAND, UK: 9-12 NOVEMBER 1987

Paper J3

ENHANCED PERFORMANCE OF IMMOBILIZED BIOREACTORS USING VARIABLE SURFACE BIOREACTORS

Paul P. Matteau
Division of Biological Sciences
National Research Council of Canada
Ottawa, Ontario, Canada K1A 0R6

ABSTRACT

The scale-up of packed-bed tubular bioreactors for fermentation processes has been impeded by two major operational phenomena -- gas build-up in the reactor; and excessive pressure drops across the packed-bed. Our approach to the solution of these problems has been to design variable surface bioreactors which would retain the desired "plug-flow" characteristics necessary for enhancing the reaction kinetics. Two geometries were chosen: a tapered packed-bed bioreactor (TB); and a radial flow packed-bed bioreactor (RFB). Theoretical analysis of bed depths for a given reactor volume were carried out as well as calculations of the expected pressure drops through the reactors. Comparative studies were then made using two reactors from each of these geometries, as well as two tubular packed-bed bioreactors. The evaluation of the reactors was made for both inert systems and active bioconversion processes. These latter included the enzymatic hydrolysis of cellobiose to glucose using calcium alginate immobilized mycelium of *Trichoderma harzianum* E58, and the gas-producing fermentation of glucose to ethanol using calcium alginate immobilized *Saccharomyces cerevisiae* NRCC 202076. The results indicate that the variable surface bioreactors do circumvent the operational difficulties normally encountered with tubular packed-bed bioreactors with no kinetic disadvantage.

INTRODUCTION

Whereas substantial efforts in the area of immobilized enzymes and cells have been reported in the literature in recent years, minimal efforts have been devoted to the design of immobilized enzyme reactors. The two most commonly used continuous reactors are the continuous flow stirred tank reactor (CSTR) and the packed-bed reactor (PBR). The reaction kinetics in the former are controlled by the effluent concentrations as a result of the reactor's well-mixed state. In the latter, the kinetics vary along the length of the reactor bed and this reactor type is normally designed in a fashion so as to achieve a fluid flow approaching "plug-flow" through the reactor bed. This plug-flow characteristic is desirable as this will often lead to reduced reactor volumes and biocatalyst requirements relative to the well-mixed reactor types [1]. In practice, plug-flow behaviour is difficult to obtain and deviations may occur owing to fluid dispersion and channelling, diffusion of reactants and temperature gradients in the reactor [2]. Another major drawback of the packed-bed reactor is that there exists a pressure drop or resistance to flow across the reactor bed. This pressure drop depends on such parameters as bed depth, feed rate, feed temperature, and composition, as well as the catalyst packing density. Additionally, if deformable particles are used, compaction will result.

During many biological conversions, an additional problem which is encountered is the formation of a gaseous product [3]. This gas results in the flotation of the biocatalyst particles (in porous materials) and leads to bed compaction and separation. Inefficient gas removal will result in liquid entrainment and thus deviation of the reactor from plug-flow behaviour. Although several reactor designs have been proposed to overcome these limitations, few designs have resulted in reactors which are truly plug-flow.

This work describes the design and testing of two types of variable-surface reactor geometries: a radial flow packed-bed bioreactor (RFB); and a tapered packed-bed bioreactor (TB). Both geometries result in an increase in bed surface area in the direction of flow and thus a decrease in fluid velocity. The increasing surface area, it was postulated, would also enhance the dispersion of any gases produced during the fermentation. The other intrinsic advantage of these reactors is that the bed depth can be substantially less than that required for an equal volume tubular reactor. Comparative studies with two reactors of each geometry as well as with tubular reactors have been carried out using two bioconversion processes: an enzymatic conversion (cellobiose to glucose using calcium alginate immobilized *Trichoderma harzianum* E58; and a gas-producing fermentation (ethanol fermentation using calcium alginate immobilized *Saccharomyces cerevisiae* NRCC 202076).

MATERIALS AND METHODS

Reactor Design: Six 1-liter reactors were constructed, two from each of the geometries (tubular, tapered and radial) and differing in their height to diameter ratio. The dimensions and material of construction of each of the reactors are present in Table 1.

TABLE 1

Description of reactors tested

Reactor	Material of Construction	Height (cm)	Diameter (cm)	
			Inlet	Outlet
Tubular (A)	Acrylic	49.3	5.1	5.1
Tubular (B)	Acrylic	87.7	3.8	3.8
Tapered (A)	ABS Plastic	16.0	1.9	15.0
Tapered (B)	Styrene-Acrylonitrile	45.0	1.3	7.6
Radial (A)	Acrylic	8.2	1.3	12.7
Radial (B)	Acrylic	24.7	1.3	7.6

Reactor Operation: All reactors were operated isothermally using either a heat-exchange jacket (tubular reactors) or by immersing the reactors in a constant temperature water bath. A schematic of the general system as used for the fermentations is presented in Figure 1. The substrate was fed to the reactor from a 20-liter reservoir using a Masterflex multi-head peristaltic pump. The use of this type of pump permitted the simultaneous feeding of three reactors at identical pump speeds. Use of different pump heads permitted the variation of the dilution rates over a wide range. The effluent was collected in a conical vessel which was connected to the outlet port of the reactor. For the fermentations, this unit also served as a gas-liquid separator. The computer (HP85) was used to monitor gas production for the fermentation and to periodically empty the gas-liquid separator. The gas production rate was used to monitor the level of activity in the fermentation reactors.

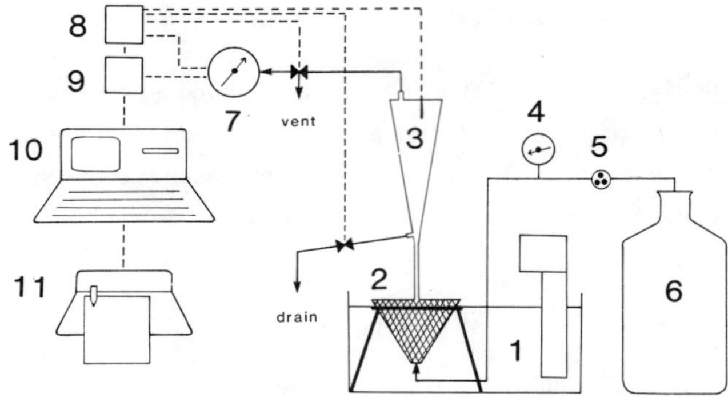

FIGURE 1. Schematic of the system used for reactor evaluation: (1) Water Bath; (2) Packed-bed Bioreactor; (3) Gas-Liquid Separator; (4) Pressure Gauge; (5) Peristaltic Pump; (6) Substrate; (7) Gas Meter; (8) Relay Actuator; (9) Digital Voltmeter; (10) HP85 Computer; (11) Printer/Plotter.

Immobilization Procedure: The general procedure used to encapsulate the organisms involved the separation of the cells from the culture filtrate and resuspension of washed solids in an aqueous 1.5% sodium alginate (Sigma) suspension. This mixture was then homogenized using a food processor and extruded under pressure through stainless steel needles into a 4% aqueous solution of calcium chloride. The size of the calcium alginate bead produced could be varied using different needle sizes. The yeast was immobilized using a 1.024 mm OD needle and the fungal mycelium was immobilized using an orifice diameter of 2.0 mm.

Microorganisms: *Trichoderma harzianum* E58 (Forintek Canada Corporation, Ottawa) was cultured at 30°C using Vogel's medium and 2% Solka Floc for the carbon source as previously described [4]. The culture solids were removed by filtration after 5 days of growth, washed with distilled water and encapsulated in calcium alginate beads. The mycelium of this fungus contains a high level of β-glucosidase (EC 3.2.1.21) and had been found effective for long-term hydrolysis of cellobiose [5]. The immobilized mycelium was then wet-packed into the test reactors.

Saccharomyces cerevisiae NRCC 202076 was grown in static culture without aeration at 30°C in a 5% glucose medium containing the following per liter of solution: 0.5 g KH_2PO_4; 0.5 g NaCl; 0.35 g $MgSO_4$; 20.0 g

$(NH_4)_2SO_4$; and 1.0 g yeast extract. This medium, with the addition of 1.5 g/l of calcium chloride, was also used for the continuous fermentation. The cells obtained from this initial growth were harvested by centrifugation, washed once with water and diluted by a factor of 4 in a solution of sodium alginate for immobilization. Once the calcium alginate beads had been formed, they were transferred into fresh aerated medium (5% glucose) to permit cell growth within the beads. The medium was changed daily over a period of 4 days and the actively-growing immobilized yeast cells were then wet-packed into the reactors.

Analytical Procedures: The performance of the reactors was determined from the degree of conversion of the substrate at different steady-state conversions. For the enzymatic hydrolysis, salicin or cellobiose were used as substrates and the glucose produced was measured as previously described [4]. For the fermentation, glucose was measured using the dinitrosalicylic acid method (DNS). Ethanol was measured using a gas chromatograph with isopropanol as an internal standard. Carbon dioxide produced during the fermentation was measured using a wet-test meter which had been connected to an HP85 computer in order to continuously calculate the rate of gas production. As equimolar amounts of carbon dioxide and ethanol are obtained during the fermentation, the carbon dioxide production rate was used as a measure of the effectiveness of the process. The productivities of the reactors were calculated based on the reactor volume. Cell concentrations in the fermentation effluent were determined as dry weights using 100 ml of sample.

RESULTS

Initial analysis of the relative velocities of liquid through various types of variable surface reactors indicated that for equal reactor volumes, the variable surface reactors would result in exit fluid velocities of less than 20% of that found in tubular reactors. A typical plot of these data is presented as Figure 2. As the pressure drop through packed beds is a function of both the bed depth and the velocity, these results indicated that either radial flow or tapered reactors would be effective in reducing the intrinsic pressure drop. Two reactors of each type were then constructed and tested and are described in Table 1 above.

Initial "inert" testing of all reactors was carried out using both calcium alginate beads and a macroporous ion exchange resin. The reactors were wet-packed with various particle sizes and pressure drops across the

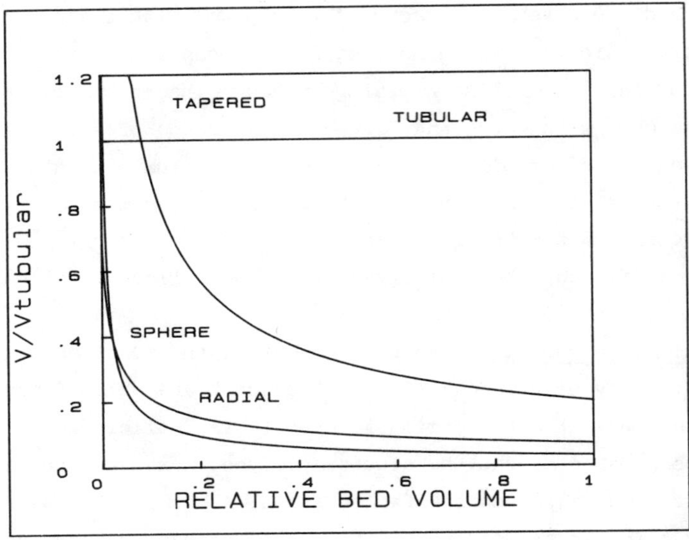

FIGURE 2. Fluid velocities through variable surface reactors relative to tubular reactors (1-liter reactors). The conditions which are presented are as follows: Tubular reactor - Diameter 5.64 cm, Height 40 cm; Tapered reactor - Inlet diameter 2.0 cm, Outlet diameter 12.7 cm, Height 10.6 cm; Radial flow reactor - Inlet diameter 1.0 cm, Outlet diameter 11.0 cm, Height 10.6 cm; Spherical reactor - Inlet diameter 1.0 cm, Outlet diameter 12.4 cm.

beds were determined at various dilution rates. The maximum dilution rate used was 6.1 h^{-1} and this resulted in a maximum pressure drop of 490 Pa in Tubular reactor (B) when it was charged with the ion exchange resin. The maximum pressure drop measured with the alginate beads (4 mm OD) was 300 Pa, again, with the tall tubular reactor. These values were significantly greater than what would have been expected from a standard calculation of the pressure drop [6], but it was noticed that in this reactor, some bed separation and compacting occurred. Tracer analysis (residence time distributions) of the reactors showed that there was very little axial dispersion in the reactors [7].

For the enzymatic reactions, the reactors which were thoroughly tested were Tubular (B), Tapered (B) and Radial (A). A typical dynamic response of a reactor (Tubular (A)) is presented as Figure 3. As can be seen, a steady state was normally achieved after approximately 4-5 hours

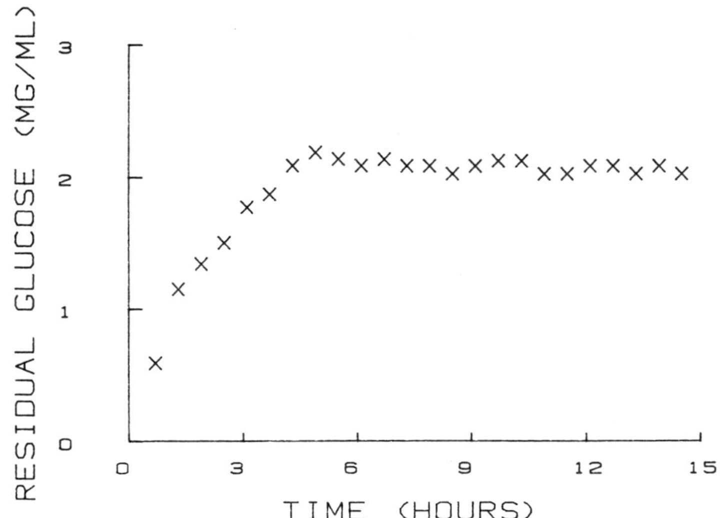

FIGURE 3. Dynamic response of Tubular (A) to 10 mM cellobiose at a dilution rate of 0.18 h^{-1}.

of operation. Using a range of substrate concentrations of between 1 and 30 mM salicin or cellobiose, the steady-state conversions were determined. A typical set of data for the conversion of 10 mM cellobiose using the three reactors indicated is presented in Figure 4. Typically, both of the

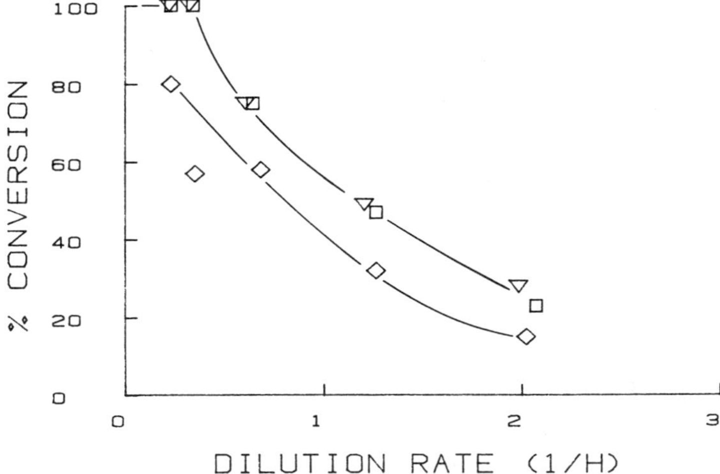

FIGURE 4. Steady-state conversion of 10 mM cellobiose as a function of dilution rate. Tubular (B) (▽); Tapered (B) (□); Radial (A) (◇).

upflow reactors performed identically, whereas the radial flow reactor consistently resulted in lower conversions. At the end of the run, it was found that the calcium alginate bed in all reactors had shrunk somewhat and as expected this would have the most serious effect on the radial reactor. The results obtained from the two upflow reactors for the range of substrate concentrations tested are presented in Figures 5 and 6. Analysis of these data (non-linear least squares) using a substrate inhibition model showed that no significant differences existed in the kinetics in either of these reactors. Values obtained for the kinetic constants were: Km = 3.2 mM; Ks = 16 mM; and the total enzyme activity, kE = 11 mmol/h. No difficulties were encountered with the operation of any of these reactors from the point of excessive pressure drops, bed separation or channelling.

Testing of the reactors with the ethanol fermentation was found to be less successful than with the enzymatic process. This was to be expected as the concommitant carbon dioxide production as well as cell growth led to less than desirable operating characteristics in the tubular reactors. Initial testing with the shorter tubular reactor (A) indicated that whereas acceptable ethanol productivities could be obtained (15-25 g ethanol/l/h) extensive bed separation and compaction occurred. During the course of the fermentation, this resulted in a continuous build-up of the pressure differential across the bed to values in excess of 150 kPa. The peristaltic feed pumps were found to be incapable of withstanding this level of backpressure and the reactor was shut down.

Further testing continued with both tapered-bed reactors and radial reactors. Fermentation runs were carried out at glucose concentrations ranging from 10 g/l to 120 g/l. A typical start-up response for the three types of reactors is presented in Figure 7. As can be seen, the radial reactor is somewhat slower to reach a steady-state value than are the two upflow reactors. In order to ensure steady-state values had been obtained, the reactors were normally run overnight at new dilution rates or substrate concentrations prior to analysis which was done during the day. In this manner, one steady-state level could be achieved in each 24-hour period. The pressure differentials across the reactors were found to decrease with bed depth as predicted. The actual pressure drops were, however, greater than predicted. This was likely a result of bed separation and compaction as was observed in reactor Tapered (B). Steady-state

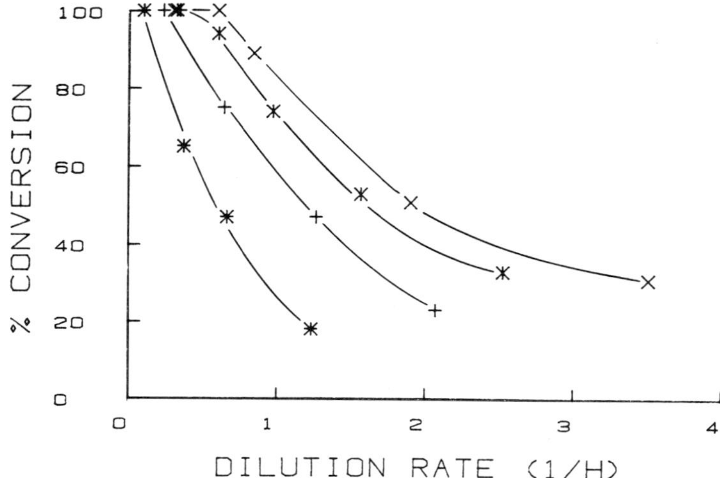

FIGURE 5. Steady-state conversion of cellobiose at various concentrations and dilution rates in Tubular (B). 5 mM (X); 7.5 mM (*); 10 mM (+); 20 mM (*).

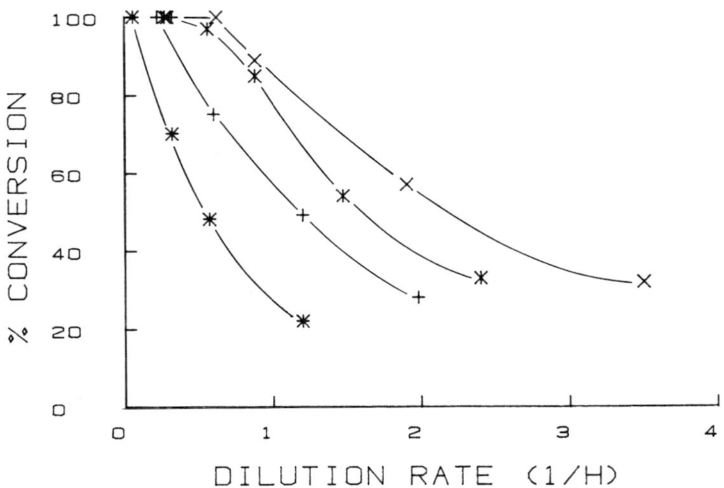

FIGURE 6. Steady-state conversions of cellobiose at various concentrations and dilution rates in Tapered (B). 5 mM (X); 7.5 mM (*); 10 mM (+); 20 mM (*).

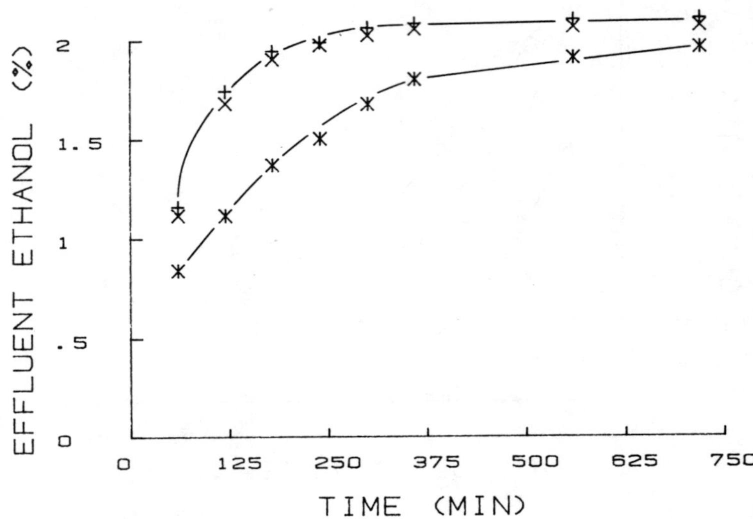

FIGURE 7. Dynamic start-up response of fermentation in reactors. Tubular (A) (+); Tapered (B) (X); Radial (A) (*). Substrate - 5% glucose at a dilution rate of 0.9 h^{-1}.

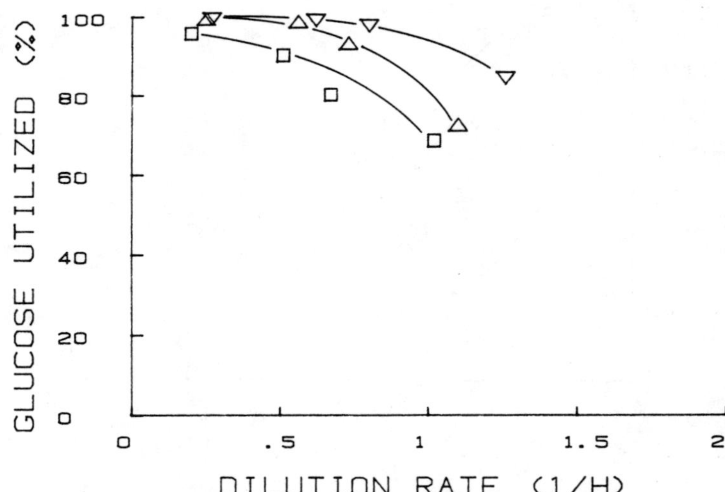

FIGURE 8. Glucose utilization at 8% feed at various dilution rates. Tapered (A) (Δ); Tapered (B) (∇); Radial (B) (□).

conversions using the two tapered bed reactors and the reactor Radial (B) are presented in Figure 8 for inlet glucose concentrations of 8%.

CONCLUSIONS

The use of variable surface reactors for bioconversion processes has demonstrated that no kinetic disadvantage is obtained relative to tubular reactors as long as no shrinkage of the bed occurs. Long-term studies with one of the Tapered reactors (A) showed that over a 22-day fermentation, the average productivity of the reactor was 17.4 g ethanol/l/h. The pressure drop across the reactor never exceeded 14 kPa and was usually in the range of 7-8 kPa. No operational difficulties were encountered. These results compare very favorably with those obtained using tubular reactors where identical kinetics were observed for both the enzymatic hydrolysis and the fermentation. On the other hand, use of the tubular reactors resulted in excessive pressure drops (>150 kPa), bed separation and obvious channelling. The superiority of the variable-surface reactors was readily observed for these types of bioconversions.

ACKNOWLEDGEMENTS

Acknowledgement is given to Ms. C. Choma for excellent technical assistance in all phases of the reactor testing and to the Division of Energy, NRCC and Department of National Defence for funding of the work.

REFERENCES

(1) Vieth, W.R., Venkatasubramanian, K., Constantinides, A. and Davidson, B., Design and Analysis of Immobilized-Enzyme Flow Reactors. In Applied Biochemistry and Bioengineering, ed. L.B. Wingard, Jr., E. Katchalski-Katzir and L. Goldstein, Academic Press, London, 1976, pp. 221-327.

(2) Levenspiel, O., Chemical Reaction Engineering, John Wiley & Sons, Toronto, 1972.

(3) Cho, G.H., Choi, C.Y., Choi, Y.D. and Han, M.H., Ethanol Production by Immobilized Yeast and its CO_2 Gas Effects in a Packed Bed Reactors. J. Chem. Tech. Biotechnol., 1982, 32, 959-967.

(4) Matteau, P.P. and Saddler, J.N., Glucose Production using Immobilized Mycelial-Associated β-Glucosidase of Trichoderma E58. Biotechnol. Lett., 1982, 4, 513-518.

(5) Matteau, P.P. and Saddler, J.N., Continuous Glucose Production using Encapsulated Mycelial-Associated β-Glucosidase from Trichoderma E58. Biotechnol. Lett., 1982, 4, 715-720.

(6) Aiba, S., Humphrey, A.E. and Hillis, N.F., Biochemical Engineering, Academic Press, New York, 1973, pp. 407.

(7) Matteau, P.P., Performance Evaluation of Variable-Surface Area Bioreactors of Cellulose Utilization. National Research Council of Canada Report 23784, 1984, 142 pp.